Sustainable Composites for Future Trends in Renewable Energy

Edited by

Sushil Kumar Verma
Department of Chemical Engineering
CoE-Suspol Indian Institute of Technology Guwahati
Guwahati 781039, Assam, India

Sonika
Department of Physics, Rajiv Gandhi University
Rono Hills, Doimukh, Itanagar 791112
Arunachal Pradesh, India

&

Arbind Prasad
Mechanical Engineering Department
Katihar Engineering College (Under Department of Science,
Technology and Technical Education, Government of Bihar)
Katihar 854109, Bihar, India

Sustainable Composites for Future Trends in Renewable Energy

Editors: Sushil Kumar Verma, Sonika Gupta and Arbind Prasad

ISBN (Online): 979-8-89881-222-5

ISBN (Print): 979-8-89881-223-2

ISBN (Paperback): 979-8-89881-224-9

© 2026, Bentham Books imprint.

Published by Bentham Science Publishers Pte. Ltd. Singapore,

in collaboration with Eureka Conferences, USA. All Rights Reserved.

First published in 2026.

BENTHAM SCIENCE PUBLISHERS LTD.
End User License Agreement (for non-institutional, personal use)

This is an agreement between you and Bentham Science Publishers Ltd. Please read this License Agreement carefully before using the ebook/echapter/ejournal (**"Work"**). Your use of the Work constitutes your agreement to the terms and conditions set forth in this License Agreement. If you do not agree to these terms and conditions then you should not use the Work.

Bentham Science Publishers agrees to grant you a non-exclusive, non-transferable limited license to use the Work subject to and in accordance with the following terms and conditions. This License Agreement is for non-library, personal use only. For a library / institutional / multi user license in respect of the Work, please contact: permission@benthamscience.org.

Usage Rules:

1. All rights reserved: The Work is the subject of copyright and Bentham Science Publishers either owns the Work (and the copyright in it) or is licensed to distribute the Work. You shall not copy, reproduce, modify, remove, delete, augment, add to, publish, transmit, sell, resell, create derivative works from, or in any way exploit the Work or make the Work available for others to do any of the same, in any form or by any means, in whole or in part, in each case without the prior written permission of Bentham Science Publishers, unless stated otherwise in this License Agreement.
2. You may download a copy of the Work on one occasion to one personal computer (including tablet, laptop, desktop, or other such devices). You may make one back-up copy of the Work to avoid losing it.
3. The unauthorised use or distribution of copyrighted or other proprietary content is illegal and could subject you to liability for substantial money damages. You will be liable for any damage resulting from your misuse of the Work or any violation of this License Agreement, including any infringement by you of copyrights or proprietary rights.

Disclaimer:

Bentham Science Publishers does not guarantee that the information in the Work is error-free, or warrant that it will meet your requirements or that access to the Work will be uninterrupted or error-free. The Work is provided "as is" without warranty of any kind, either express or implied or statutory, including, without limitation, implied warranties of merchantability and fitness for a particular purpose. The entire risk as to the results and performance of the Work is assumed by you. No responsibility is assumed by Bentham Science Publishers, its staff, editors and/or authors for any injury and/or damage to persons or property as a matter of products liability, negligence or otherwise, or from any use or operation of any methods, products instruction, advertisements or ideas contained in the Work.

Limitation of Liability:

In no event will Bentham Science Publishers, its staff, editors and/or authors, be liable for any damages, including, without limitation, special, incidental and/or consequential damages and/or damages for lost data and/or profits arising out of (whether directly or indirectly) the use or inability to use the Work. The entire liability of Bentham Science Publishers shall be limited to the amount actually paid by you for the Work.

General:

1. Any dispute or claim arising out of or in connection with this License Agreement or the Work (including non-contractual disputes or claims) will be governed by and construed in accordance with the laws of Singapore. Each party agrees that the courts of the state of Singapore shall have exclusive jurisdiction to settle any dispute or claim arising out of or in connection with this License Agreement or the Work (including non-contractual disputes or claims).
2. Your rights under this License Agreement will automatically terminate without notice and without the

need for a court order if at any point you breach any terms of this License Agreement. In no event will any delay or failure by Bentham Science Publishers in enforcing your compliance with this License Agreement constitute a waiver of any of its rights.
3. You acknowledge that you have read this License Agreement, and agree to be bound by its terms and conditions. To the extent that any other terms and conditions presented on any website of Bentham Science Publishers conflict with, or are inconsistent with, the terms and conditions set out in this License Agreement, you acknowledge that the terms and conditions set out in this License Agreement shall prevail.

Bentham Science Publishers Pte. Ltd.
No. 9 Raffles Place
Office No. 26-01
Singapore 048619
Singapore
Email: subscriptions@benthamscience.net

CONTENTS

FOREWORD	i
PREFACE	ii
LIST OF CONTRIBUTORS	iii

CHAPTER 1 BACKGROUND, HISTORY, AND INTRODUCTION TO SUSTAINABLE POLYMER NANOCOMPOSITES FOR FUTURE TRENDS IN RENEWABLE ENERGY 1
Sumit Bhowmik, Arindam Banerjee, Arbind Prasad, Bidyanand Mahto and *Sudipto Datta*

INTRODUCTION	2
SYNTHESIS TECHNIQUES OF MAGNETIC NANOCOMPOSITES	3
Co-precipitation	3
Sol-gel process	3
Sol-gel process	4
Green Synthesis	4
Microemulsion Technique	5
Chemical Vapor Deposition (CVD)	5
Flame Spray Pyrolysis (FSP)	5
Surface Functionalization and Post-synthesis Modification	5
CHARACTERIZATION TECHNIQUES	6
Structural Characterization	6
X-Ray Diffraction (XRD)	6
Fourier Transform Infrared Spectroscopy (FTIR)	6
Morphological Characterization	7
Scanning Electron Microscopy (SEM)	7
Transmission Electron Microscopy (TEM)	7
Magnetic Characterization	7
Vibrating Sample Magnetometry (VSM)	7
Superconducting Quantum Interference Device (SQUID)	7
Thermal Characterization	8
Thermogravimetric Analysis (TGA)	8
Differential Scanning Calorimetry (DSC)	8
Surface and Chemical Analysis	8
X-ray Photoelectron Spectroscopy (XPS)	8
Energy Dispersive X-ray Spectroscopy (EDS)	9
Other Characterization Techniques	9
Dynamic Light Scattering (DLS)	9
Zeta Potential Analysis	9
APPLICATIONS OF MAGNETIC NANOCOMPOSITES	9
Biomedical Applications	9
Environmental Applications	10
Catalysis	10
Energy Applications	10
Electronics	10
CHALLENGES AND LIMITATIONS	10
FUTURE DIRECTIONS	13
CONCLUSION	13
REFERENCES	14

CHAPTER 2 PROCESSING, CHARACTERIZATION, AND CLASSIFICATION OF POLYMER NANOCOMPOSITES IN SUSTAINABLE AND RENEWABLE ENERGY SECTOR 18
Ranjit Barua, Deepanjan Das, Arbind Prasad, Sachin Latiyan, Sudipto Datta and *Bidyanand Mahto*
INTRODUCTION 19
PROCESSING POLYMER NANOCOMPOSITES 20
   Processing of Polymer Nanocomposites 20
   Characterization of Polymer Nanocomposites 21
   Classification of Polymer Nanocomposites 22
APPLICATION IN HEALTHCARE 23
   Drug Delivery 23
   Tissue Engineering and Regenerative Medicine 23
   Diagnostic Tools 24
   Antimicrobial and Antiviral Applications 25
APPLICATION IN THE SUSTAINABLE RENEWABLE ENERGY SECTOR 25
   Photovoltaics (Solar Energy) 26
   Energy Storage (Batteries and Supercapacitors) 26
   Hydrogen Storage and Production 27
   Energy Harvesting from Environmental Sources 27
CHALLENGES IN POLYMERIC NANOCOMPOSITES FOR RENEWABLE ENERGY 28
FUTURE DIRECTIONS IN POLYMERIC NANOCOMPOSITES FOR RENEWABLE ENERGY 29
   Improvement in Material Performance and Stability 29
   Designing High-Performance Nanocomposite Electrodes for Energy Storage 30
   Advanced Functionalization of Polymers for Multifunctional Applications 30
   Integration with Smart Grid and IoT Systems 30
   Waste-to-Energy Technologies 31
CONCLUSION 31
REFERENCES 31

CHAPTER 3 PIEZOELECTRIC POLYMER COMPOSITES: A COMPREHENSIVE STUDY ON ENERGY-HARVESTING APPLICATIONS 36
Ratnamala Ganjir, Nitin Kumar, Nripesh Kumar, Pankaj Verma and *Sonika*
INTRODUCTION 36
   ENERGY HARVESTING 38
   PIEZOELECTRIC MATERIALS 39
   PIEZOELECTRIC POLYMER 41
   PIEZOELECTRIC ENERGY HARVESTING (PEH) 44
   CONFIGURATIONS OF PIEZOELECTRIC ENERGY HARVESTERS 46
CONCLUSION 49
REFERENCES 49

CHAPTER 4 HIGHLY STABLE NANOCOMPOSITE PVDF-BASED ENERGY HARVESTING PIEZOELECTRIC DEVICES 54
Rahutosh Ranjan, Sushama Yadav, Sujeet Kumar Chaurasia, Nitin Srivastava and *Neelabh Srivastava*
INTRODUCTION 54
HISTORY OF PIEZOELECTRIC POLYMERS 57
   Classification of Piezoelectric Polymers 58
      Bulk Piezoelectric Polymer 59
      Voided Charged Polymers (VCP) 59

Piezoelectrics Composites	60
INTRODUCTION TO PVDF	60
How is Piezoelectricity Generated?	60
Phases of PVDF	61
β-phase Enhancement	62
Piezoelectric PVDF Fabrication Method	63
Melt Blending	63
Self-poling	64
Electrospinning	64
PVDF-based cOpolymers, Composites, and Nano-composites	66
PVDF-HFP	69
PVDF-TrFE	70
PVDF NANOFIBERS	70
APPLICATION OF PIEZOELECTRIC MATERIALS	71
Application in Energy Generation	73
Piezoelectric Nanogenerator (PENG)	73
Sensors	77
Various Types of Sensors	79
Self-charging Power Cell	80
Piezoelectric Transducer	81
Application of Piezoelectric Transducers	82
Piezoelectric Actuator	83
CONCLUSION	83
REFERENCES	84
CHAPTER 5 INNOVATIVE, CUTTING-EDGE TECHNOLOGIES FOR ENERGY HARVESTING USING POLYMER NANOCOMPOSITES	90
Sonika, Manas Ranjan Nayak, Ratikanta Nayak, Varatharajan Prasanna Venkadesan and Sushil Kumar Verma	
INTRODUCTION	90
CLASSIFICATION	92
Ceramic-Matrix Nanocomposites (CMCs)	92
Metal-Matrix Nanocomposites (MMCs)	93
Polymer-Matrix Nanocomposites (PMCs)	93
LEAD-FREE MATERIALS IN THE PIEZOELECTRIC ENERGY HARVESTER	94
PIEZOELECTRIC NANOCOMPOSITE ENERGY HARVESTERS	95
INNOVATIVE STRETCHABLE PIEZOELECTRIC ENERGY HARVESTER	96
MAJOR CHALLENGES FOR ENERGY HARVESTING	98
Vibration Method	98
Radio Frequency (RF) Method	99
Thermoelectric Method	100
CONCLUSION	102
REFERENCES	102
CHAPTER 6 METHODOLOGIES EXPLOITED IN THE SYNTHESIS OF CONDUCTING POLYMER NANOCOMPOSITES: CONCEPT, STRATEGIES, AND DEVELOPMENT	109
Anju Dhillon, Manoj Kumar Srivastava, Vasudha Agarwal and Raman Sankar	
INTRODUCTION	109
Types of Conducting Polymer Nanocomposites	110
Polymer Nanocomposites With Carbon Nanotubes	110
Polymer Nanocomposites with Graphene	111
Polymer Nanocomposites with Metal Oxide Nanoparticles	111

  Polymer Nanocomposites with Nanocellulose 111
  Polymer Nanocomposites with Clay Nanoparticles 111
   Advantages of Conducting Polymer Nanocomposites 111
  Methods for the Synthesis of Conducting Polymer Nanocomposites 112
 IN SITU POLYMERIZATION 112
 SOLUTION BLENDING 117
 IN SITU REDUCTION 120
 ELECTROCHEMICAL DEPOSITION 123
 LAYER-BY-LAYER ASSEMBLY 127
 CONCLUSION 129
 REFERENCES 130

CHAPTER 7 BIODEGRADABLE NANOCOMPOSITES FOR ENERGY HARVESTING DEVICES, PIEZOELECTRIC SENSORS, AND FUEL CELLS 133
Pankaj Verma, Nitin Kumar, Nripesh Kumar, Ratnamala Ganjir, Sushil Kumar Verma and *Sonika*
 INTRODUCTION 134
  Polylactic Acid (PLA)-Based Nanocomposites 135
  Polycaprolactone Nanocomposites (PCL) 138
  Poly (p-dioxanone) (PPDO) Nanocomposites 140
  Polybutylene Succinate (PBS) Nanocomposites 141
  Poly (Hydroxyalkanoates) (PHAs) Nanocomposites 142
 METHODS FOR THE FABRICATION OF BIODEGRADABLE NANOCOMPOSITES ... 143
  Melt Mixing 143
  Solvent-based Process 144
  Electrospinning Technique 144
  3D Printing 145
  In Situ Polymerization 146
  Electrochemical Deposition 146
  Biodegradation 147
 APPLICATION OF BIODEGRADABLE NANOCOMPOSITES 148
  Piezoelectric Devices 148
  Sensors 150
  Fuel Cells 151
  Energy Harvesting 152
  Summary 153
 CONCLUSION 153
 REFERENCES 154

CHAPTER 8 ADVANCED MATERIALS WITH CARBON NANOSTRUCTURE-BASED COMPOSITES FOR ENVIRONMENTAL ENERGY HARVESTING 159
Nitin Kumar, Nripesh Kumar, Manoj Kumar Prajapati, Sushil Kumar Verma and *Sonika*
 INTRODUCTION 160
 NANOCOMPOSITES AND CARBON NANOSTRUCTURE MATERIALS 161
 AREAS OF APPLICATION OF CARBON NANOSTRUCTURED COMPOSITE MATERIALS (ENVIRONMENTAL ENERGY HARVESTING) 162
  Environmental Applications 163
  Energy Harvesting Applications 166
   Supercapacitors 166
   Solar Cells 167
   Batteries 169

| CONCLUSION | 169 |
| REFERENCE | 170 |

CHAPTER 9 EVOLUTION OF ADVANCED POLYMER-BASED NANOGENERATORS FOR ENERGY HARVESTING APPLICATIONS ... 173
Vijyendra Kumar, Sonika, Deo Karan Ram, Gamini Sahu, Nutan Kumar Sahu and *Sushil Kumar Verma*

INTRODUCTION	174
POLYMER-BASED NANOGENERATORS: MECHANISMS AND MATERIALS	175
Piezoelectric Nanogenerators (PENGs)	175
Triboelectric Nanogenerators (TENGs)	176
Pyroelectric Nanogenerators	176
Emerging Multifunctional Materials	176
Types of Nanogenerators and Their Mechanisms	177
Piezoelectric Nanogenerators (PENGs)	178
Triboelectric Nanogenerators (TENGs)	178
Pyroelectric Nanogenerators (PyNGs)	179
Emerging Techniques	181
Hybrid Nanogenerator Fabrication Advancements	181
Examples of Recent Advancements	182
APPLICATIONS	183
Environmental and Industrial Applications	183
Schematic Representation of Applications	184
CHALLENGES AND FUTURE DIRECTIONS	184
Enhancing Stability and Longevity	184
Advancing Multifunctionality	185
Efficiency	185
Stability	185
Lifetime	186
Scalability	186
CONCLUSION	187
REFERENCES	187

CHAPTER 10 POLYMERIC ENERGY MATERIALS: DEVELOPMENT, CHALLENGES, AND FUTURE BENEFITS FOR INDUSTRIALIZATION ... 191
Debasish Banerjee, Sumit Bhowmik, Arbind Prasad and *Sudipto Datt*

INTRODUCTION	192
POLYMERIC ENERGY MATERIALS: APPLICATIONS IN MODERN INDUSTRY	193
Energy Storage Systems	193
Solar Cells and Energy Conversion	194
Flexible and Wearable Electronics	194
Automotive Industry: Lightweight Energy Materials	195
Energy-Efficient Building Materials	195
POLYMERIC ENERGY MATERIALS IN HEALTHCARE APPLICATIONS	196
Wearable Health Monitoring Devices	196
Drug Delivery Systems	196
Energy Harvesting for Implantable Medical Devices	197
Biosensors for Disease Diagnosis	197
Flexible and Stretchable Electronics	198
Tissue Engineering and Regenerative Medicine	198
CHALLENGES AND FUTURE DIRECTIONS	199
CHALLENGES	199

CONCLUSION	202
REFERENCES	202

CHAPTER 11 ENHANCED ENERGY HARVESTER PERFORMANCE BY A TENSION-ANNEALED CARBON NANOTUBE YARN AT EXTREME TEMPERATURES 206
Vikas Kashyap, Shivanshu Sharma, Chandra Kumar, Anand Kumar and *Kapil Saxena*

INTRODUCTION	206
Experimental Setups	210
Methods of characterization	212
Setup for Electrochemical Analysis	212
Morphological Analysis	212
Experimental Results	213
CONCLUSION	216
REFERENCES	217

CHAPTER 12 ENERGY HARVESTING: INNOVATING ADVANCED TECHNOLOGIES THROUGH POLYMER NANOCOMPOSITES 220
Akash Ranjan, Jimli Sarma and *Sonika*

INTRODUCTION	221
PROPERTIES AND BENEFITS OF POLYMER NANOCOMPOSITES	225
SYNTHESIS TECHNIQUES FOR POLYMER NANOCOMPOSITES	227
USES IN ENERGY HARVESTING DEVICES	230
CONCLUSION	235
REFERENCES	235

CHAPTER 13 A BRIEF OVERVIEW OF ENERGY HARVESTING IN ADVANCED SUSTAINABLE POLYMERS 238
Akash Ranjan, Sabira Sultana Khadim and *Sonika*

INTRODUCTION	239
SIGNIFICANCE OF SUSTAINABLE POLYMERS IN MODERN ENERGY APPLICATIONS	240
Renewable Energy Technologies	242
Energy Storage Systems	242
Energy Efficiency	242
Waste-to-Energy Conversion	242
Environmental Benefits	243
ROLE OF POLYMERS IN ENERGY HARVESTING	243
Photovoltaics (Solar Energy Harvesting)	243
Piezoelectric Energy Harvesting	243
Thermoelectric Energy Harvesting	244
Bioenergy Harvesting	244
Triboelectric Energy Harvesting	244
Wind and Wave Energy	244
Energy Storage Integration	244
SUSTAINABLE POLYMERS	244
TYPES OF SUSTAINABLE POLYMERS	245
Bio-based Polymers	245
Biodegradable Polymers	247
Recycled Polymers	247
Hybrid Polymers	247
BENEFITS OVER TRADITIONAL POLYMERS	248

MECHANISMS OF ENERGY HARVESTING IN POLYMERS ... 248
    Piezoelectric Polymers ... 249
    Thermoelectric Polymers ... 250
    Photovoltaic Polymers ... 251
RECENT ADVANCES IN POLYMER-BASED ENERGY HARVESTING ... 251
APPLICATIONS OF ENERGY HARVESTING POLYMERS ... 254
    Electronics that are worn ... 254
    Medical Equipment ... 254
    Systems of Renewable Energy ... 254
    Electronics for consumers ... 255
    Uses in Automobiles ... 255
    Internet of Things (IoT) ... 255
    Monitoring of the Environment ... 255
    Intelligent Textiles ... 255
    Use in Industry ... 255
    Applications in Aerospace ... 256
    Agriculture ... 256
CHALLENGES AND FUTURE DIRECTIONS IN POLYMER-BASED ENERGY HARVESTING ... 256
    Low Efficiency of Energy Conversion ... 256
    Stability and Durability ... 256
    Manufacturing and Scalability ... 257
    Integration of Mechanical and Electrical ... 257
    Density of Power ... 257
    Availability and Cost of Materials ... 257
FUTURE DIRECTIONS ... 258
    High-Performance Hybrid Materials:- ... 258
    Flexible and Transparent Solar Cells ... 258
    Multipurpose Equipment ... 259
    Smart Textiles ... 259
    Better Thermoelectric Materials ... 259
    Sustainable Energy Harvesting Polymers ... 259
    Self-Healing Polymers ... 259
    Implantable and Wearable Medical Equipment ... 259
    Economic Feasibility and Market Readiness ... 259
CONCLUSION ... 260
REFERENCES ... 260

CHAPTER 14 BIOBASED RESORBABLE POLYMERIC NANOCOMPOSITES FOR SUSTAINABLE HEALTHCARE APPLICATIONS ... 263
Sumit Bhowmik, Debasish Banerjee, Arbind Prasad, Souvik Debnath and *Sudipto Datta*
INTRODUCTION ... 263
BIOBASED RESORBABLE POLYMERIC NANOCOMPOSITES ... 264
APPLICATIONS IN HEALTHCARE ... 265
    Advanced Drug Delivery Systems ... 266
    Tissue Engineering Scaffolds ... 266
    Resorbable Medical Implants ... 266
    Wound Healing and Skin Regeneration ... 267
    Biodegradable Biosensors ... 267
    Dental Applications ... 267

    Antimicrobial Coatings for Medical Devices 267
    Controlled Release of Growth Factors 267
    Injectable Nanocomposites 268
    Anticancer Applications 268
CHALLENGES AND LIMITATIONS 268
    Degradation Control 268
    Cost and Scalability 268
    Toxicological Concerns 268
FUTURE DIRECTIONS OF BIOBASED RESORBABLE POLYMERIC NANOCOMPOSITES IN HEALTHCARE 268
    Advanced Material Design with Artificial Intelligence 269
    Development of Smart Nanocomposites 269
    Bioinspired Nanocomposites 269
    Green Manufacturing Techniques 270
    Multifunctional Healthcare Devices 270
    Exploration of Novel Biopolymers and Nanofillers 270
PERSONALIZATION THROUGH 3D PRINTING 271
    Long-Term Biocompatibility and Toxicity Studies 271
    Applications in Emerging Fields 271
    Global Collaboration and Standardization 271
CONCLUSION 271
REFERENCES 272

CHAPTER 15 CARBON-BASED NANOCOMPOSITES (CBNS) FOR ENVIRONMENTAL ENERGY HARVESTING APPLICATIONS 276
Sagar Vikal, Durvesh Gautam, Ajay Kumar, Rizwan Khan, Yogendra K. Gautam, Ashwani Kumar and Neetu Singh

INTRODUCTION 277
FABRICATION AND CHARACTERIZATION OF CBNS 279
CARBON-BASED NANOCOMPOSITES FOR ENVIRONMENTAL APPLICATIONS 281
    Detection and Removal of Heavy Metals 281
    Environmental Gas Sensor for Inorganic and Organic Vapor 281
    Bacterial Detection and Eradication 282
    Degradation and Removal of Organic Species 284
    Environmental and Health Implications of CBNs 284
CARBON-BASED NANOCOMPOSITES FOR ENERGY HARVESTING APPLICATIONS 285
    Photovoltaics and Solar Energy Harvesting 286
        CNTs in Solar Cells 287
        Graphene-Based Photovoltaic Devices 288
        Hybrid Nanocomposites for Enhanced Solar Energy Conversion 288
    Energy Storage Systems 289
        CBNs for Batteries 291
        Carbon Nanomaterials in Supercapacitors 293
        Fuel Cells 295
        Thermoelectric Energy Conversion 299
        Piezoelectric and Triboelectric Energy Harvesting 302
CONCLUSION AND FUTURE PERSPECTIVES 305
REFERENCES 306

CHAPTER 16 RECENT ADVANCEMENTS IN METAL OXIDE NANOCOMPOSITES FOR ENERGY HARVESTING APPLICATIONS 316

Akash Ranjan, Shaistah Tabassum and *Sonika*

INTRODUCTION	317
INTRODUCTION TO NANOCOMPOSITES IN ENERGY HARVESTING	318
NANOCOMPOSITES: STRUCTURE, PROPERTIES, AND SYNTHESIS	320
ROLE OF METAL OXIDES IN ENERGY HARVESTING	323
RECENT DEVELOPMENTS IN METAL OXIDE NANOCOMPOSITES FOR ENERGY HARVESTING	325
NANOCOMPOSITE METAL OXIDES FOR SOLAR ENERGY HARVESTING	326
THERMOELECTRIC APPLICATIONS OF METAL OXIDE NANOCOMPOSITES	329
PIEZOELECTRIC AND TRIBOELECTRIC ENERGY HARVESTING WITH METAL OXIDE NANOCOMPOSITES	332
ELECTROCHEMICAL ENERGY HARVESTING: METAL OXIDE NANOCOMPOSITES IN SUPERCAPACITORS AND BATTERIES	336
CASE STUDIES AND REAL-WORLD APPLICATIONS	339
Solar Energy Harvesting with Metal Oxide Nanocomposites	340
Thermoelectric Applications of Metal Oxide Nanocomposites	340
Piezoelectric Energy Harvesting with Metal Oxide Nanocomposites	341
Triboelectric Energy Harvesting with Metal Oxide Nanocomposites	342
Electrochemical Energy Harvesting with Metal Oxide Nanocomposites	342
CHALLENGES AND LIMITATIONS IN METAL OXIDE NANOCOMPOSITE DEVELOPMENT	343
Synthesis and Fabrication Challenges	343
Material Stability and Durability	344
Limited Conductivity and Performance	344
Compatibility with Existing Technologies and Integration	345
High Costs and Environmental Impact	345
FUTURE PROSPECTS OF METAL OXIDE NANOCOMPOSITES IN ENERGY HARVESTING	346
CONCLUSION	350
REFERENCES	350
CHAPTER 17 CURRENT DEVELOPMENTS AND FUTURE PERSPECTIVES OF MAGNETIC NANOCOMPOSITES FOR ADVANCED APPLICATIONS	352
Debasish Banerjee, Arindam Banerjee, Arbind Prasad, Bidyanand Mahto and *Sudipto Datta*	
INTRODUCTION	353
MAGNETIC NANOCOMPOSITES IN TARGETED DRUG DELIVERY	354
MAGNETIC NANOCOMPOSITES IN ENERGY STORAGE AND ENVIRONMENTAL REMEDIATION	355
Magnetic Nanocomposites in Energy Storage	355
Magnetic Nanocomposites in Environmental Remediation	356
MAGNETIC HYPERTHERMIA: A REVOLUTIONARY CANCER TREATMENT APPROACH	357
Mechanism of Magnetic Hyperthermia	357
Challenges in Magnetic Hyperthermia	357
Recent Advances in Magnetic Hyperthermia	358
MAGNETIC NANOCOMPOSITES AND SUSTAINABILITY	358
Magnetic Nanocomposites in Energy Storage	358
Biodegradable and Green Synthesis Approaches	358
CHALLENGES AND FUTURE DIRECTIONS	359
CONCLUSION	360
REFERENCES	360
SUBJECT INDEX	364

FOREWORD

In an era where sustainability is the primary goal for scientific research and international policy, materials science is essential for advancing progress. The demand for sustainable, cost-efficient, and eco-friendly materials for renewable energy technologies has become increasingly urgent. ***Sustainable Composites for Future Trends in Renewable Energy*** addresses this pressing issue, offering an in-depth examination of advanced composite materials and their transformative potential in the renewable energy sector.

An important contribution to the growing field of sustainable energy materials is this publication. It covers topics like the design, manufacturing, and performance improvement of composites for energy generation, storage, and conversion, thereby bridging the gap between basic research and practical applications. The book offers an in-depth overview of the prospects and difficulties associated with developing a sustainable future by combining perspectives from a wide range of areas, such as energy systems engineering, nanotechnology, and polymer science.

In order to cover disciplines like bio-based composites and new materials for use in photovoltaics, fuel cells, and wind energy systems, the editors have expertly arranged chapters written by prominent experts. This book stands out for its emphasis on sustainability and dedication to promoting cutting-edge scientific research, which makes it a priceless tool for academics, business leaders, and legislators. This book provides direction and motivation as the world transitions to renewable energy. It highlights the revolutionary potential of sustainable composites and the vital role that collaboration, innovation, and moral responsibility have in addressing the world's energy problems.

The dedication and foresight of the editors and contributors in creating this work are commendable. I hope that ***Sustainable Composites for Future Trends in Renewable Energy*** will motivate future researchers and practitioners to contribute to a cleaner and more sustainable world.

Anoop Kumar
Delhi Pharmaceutical Sciences and Research University (DPSRU)
Govt. of NCT of Delhi
India

PREFACE

Our modern world's persistent demand for energy has elevated sustainability to the forefront of industrial and scientific innovation. The development of materials that adhere to sustainability principles has become crucial as we struggle with the critical issues of climate change and the depletion of natural resources. A timely response to these issues is this book, Sustainable Composites for Future Trends in Renewable Energy, which explores the potential of composites to transform the renewable energy industry. Composites are becoming increasingly essential in renewable energy applications because of their adaptable qualities, lightweight design, and variable performance. Sustainable composite materials are essential in challenging the limits of what is feasible in a variety of applications, including wind turbine blades, sophisticated solar systems, and next-generation energy storage devices. We achieve the demands of performance while also addressing environmental imperatives by emphasizing materials produced from renewable resources and environmentally friendly manufacturing techniques. The objective of the book is to present an extensive overview of the most important recent developments, discoveries, and studies in the field of Sustainable Composites for Future Trends in Renewable Energy. Numerous subjects are covered in the chapters, such as life cycle evaluation, application-specific optimization, fabrication techniques, and material design. In order to ensure that the insights provided here are both innovative and useful, the goal is to bridge the gap between scholarly research and industrial applications. Without the diligent efforts of researchers, professionals, and entrepreneurs who are dedicated to establishing a sustainable future, this book would not have been possible.

I sincerely appreciate everyone who contributed by sharing their expertise and perspectives. Their endeavors give us hope that we can overcome the obstacles of the twenty-first century and achieve an environmentally friendly and more sustainable future by working together and using our creativity. For students, researchers, scientists, and government officials interested in the nexus between materials science and renewable energy, I hope this book will be a useful resource. We can make the idea of a greener world a reality by utilizing the potential of sustainable composites.

Sushil Kumar Verma
Department of Chemical Engineering
CoE-Suspol Indian Institute of Technology Guwahati
Guwahati 781039, Assam, India

Sonika
Department of Physics, Rajiv Gandhi University
Rono Hills, Doimukh
Itanagar 791112, Arunachal Pradesh, India

&

Arbind Prasad
Mechanical Engineering Department
Katihar Engineering College (Under Department of Science Technology and Technical Education, Government of Bihar)
Katihar 854109, Bihar, India

List of Contributors

Anju Dhillon	Department of Applied Sciences, Maharaja Surajmal Institute of Technology affiliated to GGSIP University, New Delhi 110058, India
Anand Kumar	Department of Applied Sciences and Humanities, Invertis University, Bareilly 243001, Uttar Pradesh, India
Ajay Kumar	Department of Biotechnology, Mewar University, Chittorgarh 312901, Rajasthan, India
Ashwani Kumar	Department of Physics, Regional Institute of Education (NCERT), Bhubaneswar 751022, Odisha, India
Akash Ranjan	Faculty of Education, Banaras Hindu University, Varanasi, Uttar Pradesh, India
Arindam Banerjee	Mechanical Engineering Department, Omdayal Group of Institutions, Howrah 711316, West Bengal, India
Arbind Prasad	Mechanical Engineering Department, Katihar Engineering College (Under Department of Science, Technology and Technical Education, Government of Bihar), Katihar 854109, Bihar, India
Bidyanand Mahto	Government Engineering College, Vaishali (Under Department of Science, Technology and Technical Education, Government of Bihar), Vaishali 844115, Bihar, India
Chandra Kumar	Escuela de Ingeniería, Facultad de Ciencias, Ingeniería y Tecnología, Universidad Mayor, Santiago 7500994, Chile
Deepanjan Das	Mechanical Engineering Department, Omdayal Group of Institutions, Howrah 711316, West Bengal, India
Deo Karan Ram	Department of Petroleum & Chemical Engg., NIMS University, Jaipur 303121, Rajasthan, India
Durvesh Gautam	Smart Materials and Sensor Laboratory, Department of Physics, Ch. Charan Singh University, Meerut 250004, Uttar Pradesh, India
Debasish Banerjee	Mechanical Engineering Department, Omdayal Group of Institutions, Howrah 711316, West Bengal, India
Gamini Sahu	School of Life Science, Pt. Ravishankar Shukla University, Raipur 492001, Chhattisgarh, India
Jimli Sarma	Centre for Multidisciplinary Research, Tezpur University, Tezpur 784028, Assam, India
Kapil Saxena	Department of Applied Sciences, Kamla Nehru Institute of Technology, Sultanpur 228118, Uttar Pradesh, India
Manas Ranjan Nayak	Novel Material Research Laboratory, NIST University, Berhampur 761008, Odisha, India
Manoj Kumar Prajapati	Department of Physics, SSSVS Government Post Graduate College Chunar Mirzapur, Mirzapur 231304, Uttar Pradesh, India
Manoj Kumar Srivastava	Department of Physics, DAV PG College, DDU Gorakhpur University, Gorakhpur 273001, Uttar Pradesh, India

Nripesh Kumar	Department of Physics, National Institute of Technology Mizoram, Aizawl 796012, Mizoram, India Department of Physics, Bharat Institute of Engineering and Technology, Hyderabad 501510, India
Nitin Srivastava	Department of Physics, Sri Chitragupt P. G. College, Mainpuri 205001, Uttar Pradesh, India
Neelabh Srivastava	Department of Physics, School of Physical Sciences, Mahatma Gandhi Central University, Motihari, East Champaran 845401, Bihar, India
Nitin Kumar	Department of Physics, National Institute of Technology Mizoram, Aizawl 796012, Mizoram, India
Nutan Kumar Sahu	Department of Chemical Engineering, Raipur Institute of Technology, Raipur 49001, Chhattisgarh, India
Neetu Singh	Department of Biotechnology, Mewar University, Chittorgarh 312901, Rajasthan, India
Pankaj Verma	Department of Applied Sciences, Galgotias College of Engineering and Technology, Knowledge Park-II, Greater Noida, Uttar Pradesh, India
Ranjit Barua	Mechanical Engineering Department, Omdayal Group of Institutions, Howrah 711316, West Bengal, India
Rahutosh Ranjan	Department of Physics, School of Physical Sciences, Mahatma Gandhi Central University, Motihari, East Champaran 845401, Bihar, India
Ratikanta Nayak	Novel Material Research Laboratory, NIST University, Berhampur 761008, Odisha, India
Raman Sankar	Institute of Physics, Academia Sinica, Nankang 11529, Taipei, Taiwan
Ratnamala Ganjir	Department of Physics, Government J. Yojanandam Chhatisgarh College, Raipur 492001, Chhattisgarh, India
Rizwan Khan	Smart Materials and Sensor Laboratory, Department of Physics, Ch. Charan Singh University, Meerut 250004, Uttar Pradesh, India
Sachin Latiyan	Department of Materials Engineering, Indian Institute of Science, Bangalore 560012, Karnataka, India
Sudipto Datta	Department of Materials Engineering, Indian Institute of Science, Bangalore 560012, Karnataka, India
Sushama Yadav	Centre for Nanoscience & Technology, Veer Bahadur Purvanchal University, Jaunpur 222003, Uttar Pradesh, India
Sujeet Kumar Chaurasia	Centre for Nanoscience & Technology, Veer Bahadur Purvanchal University, Jaunpur 222003, Uttar Pradesh, India
Sushil Kumar Verma	Department of Chemical Engineering, CoE-Suspol, Indian Institute of Technology Guwahati, Guwahati 781039, Assam, India
Shivanshu Sharma	Department of Physics, Panjab University, Chandigarh 16001, Punjab, India
Sonika	Department of Physics, Rajiv Gandhi University, Rono Hills, Doimukh, Itanagar 791112, Arunachal Pradesh, India
Sabira Sultana Khadim	Department of Education, Lima Aier Higher Secondary School, Dimapur, Nagaland, India

Sonika	Department of Physics, Rajiv Gandhi University, Itanagar, Rono Hills, Doimukh, Papumpare 791112, Arunachal Pradesh, India
Sumit Bhowmik	Mechanical Engineering Department, Omdayal Group of Institutions, Howrah 711316, West Bengal, India
Souvik Debnath	Department of Materials Engineering, Indian Institute of Science, Bangalore 560012, Karnataka, India
Sagar Vikal	Smart Materials and Sensor Laboratory, Department of Physics, Ch. Charan Singh University, Meerut 250004, Uttar Pradesh, India
Shaistah Tabassum	Department of Education, Tezpur University, Tezpur 784028, Assam, India
Varatharajan Prasanna Venkadesan	School of Mechanical and Aerospace Engineering, Queens University of Belfast, Belfast, Northern Ireland, United Kingdom
Vasudha Agarwal	Department of Physics, Maitreyi College (University of Delhi), Chanakyapuri, Delhi, India
Vijyendra Kumar	Department of Chemical Engineering, Raipur Institute of Technology, Raipur 49001, Chhattisgarh, India
Vikas Kashyap	Department of Physics, Panjab University, Chandigarh 16001, Punjab, India
Yogendra K. Gautam	Smart Materials and Sensor Laboratory, Department of Physics, Ch. Charan Singh University, Meerut 250004, Uttar Pradesh, India

CHAPTER 1

Background, History, and Introduction to Sustainable Polymer Nanocomposites for Future Trends in Renewable Energy

Sumit Bhowmik[1], **Arindam Banerjee**[1], **Arbind Prasad**[2,*], **Bidyanand Mahto**[3] and **Sudipto Datta**[4,*]

[1] *Mechanical Engineering Department, Omdayal Group of Institutions, Howrah 711316, West Bengal, India*

[2] *Mechanical Engineering Department, Katihar Engineering College (Under Department of Science, Technology and Technical Education, Government of Bihar), Katihar 854109, Bihar, India*

[3] *Government Engineering College, Vaishali (Under Department of Science, Technology and Technical Education, Government of Bihar), Vaishali 844115, Bihar, India*

[4] *Department of Materials Engineering, Indian Institute of Science, Bangalore 560012, Karnataka, India*

Abstract: Magnetic nanocomposites are among the most significant categories of materials because they have excellent magnetic, thermal, and mechanical characteristics for use in various professions. The progress made in the last half decade has stemmed from the preparation of new nanocomposites that display enhanced performance in healthcare, environmental treatment, the energy industry, and electronics. Various synthesis methods for producing M-TiO_2 have been enhanced effectively, including co-precipitation, sol-gel processes, and hydrothermal synthesis to possess the required structural and functional characteristics. Magnetic nanoparticles, which have been approved for targeted drug delivery combined with polymers, ceramics, and metal matrices, offer expanded functionality in applications such as MRI and catalytic systems. Nevertheless, some problems of the approach include the problems of scalability, stability, and environmental impact that may require further inquiries. In this chapter, we focus on the recent progress, including computational modeling, artificial intelligence, and green chemistry. The last one is devoted to present trends, where the roles of bioinspired and hybrid magnetic nanocomposites in shaping future technologies are described.

---

[*] **Corresponding authors Arbind Prasad and Sudipto Datta:** Mechanical Engineering Department, Katihar Engineering College (Under Department of Science, Technology and Technical Education, Government of Bihar), Katihar-854109, Bihar, India; Department of Materials Engineering, Indian Institute of Science, Bangalore 560012, Karnataka, India; E-mail: sudiptodatta1990@gmail.com; arbind.iitg@gmail.com

Keywords: Advanced materials, Biomedical applications, Energy storage, Environmental remediation, Green chemistry, Hybrid composites, Magnetic nanocomposites.

INTRODUCTION

Magnetic nanocomposites are a relatively new generation of multifunctional advanced materials possessing both magnetic characteristics and nanoscale size, as well as the advantages of composites. These materials have recently attracted the interest of researchers in both scientific studies and industries because of their unique physical, chemical, and mechanical characteristics [1]. Due to their exciting ability to selectively control the magnetic properties on the nanoscale, they are potential candidates for multifaceted applications such as drug delivery, pollutant removal, energy storage, and nanoscale electronics [2]. Recent developments in nanotechnology have seen improvements in the synthesis, as well as the manipulation of properties of numerous magnetic nanocomposites, leading to improvements in their properties and applications [3]. The inclusion of magnetic nanoparticles with polymers, metals, and ceramics has enabled the design of materials with specific characteristics of use, for example, superparamagnetism, high thermal stability, and biocompatibility [4]. These materials are especially important in new scientific disciplines, including biomedicine, where they are used as diagnostic imaging tools or components of drug delivery systems and hyperthermia therapy devices. The natural polymers and derivatives that the BioPol4fun research group uses include chitosan, agarose, alginate, cellulose, gelatin, *etc.*, to design materials [5].

Magnetic nanocomposite synthesis has advanced towards techniques such as co-precipitation, sol-gel processing, and hydrothermal techniques, which provide the adaptability for larger-scale production and result in lower costs [6]. Additionally, there has been a development of innovative remedies in green-synthesis engineering that allow for operation under environmentally sensitive conditions and offer high efficacy. These developments have provided opportunities for the green synthesis of magnetic nanocomposites for global sustainable production and proper utilization of natural resources [7]. Characterization techniques such as X-Ray Diffraction (XRD), Scanning Electron Microscopy (SEM), TEM, and Vibrating Sample Magnetometry (VSM) have played a vital role in the characterization and manipulation of properties in these materials [8]. These techniques have proved helpful to researchers in acquiring information on the structural, morphological, and magnetic characteristics of nanocomposites and their enhancement for particular applications [9]. However, magnetic nanocomposites are not exempted from some issues like scalability, stability during operation, and toxicity effects that may be incurred on the environment and human beings [10]. These are problems that will need to be solved at the interface of materials, engineering, and computation. AI and ML in the recent past have come up as essential tools in enhancing the identification and designing of magnetic nanocomposites by providing behavioral and performance forecasts.

This chapter is intended to give a brief account of some recent progress made in the synthesis of magnetic nanocomposites, their characterization, and multifarious fields of applications. It also highlights the challenges associated with these materials and discusses the prospects and directions for development, including trends such as bioinspired designs, hybrid nanocomposites, and knowledge of the use of AI-driven approaches. In targeting a parallel between research and application, this chapter aims to encourage more research and application of magnetic nanocomposite materials in innovative technologies.

SYNTHESIS TECHNIQUES OF MAGNETIC NANOCOMPOSITES

Nanocomposites are widely used in the medical sector for various applications, such as in orthopedics, cardiovascular implants, and the fabrication of implants. Recent progress in synthesis techniques has been widely seen to have a large impact on the development of magnetic nanocomposites. These methods seek to provide accurate and reproducible control over aspects such as size, shape, magnetic characteristics, and distribution of nanoparticles in the host matrix. All of the synthesis techniques present certain advantages and are suitable for solving certain problems; this is why the choice of the synthesis method is critical.

Co-precipitation

Magnetic nanocomposites prepared by the co-precipitation method are discussed in this section because this technique is quite general and inexpensive. In this technique, magnetic nanoparticles are separated from a solution containing metal salts at a specific pH and temperature [11]. Normally, Fe^{2+} and Fe^{3+} salts are precipitated by using a base, especially ammonia solution or ammonium hydroxide. Although this technique offers the benefits of large-volume production and precise control over particle size and size distribution on the nanoscale, it has limitations in that the crystallinity and, conversely, uniformity of the produced nanoparticles may be low and require additional post-synthesis treatments like annealing.

Sol-gel process

Another general approach to obtaining magnetic nanocomposites is the sol-gel process, which becomes crucial for obtaining target materials with high purity and homogeneity [12]. This process involves the reaction of metal alkoxides or metal salts through hydrolysis and condensation to form a gel that, upon drying and calcination, yields the prepared material. The incorporation of magnetic nanoparticles during this process helps in getting a uniform distribution of the nanoparticles within the composite matrix [12]. Coating synthesis used in sol-gel synthesis offers the advantage of producing thin films with specific compositions and morphologies that are desirable in particular applications with specific desired magnetic properties. However, the process consumes a lot of time and depends on the environmental conditions, including humidity and temperature levels. The effects of plastic on the environment and the sources that correspond to them can be widely studied in the literature [13].

Sol-gel process

Hydrothermal and solvothermal methods are incredibly suitable for preparing largely crystalline magnetic nanocomposites [14]. These techniques include comprehending reactions that take place at high temperatures and pressures in an autoclave. It is possible to achieve very good magnetic characteristics of nanocomposites by employing metal precursors dissolved in water or organic solvents [15]. The prospects of subtle variations in the behaviors of therapeutically relevant molecules open doors to the development of drugs with enhanced efficacies by achieving precise sub-cellular targeting and evading the defense mechanisms of pathogenic organisms or cells with the help of externally applicable magnetic or electric fields, while the presence of surfactants or stabilizers in the formulations also allows for direct regulation of nanoparticle morphology. Although these methods yield high-quality materials, their drawbacks are the general need for the use of special equipment and longer reaction periods, which can lead to high costs.

Green Synthesis

Green synthesis is now being developed as a green approach to fabricating magnetic nanocomposites [16]. These methods use biological agents, which may be plant products or microorganisms, to bring about the reduction of metal ions into nanoparticles. These compounds, *viz.* phenolics and alkaloids, in these agents have the role of natural reducers and stabilizers. Green synthesis not only involves the absence of dangerous chemicals but is also consistent with sustainable development by minimizing energy [17]. However, the use of biological sources introduces some uncertainty in the size and characteristics of resulting particles, which can be a serious issue for mass production. It has therefore secured an enviable place as an eco-friendly green synthesis methodology, utilizing plant extracts, microorganisms, and biodegradable polymers for the synthesis of magnetic nanocomposites [18].

Microemulsion Technique

The microemulsion technique is another method of preparing magnetic nanocomposites. This strategy involves the development of stable emulsions in which the use of surfactants allows the formation of magnetic nanoparticles in the nanoscale reaction vessel. The advantage of this method is that it is possible to exert exact control on the size and the shape of the nanoparticles, which is useful in applications where homogeneity is expected. However, problems like contamination that arise with the addition of surfactants need a lot of purification steps to clear up.

Chemical Vapor Deposition (CVD)

Thin films and coatings are well produced using the Chemical Vapor Deposition (CVD) technique. This gas-phase technique also results in products of enhanced purity and crystallinity of the nanocomposites. An essential precursor gas breaks down on a hot surface, which leads to the formation of magnetic nanoparticles. CVD, therefore, provides a thin layer with good control of its thickness, a property that makes CVD especially suitable in electronic and magnetic devices. However, it costs dramatically more than other methods, as well as requires highly specialized equipment.

Flame Spray Pyrolysis (FSP)

Flame Spray Pyrolysis, or FSP, is a fast and easy method of preparing magnetic nanocomposites. This is a method in which a focused solution of metal precursors is burned in a high-temperature flame, decomposing into nanoparticles. It was also reported that FSP is highly efficient in making nanoparticles with good control over size and composition. However, its high energy demand and other issues, such as difficulty in preventing particle agglomeration, can also affect its general viability at the industrial scale.

Surface Functionalization and Post-synthesis Modification

Secondary synthesis processes, including surface modification of magnetic nanoparticles, are instrumental in controlling the properties of the resultant nanocomposites. They include salinization, polymer coating, and ligand exchange that enhance stability, biocompatibility, and functionality and expand the use of these materials in biomedical engineering and environmental conservation.

Thus, the available synthesis techniques for magnetic nanocomposites have improved greatly, and their application potential has expanded even further due to tunable material characteristics. Each of the three methods has its strengths and weaknesses, but further development of the technology and increased interdisciplinary collaboration have helped to work through these problems. The sustainability of nanocomposites and the application of artificial intelligence models may be used for the subsequent generation of magnetic nanocomposites.

CHARACTERIZATION TECHNIQUES

Evaluation and identification of the characteristics of magnetic nanocomposites are important to determine the structural, morphological, magnetic, thermal, and chemical properties, controlling the utilization of nanocomposites in numerous applications. Current scientific advances offer a much deeper understanding of these properties, and nanocomposites can be designed to fit certain requirements. We hereby describe the commonly employed characterization techniques and their relevance in the study of magnetic nanocomposites.

Structural Characterization

X-Ray Diffraction (XRD)

Crystal structure, phase composition, and crystallinity, which are important characteristics of magnetic nanocomposites, are determined using X-ray diffraction. The forthcoming diffraction patterns are used to ascertain the crystal structure and lattice constants.

Application: XRD is especially important for the identification of certain phases like magnetite Fe_3O_4 or maghemite $-\gamma$-Fe_2O_3 and to ensure proper incorporation of magnetic nanoparticles into the composite material [19].

Significance: It also enables the determination of the purity and stability of nanocomposites using different synthesis parameters.

Fourier Transform Infrared Spectroscopy (FTIR)

Through vibrational mode analysis, FTIR spectroscopy determines functional groups and chemical bonds in the nanocomposites that produce the spectra.

Application: The analytical applications of this technique mainly have their advantages in verifying the existence of organic coatings, polymer matrices, or even surface functionalization on magnetic nanoparticles [20].

Significance: This technique is very useful in explaining how nanoparticles respond to composite matrices.

Morphological Characterization

Scanning Electron Microscopy (SEM)

SEM produces clear images of the surface features of Mg-Al Co nanostructures, including surface roughness and topography.

Application: It is used extensively in the analysis of the distribution of nanoparticles within the polymer matrix as well as in the general surface morphology of the composite.

Significance: SEM is useful in examining defects, aggregation, and the distribution of nanoparticles and how these are likely to affect the performance of the material [21].

Transmission Electron Microscopy (TEM)

A detailed understanding of the internal structures, size, and shape of the nanoparticles can be done with the help of TEM.

Application: This technique directly gives proof of the dispersion of nanoparticles, interaction on the interface, and the core-shell arrangement in magnetic nanocomposite materials.

Significance: In volumetric and uniformity aspects, especially involving substructures such as nanoscale dimensions of magnetic nanoparticles within the composite matrix, the role of TEM is significant.

Magnetic Characterization
Vibrating Sample Magnetometry (VSM)

VSM determines the saturation magnetization, coercivity, and remanence of the nanocomposites subjected to an applied field of varying magnitude.

Application: This technique assesses the superparamagnetic properties of nanoparticles and how they engage with the external fields.

Significance: $Na0.5Bi0.5TiO_3$ and $BaTiO_3$-XNx thin films are fabricated and characterized by *vs*M, which aids in enhancing the applicability of nanocomposites in magnetic storage, drug delivery, and hyperthermia [22].

Superconducting Quantum Interference Device (SQUID)

SQUID is an excellent instrument for recognizing weak magnetic fields in the nanocomposite system.

Application: They are widely applied in revealing low-temperature magnetic characteristics of a material, such as the blocking temperature and spin-glass effects.

Significance: SQUID gives additional information concerning the magnetic coupling at the nanoscale level, which is inevitable for most applications.

Thermal Characterization

Thermogravimetric Analysis (TGA)

TGA is used to determine weight changes in nanocomposites as a function of temperature, giving information on thermal stability and composition.

Application: It is widely employed to study the thermal degradation of polymer matrices and the stability of surface coatings on nanoparticles.

Significance: TGA assists in the determination of the operating temperature range for the nanocomposites [23].

Differential Scanning Calorimetry (DSC)

DSC monitors mass-temperature data during heating and cooling, allowing one to observe phase transformations, such as melting and crystallization, in nanocomposites.

Application: It is used to investigate thermal changes in the polymer matrix and the effect of nanofillers on these changes.

Significance: Thermal characterization of nanocomposites under operational conditions of DSC helps to identify their thermal characteristics.

Surface and Chemical Analysis

X-ray Photoelectron Spectroscopy (XPS)

XPS is useful for determining the elemental and chemical constitution of the surface and depth profile of nanocomposites.

Application: XPS is used to determine the oxidation state of the magnetic nanoparticles as well as the type of bonding that is present between the nanoparticles and the matrix [24].

Significance: XPS provides clearer insight into the nature of interactions at the surface, which is useful in catalysis and drug delivery.

Energy Dispersive X-ray Spectroscopy (EDS)

EDS used in conjunction with SEM or TEM provides elemental information about nanocomposites.

Application: EDS can be employed to verify the elemental constitution of magnetic nanoparticles as well as the configuration of a component.

Significance: EDS confirms that the number concentration of the elements matches the design of the material as expected.

Other Characterization Techniques

Dynamic Light Scattering (DLS)

DLS size distribution concentrates on the hydrodynamic size of particles in a solution.

Application: It is particularly applied to assess the dispersion stability of nanocomposites in the liquid phase.

Significance: DLS is highly useful when applied to biomedicine because colloidal stability affects application properties [25].

Zeta Potential Analysis

This technique determines the surface charge of nanoparticles, which gives an understanding of their stability in a colloidal system.

Application: It is employed for surface properties enhancement of nanocomposites for biosensory applications and environments.

Significance: Zeta potential analysis plays an important role in avoiding the problem of particle clumping and achieving proper distribution.

APPLICATIONS OF MAGNETIC NANOCOMPOSITES

Biomedical Applications

Magnetic nanoparticles such as Fe_3O_4 are effective drug delivery agents; however, when they are coated with a biocompatible polymer, they can selectively deliver drugs to target tissues using an external magnetic field [26]. Various resorbable composites, along with magnetic particles, are also used for targeted drug deliveries [36 - 40]. Superparamagnetic nanocomposites increase MRI sensitivity by changing the relaxation time of hydrogen protons nearby [27].

Environmental Applications

Magnetic nanocomposites are suitable for the removal of heavy metals and organic pollutants because of the large surface area of the nanocomposites and their magnetic nature that enables their recovery easily [28].

Catalysis

Surface-modified magnetic nanoparticles are immensely useful in the catalysis of environmental remediation procedures.

Energy Applications

In lithium-ion batteries and supercapacitors, magnetic nanocomposites have been incorporated to improve energy density and stability [29, 30]. The use of magnetic materials in thermomagnetic appliances is seen as having a promising future for efficient cooling apparatus.

Electronics

Small sensors and data storage components, in terms of magnetic field and data storage, make use of the nano characteristics of magnetic nanocomposites [31].

CHALLENGES AND LIMITATIONS

Although magnetic nanocomposites has promising potential in various fields, their synthesis, characterization, and application also have many challenges and limitations [32 - 35]. Incidentally, these limitations prevent them from being deployed more often and optimizing their performance in cutting-edge domains, such as biomedicine, decontamination, and energy storage [36 - 38]. Meeting these demands requires research and development in materials science, engineering, and computational modeling to work on prevailing hurdles. A major concern that often arises during the synthesis of magnetic nanocomposites is how to guarantee proper dispersion of the magnetic nanoparticles within the host matrix [39, 40]. Magnetic nanoparticles require coordination and dispersion because their large dipole-dipole interaction forces and van der Waals interactions cause them to clump together, leading to a nonhomogeneous response and inferior overall functionality. These agglomeration states are also characterized by low magnetic properties and mechanical strength, which is important in applications that call for fine accuracy [41]. Another recurrent problem is the control of the particle size and morphology during synthesis. Superparamagnetism, coercivity, and saturation magnetization are thus highly dependent on the size and shape of the nanoparticles. Reproducibility of the method is critical as any changes in synthesis parameters, such as temperature, pH, concentration of the precursor, and time taken for the reaction, are likely to have an impact on the properties of the resulting material. However, the dependence of the magnetic properties of magnetic nanoparticles on their degree of crystallinity is a critical factor in determining the value of this parameter. However, some of the synthesis methods introduced here involve post-synthesis treatments such as annealing, which additionally lead to changes in the material morphology or the appearance of unwanted phase transformations. The limitability problem related to scalability is another major limitation, especially for complex synthesis procedures like hydrothermal ones, sol-gel methods, or Chemical Vapor Deposition (CVD). Although these methods yield good quality material, the scalability is constrained by high costs, longer reaction time, and the requirement for sophisticated equipment. While green synthesis methods are environmentally friendly, it is limited by high variance in biological precursors, resulting in unequal particle size and characteristics.

Comprehensive studies on the structure and properties of magnetic nanocomposites are therefore required to fill this huge gap. Nevertheless, many characterization techniques analyzed in the frame of classical systems show some constraints when it comes to the equating of nanoscale systems. For example, methods such as X-Ray Diffraction (XRD) and Fourier transform infrared spectroscopy (FTIR) may not provide sufficient capabilities to distinguish differences in surface chemistry or the interface between nanoparticles and the matrix. It is even more difficult to characterize magnetic properties at the nanoscale. Measuring techniques like vibrated sample magnetometers and superconducting quantum interference devices yield additional insight, but the tools employed in exercises like these are only available to a select group of people with knowledge in this area. Furthermore, in cases involving multi-phase samples or multi-modal particle size distributions, the task of analyzing the data for the correlation of experimental work with the theoretical models is quite complicated [42]. One disadvantage of magnetic nanocomposites is that many of them undergo stability problems during synthesis, storage, and use. Actually, with time, particle degradation through physical oxidation or any other change in the particle surface in aggressive environments negatively influences the magnetic characteristics and the structure of the particles. For instance, the stability of iron-based nanoparticles is not acceptable due to their tendency to oxidize, resulting in reduced saturation magnetization and weak results in magnetic separation or biomedical applications [43]. Dispersed in aqueous settings, magnetic nanoparticles may coagulate and release toxic ions, threatening both biocompatibility and the environment. Stabilization at the surface through functionalization and coatings helps increase stability, but at the same time, it also increases the difficulty of synthesis, and properties such as thermal conductivity or magnetic character may suffer.

In various applications of magnetic nanocomposites, each application presents its own challenges that must be addressed to achieve the optimal value. Key issues when working with biopolymers in biomedical applications are biocompatibility and toxicity. However, such magnetic nanoparticles, including magnetite (Fe_3O_4), are generally considered biocompatible but contain PAH cytotoxicity if synthesized or surface functionalized improperly. The challenges must be addressed to ensure quality as well as safe and constant delivery for imperative uses such as targeted medication delivery methods [42]. Several issues exist in the environmental remediation of MMMs; for instance, to ensure the long-term usage of magnetic nanocomposites for the said purposes, the efficiency of such a medium should not diminish with the number of used/reused cycles. Magnetic nanocomposites employed for the elimination of pollutants from wastewater may have reduced magnetic and adsorptive efficiency after several regeneration cycles. Further, collecting nanocomposites after treating water or soil requires efficient magnetic removal, which often involves expensive and energy-consuming equipment. In energy storage and conversion, including lithium-ion batteries or magnetic sensors, some of the constraints include the heat resistance of the composite and the stability of the magnetic phase in cycling. These materials, therefore, have to experience high-temperature areas and mechanical stresses during use without degrading; this is a difficult prospect because of issues of balance between performance and stability of the nanocomposites. It is difficult to obtain magnetic nanocomposites due to the high cost of raw materials and synthesis processes, as well as expensive characterization protocols, hence not ideal for broad use in industries. Most modern techniques, including chemical vapor deposition or flame spray pyrolysis, use expensive chemicals and equipment and hence are not feasible for large-scale production. Besides, the consequences of synthesizing magnetic nanocomposites are now having important effects on the environment. The cheapest sources of these materials involve the utilization of hazardous chemicals and energy-intensive processes, accompanied by the production of waste in the synthesis process. However, green synthesis methods aim to provide a sustainable solution, but the problems of scalability and yielding consistent results have not been fully addressed.

The synthesis of magnetic nanocomposites sometimes involves the cooperation of material scientists and engineers, chemists, and biologists. Nevertheless, a lack of overall regulation, primary protocols, and inter-disciplinary communication difficulties can negatively influence the process. By way of an example, a material may generate high-performance indices in a laboratory setting and be a candidate for an ideal material; nevertheless, the actual commercialization of the material into a product entails a long process of extensive testing, receiving permits, and enhancement of cost-effective manufacturing strategies, all of which are not only costly but also take time.

FUTURE DIRECTIONS

These challenges need to be addressed through intervention from many sectors or disciplines. Further advancements in computational modeling and artificial intelligence can improve the design and optimization of the magnetic nanocomposites, forecasting their behavior in various conditions. Moreover, it is also possible to mention that the synthesis is still in progress to achieve the best methods for the different tasks, and the development of a hybrid synthesis that combines the beneficial features of several approaches might also cope with shortcomings like scalability or reproducibility.

Reduced usage of hazardous chemicals during synthesis and the reuse of raw materials should be valued with an eye towards environmental friendliness. Studies in multiple disciplines, integrating laboratory findings with practices and standardization of procedures, can help advance the conversion of laboratory developments to industrial use. Last but not least, the integration and cooperation among academia, industry, and governmental bodies to jointly strive to improve the issues of regulations, safety, and cost will create a more effective pathway for the application of magnetic nanocomposites in a variety of applications.

CONCLUSION

Magnetic nanocomposites are the next generation of material science innovations that hold the promise of creating a revolution in multiple domains, ranging from biomedicine and environmental cleanup to energy storage and sensors. A combination of magnetic properties and adjustable functionality has rendered the latter solution indispensable for modern scientific and technological problems. Nevertheless, several emerging problems and research limitations are crucial issues that call for intentional approaches for the further evolution and use of these materials. Although the synthesis of magnetic nanocomposites is general, several challenges limit the well-dispersed nanocomposites, desired size, and batch-to-batch reproducibility. High-quality materials are obtained due to an improved synthesis process; however, scalability as well as environmental efficiency is an issue. There are also drawbacks to characterization techniques, even though they have been developedtoexcel in terms of resolution, accuracy, and ability to analyze structures at the nanoscale. These inherent limitations call for new avenues in both synthesis and characterization to address the gap between such experimental works and practical applications.

Moreover, the functional issues that may occur in each application area, including the biocompatibility of materials in the biomedical field, stability in the environment, and durability in energy storage devices, indicate the necessity of controllability and designability of magnetic nanocomposites. Still, the cost of production together with the environmental impact of synthesis processes also motivates the need for green and cheaper solutions. The above challenges need to be addressed through interdisciplinary approaches that reflect the advancements in computational modeling, machine learning, and the synthesis method hybridization. Efforts made by material scientists, engineers, chemists, and biologists shall therefore be geared towards new understandings and solutions for the challenges. Furthermore, the process of harmonizing protocols and promoting collaborations will make it possible to transfer magnetic nanocomposites from laboratories to commercial and industrial applications.

REFERENCES

[1] Nivetha, P., Siranjeevi, R., Susmitha, R., Shabnum, S.S., Raj, C.K., Benazir, K., Saravanan, A. and Vickram, A.S., 2025. A comprehensive review on advances in synthesis and characterization of nanocomposite: current status and emerging applications. Environmental Quality Management, 35(1), p.e70108.
[http://dx.doi.org/10.1016/j.jmst.2020.11.031]

[2] M. B. Gawande, A. Goswami, and T. Asefa, "Advances in magnetic nanocomposite synthesis", In: *ACS Nano* vol. 14. , 2020, no. 4, pp. 2873-2894.
[http://dx.doi.org/10.1021/acsnano.0c04563]

[3] R. Gupta, A. Jain, and A. Shrivastava, "Biomedical applications of magnetic nanocomposites: a review", In: *Adv. Biomed.* vol. 5. , 2019, no. 2, pp. 125-139.
[http://dx.doi.org/10.1016/j.ab.2018.09.011]

[4] Hosny, S., Mohamed, L.Z., Ragab, M.S., Alomoush, Q.K., Abdalla, E.M. and Aly, S.A., 2025. Nanomaterials in biomedical applications: opportunities and challenges—a review. Chemical Papers, pp.1-22.
[http://dx.doi.org/10.1201/9781003569817-3]

[5] Amponsah, O., Nopuo, P.S.A., Manga, F.A., Catli, N.B. and Labus, K., 2025. Future-oriented biomaterials based on natural polymer resources: characteristics, application innovations, and development trends. International Journal of Molecular Sciences, 26(12), p.5518.

[6] Rohith, S., Radhakrishnan, K., Dinesh, A., Sakthivel, S., Patil, R.P., Gnanasekaran, L., Mohanavel, V., Ayyar, M., Iqbal, M., Santhamoorthy, M. and Jaganathan, S.K., 2025. Review on the recent developments in magnetic nanocomposites for energy storage applications. Semiconductors, 59(1), pp.91-114.
[http://dx.doi.org/10.1016/j.matpr.2021.12.041]

[7] Urrea, C., 2025. Artificial intelligence-driven and bio-inspired control strategies for industrial robotics: a systematic review of trends, challenges, and sustainable innovations toward industry 5.0. Machines, 13(8), p.666.
[http://dx.doi.org/10.4018/979-8-3693-1277-3.ch003]

[8] Kuppusamy, S., Bhattacharjee, B., Ghose, S., Tamilanban, T., Barman, D., Ahmed, A.B. and Sahu, R.K., 2025. Current status and future prospect of bioremediation using green synthesis of nanoparticle/nanomaterials. in nanomaterials as a catalyst for biofuel production (pp. 295-327). Singapore: Springer Nature Singapore.
[http://dx.doi.org/10.1080/17518253.2022.204568]

[9] Fu, S., Dong, S., Shen, H., Chen, Z., Ma, G., Cai, M., Huang, C., Peng, Q., Bai, C., Dong, Y. and Liu, H., 2025. Multifunctional magnetic catheter robot with triaxial force sensing capability for minimally invasive surgery. Research, 8, p.0681.
[http://dx.doi.org/10.1177/09544119221143860]

[10] Skosana, S.J., Khoathane, C. and Malwela, T., 2025. Driving towards sustainability: A review of natural fiber reinforced polymer composites for eco-friendly automotive light-weighting. Journal of Thermoplastic Composite Materials, 38(2), pp.754-780.

[11] Nassar, K.I., Teixeira, S.S. and Graça, M.P., 2025. Sol–gel-synthesized metal oxide nanostructures: advancements and prospects for spintronic applications—a comprehensive review. Gels, 11(8), p.657.

[12] Y. O. Waidi, R. Barua, and S. Datta, "Metals, polymers, ceramics, composites biomaterials used in additive manufacturing for biomedical applications", In: *Adv. Chem. Mater. Eng.*, 2023, pp. 165-184.
[http://dx.doi.org/10.4018/978-1-6684-9224-6.ch008]

[13] Dias, J.C., Marques, S., Branco, P.C., Rodrigues, T., Torres, C.A., Freitas, F., Evtyugin, D.V. and Silva, C.J., 2025. Biopolymers derived from forest biomass for the sustainable textile industry. Forests, 16(1), p.163.
[http://dx.doi.org/10.3390/polym14040692]

[14] Nakao, S., Tadano, K. and Sonoda, K.H., 2025. Advancements in robotic surgery for vitreoretinal diseases: current trends and the future. Japanese Journal of Ophthalmology, pp.1-12.
[http://dx.doi.org/10.4018/979-8-3693-3218-4.ch015]

[15] Pereda, E.D.C., Rojas, H.A.G. and Egea, A.J.S., 2025. Dynamic modelling of needle-tissue interaction applied to soft tissue damage during needle extraction. IEEE access. for minimally invasive surgery (MIS)., *Mater. Today Proc.*, vol. 57, pp. 259-264, 2022.
[http://dx.doi.org/10.1016/j.matpr.2022.02.498]

[16] T. Zhang, Y. Zhao, and L. Wang, "Magnetic nanocomposites for high-energy density storage devices", In: *Energy Storage Mater.* vol. 34., 2021, no. 5, pp. 265-279.
[http://dx.doi.org/10.1016/j.ensm.2021.05.019]

[17] Segneanu, A.E., Bejenaru, L.E., Bejenaru, C., Blendea, A., Mogoşanu, G.D., Biţă, A. and Boia, E.R., 2025. Advancements in hydrogels: a comprehensive review of natural and synthetic innovations for biomedical applications. Polymers, 17(15), p.2026,

[18] Xue, Y., Liu, X., Cui, X., Zhao, Y. and Lu, X., 2025. Precise magnetic tailoring in ZrO_2/Fe_3O_4-Fe/C nanocomposites via electron transfer modulation for enhanced electromagnetic wave absorption. Nano Research, 18(6), p.94907387.

[19] P. Kumar, M. Sharma, and R. Gupta, "Functional analysis of magnetic nanocomposites using spectroscopic tools", *Appl. Nanosci.*, vol. 7, no. 6, pp. 456-472, 2021.

[20] Tariq, A., Mehmood, K., Shahzad, A., Asim, H., Tariq, S. and Asif, Q.U.A., 2025. Role of Cr-dopant in tuning the properties of Mg-Ni ferrites nanoparticles synthesized by sol-gel auto combustion method for applications in electronic device. Materials Today Communications, 42, p.111504.

[21] Aksoy, Y.T., 2025. Nanofluids for sustainable heat transfer enhancement: beyond thermal conductivity. Sustainability, 17(17), p.8006,
[http://dx.doi.org/10.4018/979-8-3693-3625-0.ch008]

[22] R. Barua, "An in-depth exploration of AI and humanoid robotics' role in contemporary healthcare", In: *Adv. Med. Technol. Clin. Pract.*, 2024, pp. 42-61.
[http://dx.doi.org/10.4018/979-8-3693-2238-3.ch003]

[23] A. Mandal, T. Biswas, and R. Sinha, "Thermal and magnetic stability of polymer-based nanocomposites", *Int. J. Polym. Sci.*, vol. 12, no. 9, pp. 823-836, 2023.

[24] F. Rahman, P. Singh, and K. Banerjee, "Advanced microscopy for magnetic nanocomposites", *Nano Imaging Anal.*, vol. 8, no. 7, pp. 201-218, 2020.

[25] R. Barua, N. Biswas, and D. Das, "Emergent applications of organ-on-a-chip (OOAC) technologies with artificial vascular networks in the 21st century", In: *Adv. Healthc. Inf. Syst. Admin.*, 2024, pp. 198-219.
[http://dx.doi.org/10.4018/979-8-3693-1214-8.ch010]

[26] González Burgos, E.M., Serrano López, D.R. and Susana, Y.J., 2025. Fabrication of Organ-on-a-Chip using Microfluidics.

[27] R. Barua, Advanced biomimetic compound continuum robot for minimally invasive surgical applications. Modeling, simulation, and control of AI robotics and autonomous systems.

[28] Dutta, S., Sarma, D.K., Vora, J., Chaudhari, R., Bhowmik, A., Samal, P. and Khanna, S., 2025. A State-of-the-Art review on Micro-Machining of nitinol shape memory alloys and optimization of process variables considering the future trends of research. Journal of Manufacturing and Materials Processing, 9(6), p.183.
[http://dx.doi.org/10.4018/978-1-6684-9385-4.ch006]

[29] Z. Zhou, F. Liu, and W. Chen, "Surface analysis techniques for next-generation magnetic nanocomposites", *Surf. Sci. Adv.*, vol. 6, no. 4, pp. 321-338, 2021.

[30] Li, Y., Ling, W., Yang, J. and Xing, Y., 2025. Risk assessment of microplastics in humans: distribution, exposure, and toxicological effects. Polymers, 17(12), p.1699.
[http://dx.doi.org/10.4018/978-1-7998-9723-1.ch007]

[31] S. K. Verma, A. Prasad, and V. Katiyar, "State of art review on sustainable biodegradable polymers with a market overview for sustainability packaging", *Mater. Today Sustain.*, p. 100776, 2024.

[32] Jiang, Y., Kyeremeh, J., Luo, X., Wang, Z., Zhang, K., Cao, F., Asciak, L., Kazakidi, A., Stewart, G.D. and Shu, W., 2025. A numerical simulation study of soft tissue resection for low-damage precision cancer surgery. Computer Methods and Programs in Biomedicine, p.108937.
[http://dx.doi.org/10.1177/09544119221122024]

[33] Rumon, M.M.H., Rahman, M.S., Akib, A.A., Sohag, M.S., Rakib, M.R.A., Khan, M.A.R., Yesmin, F., Shakil, M.S. and Rahman Khan, M.M., 2025. Progress in hydrogel toughening: Addressing structural and crosslinking challenges for biomedical applications. Discover Materials, 5(1), p.5.

[34] Samaranayake, P., Nagalingam, S., Parkes, J. and Jagoda, K., 2025. Advancing sustainable manufacturing with industry 4.0 for enhanced performance: a conceptual framework and empirical validation. Available at SSRN 5219287.

[35] Wang, F., Long, X. and Han, Y.C., Hot-press molding preparation and properties of poly (lactic acid)/nano-hydroxyapatite composites. Polym. Bull, pp.9-11.

[36] Plaass, C., Reifenrath, J. and Richter, A., 2025. Innovative Fixation methods of osteotomies for hallux valgus correction. Foot and Ankle Clinics, 30(2), pp.269-283.

[37] Yakoubi, S., 2025. Sustainable revolution: AI-driven enhancements for composite polymer processing and optimization in intelligent food packaging. Food and Bioprocess Technology, 18(1), pp.82-107.

[38] Madhu, K., Dhal, M.K., Banerjee, A., Katiyar, V. and Kumar, A., 2025. Melt-processed cast films of calcite reinforced starch/guar-gum biopolymer composites for packaging applications. Journal of Materials Science, 60(5), pp.2689-2708.

[39] A. Prasad, S. Datta, S. De, P. Singh, and B. Mahto, Bioresorbable composite for orthopedics and drug delivery applications.*Applications of Biotribology in Biomedical Systems*. Springer Nature Switzerland: Cham, 2024, pp. 327-344.

[40] Zhu, Y., Yin, Y., Shen, Z., Zhao, Z., Song, H., Wang, S., Shen, D. and Wang, Q., 2025. UniCAD: Efficient and extendable architecture for multi-task computer-aided diagnosis system. arXiv preprint arXiv:2505.09178.
[http://dx.doi.org/10.4018/978-1-6684-4671-3.ch002]

[41] Sharma, R., Kumar, H., Kumari, R., Kumar, G., Swami, B., Kumar, A., Rani, G. and Kumar, R., 2025.

Next-generation nanocomposites: Optimizing Al$_2$O$_3$-CuO-ZnO and reduced graphene oxide for enhanced performance. Next Nanotechnology, 7, p.100119.

[42] Lutz, T.M., De Breuck, J., Salehi, S. and Leiske, M.N., 2025. Smart, bio-inspired polymers and bio-based molecules modified by zwitterionic motifs to design next-generation materials for medical applications. Advanced Functional Materials, p.e13765.
[http://dx.doi.org/10.4018/978-1-6684-9224-6.ch003]

CHAPTER 2

Processing, Characterization, and Classification of Polymer Nanocomposites in Sustainable and Renewable Energy Sector

Ranjit Barua[1]**, Deepanjan Das**[1]**, Arbind Prasad**[2,*]**, Sachin Latiyan**[3]**, Sudipto Datta**[3,*] **and Bidyanand Mahto**[4]

[1] *Mechanical Engineering Department, Omdayal Group of Institutions, Howrah 711316, West Bengal, India*

[2] *Mechanical Engineering Department, Katihar Engineering College (Under Department of Science, Technology and Technical Education, Government of Bihar), Katihar 854109, Bihar, India*

[3] *Department of Materials Engineering, Indian Institute of Science, Bangalore 560012, Karnataka, India*

[4] *Government Engineering College, Vaishali (Under Department of Science, Technology and Technical Education, Government of Bihar), Vaishali 844115, Bihar, India*

Abstract: The potential of Polymer Nanocomposites (PNCs) to enhance energy systems' performance and efficiency while providing sustainability advantages has drawn a lot of interest in the field of renewable energy. These materials' mechanical, thermal, and electrical properties are improved by the integration of nanoparticles into a polymer matrix. The usefulness of polymer nanocomposites is greatly influenced by their processing, with methods like melt mixing, solution casting, and in situ polymerization being often employed. The structural, morphological, and mechanical characteristics of these materials are evaluated using characterization techniques such as spectroscopy, microscopy, and thermal analysis. PNCs are also categorized according to their composition and use, with an emphasis on technology for energy conversion, storage, and harvesting. PNCs have been investigated for use in solar cells, batteries, fuel cells, and supercapacitors in the field of renewable energy, which will improve durability, energy efficiency, and environmental sustainability. The production, characterization, and classification of polymer nanocomposites are covered in this paper, along with how they contribute to the development of sustainable and renewable energy systems. PNCs' success in the renewable energy industry depends on resolving issues of cost-effectiveness, reliability, and scalability.

---

[*] **Corresponding authors Arbind Prasad:** Mechanical Engineering Department, Katihar Engineering College (Under Department of Science, Technology and Technical Education, Government of Bihar), Katihar 854109, Bihar, India; E-mail: arbind.iitg@gmail.com;
Sudipto Datta: Department of Materials Engineering, Indian Institute of Science, Bangalore 560012, Karnataka, India; E-mail: sudiptodatta1990@gmail.com

Sushil Kumar Verma, Sonika & Arbind Prasad (Eds.)
All rights reserved-© 2026 Bentham Science Publishers

Keywords: Characterization methods, Energy conversion, Energy storage, Nanomaterials, Polymer nanocomposites, Processing techniques, Renewable energy applications, Renewable energy, Sustainable energy, Sustainable materials.

INTRODUCTION

The mechanical, thermal, and electrical properties of Polymer Nanocomposites (PNCs), a class of sophisticated materials, are greatly improved by the addition of nanoscale fillers to a polymer matrix [1]. Because of their potential to enhance the sustainability and efficacy of energy systems, these materials have garnered a lot of interest across several industries, including renewable energy [2]. PNCs offer improved efficiency, lightweight constructions, and ecologically favorable features in a variety of renewable energy applications, including energy storage, energy conversion, and energy harvesting technologies [2]. To modify polymer nanocomposites' characteristics to satisfy the unique requirements of renewable energy applications, processing is essential [3]. To successfully incorporate nanoparticles into polymer matrices, a variety of processing methods have been investigated, including melt blending, solution casting, and *in situ* polymerization [4]. These techniques have an impact on how nanofillers disperse, which has a direct impact on the material's overall performance [5]. For example, the mechanical strength and electrical conductivity of the composites—both crucial for energy applications, including solar cells, batteries, and supercapacitors—can be greatly impacted by the shape and distribution of nanofillers inside the polymer matrix [6].

Understanding the structure-property relationship of polymer nanocomposites is crucial for characterization, which enables researchers to evaluate the materials' suitability for a range of applications [7]. The morphology, crystallinity, and thermal stability of the nanocomposites are frequently examined using methods including Scanning Electron Microscopy (SEM), Transmission Electron Microscopy (TEM), X-Ray Diffraction (XRD), and Thermogravimetric Analysis (TGA) [8]. The accurate assessment of PNCs' mechanical, chemical, and physical characteristics is made possible by these characterization technologies, guaranteeing their efficient deployment in energy-related applications [9]. The classification of polymer nanocomposites is often based on the composition, the kind of nanofillers, and the application for which the nanocomposite is intended [10]. PNCs have been studied for application in energy harvesting gadgets, energy conversion technologies like organic photovoltaics, and energy storage systems such as lithium-ion batteries and supercapacitors in the field of renewable energy [11]. It is established that the introduction of nanoparticles such as metal oxides, inorganic nanoparticles, and carbon structures (graphene, carbon nanotubes) into the polymer matrix increases the strength, conductivity, and thermal stability of the material, whichis essential for enhancing the relevant device performance and energy conversion efficiency [12]. With an emphasis on their uses in energy storage, conversion, and harvesting technologies, this study attempts to investigate the processing, characterization, and classification of polymer nanocomposites in the renewable energy field. The difficulties with PNCs' scalability, stability, and cost-effectiveness will also be covered, along with potential future research and development avenues. Several kinds of nanofillers for nanocomposites are made of polymers, in which nanofillers are mainly classified into organic and inorganic composites [13].

PROCESSING POLYMER NANOCOMPOSITES

Polymer Nanocomposites (PNCs) are a class of sophisticated materials that improve the properties of polymers by combining them with nanoscale fillers. Their processing, characterization, and classification are critical to their use in renewable energy systems [14]. Optimizing their performance for particular uses, such as energy harvesting, energy conversion, and storage, requires an understanding of these factors [15]. A discussion of several processing, characterization, and classification-related topics is provided here, along with the relevant references [16].

Processing of Polymer Nanocomposites

The structural and functional characteristics of polymer nanocomposites are significantly influenced by their processing [17]. Nanomaterials are frequently added to polymer matrices using a variety of processing techniques, including melt mixing, solution casting, and *in situ* polymerization [18]. The performance of the composite material is directly impacted by each method's effects on the homogeneity and dispersion of the nanofillers [19]. Green polymeric nanocomposites for supercapacitors can be seen widely in the literature [20].

a. **Melt Blending:** By combining the polymer with molten nanoparticles, this method enables the filler to be evenly distributed throughout the polymer matrix [19]. Its ease of use and scalability make it one of the most used techniques for processing PNCs. However, controlling the dispersion quality of nanofillers can be difficult [21].

b. **Solution Casting:** This entails adding the nanomaterials after the polymer has been dissolved in an appropriate solvent. After that, the solution is dried and cast into a thin layer. For the production of thin, flexible films for energy devices like solar cells and supercapacitors, this technique offers improved control over filler dispersion [22].

c. **In-Situ Polymerization:** This approach can also provide strong interface adhesion between the filler and the matrix, where the monomer may polymerize in the presence of nanofillers [23]. It leads to good dispersion of nanofillers and enhances the composite's mechanical and thermal properties, which makes them suitable for high performance in energy storage systems. Meanwhile, the following technique ensures that the monomer is polymerized in the presence of a nanofiller and a strong interaction between the filler and the matrix [24]. It results in self-exfoliated nanofillers and improves the composites' mechanical and thermal performance, thereby finding applications in high-end energy storage systems. Table 1 shows the summary of processing, characterization, and classification of polymer nanocomposites.

Table 1. Aspects, description, techniques, and applications of polymeric nanocomposites.

Aspect	Description	Techniques/Materials	Application
Processing	Techniques to incorporate nanofillers into polymer matrices.	Melt blending, solution casting, in situ polymerization.	Energy storage, energy conversion.
Characterization	Methods used to assess the structure-property relationship of PNCs.	SEM, TEM, XRD, TGA.	Morphology, thermal stability, conductivity.
Classification	Categorization of PNCs based on composition, nanofillers, and application.	Organic/organic, organic/inorganic, carbon-based, metal oxides, clay.	Energy storage, energy harvesting, fuel cells.

Characterization of Polymer Nanocomposites

Characterization is essential to understanding the structure-property relationships of polymer nanocomposites, ensuring that they meet the required specifications for renewable energy applications. Several techniques are employed for the characterization of PNCs, including:

a. **Scanning Electron Microscopy (SEM):** SEM is commonly used to examine the morphology and dispersion of nanofillers within the polymer matrix. It provides high-resolution images that allow for the assessment of filler distribution and particle aggregation [25]. The graphene quantum dots from the basis of green nanocomposites can be studied widely in the literature [23].
b. **Transmission Electron Microscopy (TEM):** TEM offers higher resolution compared to SEM and is used for observing the internal structure of PNCs at the nanoscale. It can provide insights into the interaction between the nanofillers and the polymer matrix [2].
c. **X-Ray Diffraction (XRD):** XRD is used to analyze the crystallinity of the polymer matrix and the nanofillers. The diffraction patterns help determine the

degree of crystalline order in the composite and how the nanofillers influence the polymer's crystallization [26].

d. **Thermogravimetric Analysis (TGA):** TGA measures the thermal stability of PNCs by monitoring the weight loss as the material is heated [27]. This technique is essential for assessing the thermal properties of PNCs, which are particularly important for applications in energy storage and conversion devices [3].

Classification of Polymer Nanocomposites

Polymer nanocomposites can be classified based on their composition, the type of nanofillers used, and the intended application. The classification is important for selecting the right type of nanocomposite for specific renewable energy applications.

a. Based on Composition:
  i. **Organic/Organic PNCs:** These composites use both polymer matrices and organic nanofillers such as graphene or carbon nanotubes. These are mainly used for applications requiring good electrical conductivity [28].
  ii. **Organic/Inorganic PNCs:** In these composites, organic polymers are combined with inorganic fillers like metal oxides or clay nanoparticles. These are often used in energy storage applications [29].
b. Based on Nanofiller Type:
  a. **Carbon-Based Nanomaterials:** Graphene, carbon nanotubes, and carbon black are used in polymer nanocomposites to improve electrical conductivity and mechanical strength [30].
  b. **Metal Oxides:** Metal oxide nanoparticles, such as titanium dioxide (TiO_2) and zinc oxide (ZnO), are incorporated into polymer matrices for applications in energy conversion and storage [31].
  c. **Clay Nanocomposites:** Clay nanoparticles are often used in PNCs due to their ability to enhance barrier properties and thermal stability, making them useful in energy applications like fuel cells [3].
c. Based on Application:
  a. **Energy Storage:** PNCs are often integrated into energy storage systems such as lithium-ion batteries and supercapacitors because of enhanced conductivity and mechanical strength [32].
  b. **Energy Conversion:** PNCs are also used in energy conversion technologies like solar cells and fuel cells, where their enhanced thermal stability and electrical properties improve device efficiency [33].

APPLICATION IN HEALTHCARE

Because of its special qualities, including increased mechanical strength, biocompatibility, and the capacity to incorporate functional nanomaterials, polymeric nanocomposites, or PNCs, have drawn a lot of interest in the healthcare industry. PNCs are being used in a wide range of healthcare applications, such as wound healing, tissue engineering, medication administration, and diagnostic tools. These materials offer creative ways to address the intricate requirements of medical applications.

Drug Delivery

Polymeric nanocomposites have been widely explored for drug delivery systems due to their ability to enhance the solubility, stability, and bioavailability of therapeutic agents. The incorporation of nanomaterials such as nanoparticles, nanorods, and nanotubes into polymer matrices allows for the controlled and sustained release of drugs, targeting specific tissues or organs.

Controlled Release and Targeted Drug Delivery: PNCs can provide controlled drug release, minimizing side effects and improving patient compliance. For example, by functionalizing nanofillers (*e.g.*, silica or carbon-based materials), it is possible to target drugs specifically to diseased tissues [34]. Nanocomposites loaded with anticancer drugs have shown significant promise in targeting tumor cells with reduced toxicity to surrounding healthy cells.

Polymeric Nanocomposites for Biodegradable Drug Delivery Systems: Polymers such as Poly(Lactic-co-Glycolic Acid) (PLGA) and polycaprolactone (PCL) are frequently used as matrices for drug delivery systems. When combined with nanofillers, these systems exhibit enhanced mechanical strength and controlled degradation rates, making them ideal for sustained drug release in cancer therapy and chronic disease treatment [35].

Magnetic Nanocomposites for Targeted Drug Delivery: Magnetic nanoparticles embedded in polymer matrices can be used for the targeted delivery of drugs, using an external magnetic field to guide drug-loaded carriers to the target site. This approach has been applied to cancer therapy, where magnetic nanocomposites are employed to deliver chemotherapeutic agents directly to tumors, improving therapeutic outcomes while reducing systemic toxicity [36].

Tissue Engineering and Regenerative Medicine

Polymeric nanocomposites are also a focus of attention concerning tissue engineering applications, where they may offer support structures for cell proliferation and tissue repair. The advantages of this concept are that the qualities of the nanocomposites can be adjusted to resemble the mechanical and structural characteristics of the native tissue and thus become worthy candidates for regenerative medicine.

a. **Bone Tissue Engineering**: In bone tissue engineering, polymer nanocomposites are used as scaffolds for bone regeneration. By incorporating nanofillers such as Hydroxyapatite (HA) or bioactive glass into polymer matrices, the mechanical properties and bioactivity of the composites can be enhanced, promoting osteogenesis [15]. This approach is particularly promising for orthopedic implants and bone defect healing.
b. **Cartilage Repair**: Similarly, PNCs have shown promise in cartilage repair applications. The incorporation of nanoparticles such as silica, titanium dioxide (TiO_2), or carbon nanotubes into biodegradable polymer matrices can improve the mechanical and biochemical properties of the scaffold, supporting the growth and differentiation of chondrocytes and enhancing the repair of damaged cartilage [25].
c. **Wound Healing**: The combination of polymeric matrices and nanofillers has also been applied in wound healing. Nanocomposites are used to create wound dressings that promote faster healing, prevent infection, and provide a controlled release of antibiotics. For instance, silver nanoparticles incorporated into polymer matrices exhibit antibacterial properties, making them suitable for use in burn treatments and chronic wound care [27].

Diagnostic Tools

Polymeric nanocomposites are also widely used in diagnostics, particularly for biosensing and imaging. The inclusion of nanomaterials in polymers enhances their sensitivity and selectivity, making them useful for detecting disease biomarkers, pathogens, or other diagnostic indicators.

a. **Biosensors for Early Disease Detection**: PNCs have been incorporated into biosensors for the detection of cancer markers, pathogens, and other diseases. The use of nanocomposites in biosensors allows for rapid, sensitive, and cost-effective detection of disease biomarkers, which is essential for early diagnosis [22].
b. **Magnetic Nanocomposites in Imaging**: Magnetic nanoparticles embedded in polymeric matrices are utilized in Magnetic Resonance Imaging (MRI) and other imaging techniques. These magnetic nanocomposites provide better contrast and resolution in imaging, aiding in the detection and diagnosis of diseases such as cancer and neurological disorders [25].

Antimicrobial and Antiviral Applications

Polymeric nanocomposites have also been reported to show excellent antimicrobial and antiviral activity, and as such, this research has found these nanocomposite materials to have several medical applications, such as in drug delivery and tissue engineering. The antibacterial and antiviral effectiveness of polymers can be further improved when antimicrobial agents like silver, copper, or zinc oxide are embedded within their matrix, which is useful in avoiding microbial infections in implanted devices [31 - 35].

a. **Antimicrobial Coatings**: Polymeric nanocomposites with antimicrobial agents are widely used for coating medical devices such as catheters, prosthetics, and wound dressings to prevent infection. The sustained release of antimicrobial agents from the nanocomposites ensures long-lasting protection against bacterial growth [27].
b. **Antiviral Polymeric Nanocomposites**: With the rise of viral infections, polymeric nanocomposites have been explored as antiviral agents. The incorporation of antiviral materials into polymeric matrices enhances their effectiveness in preventing the spread of viruses, such as influenza or the novel coronavirus [21].

Polymeric nanocomposites hold significant promise in a wide range of healthcare applications due to their unique combination of polymer properties and nanomaterial enhancements. Their ability to improve drug delivery systems, facilitate tissue engineering, provide diagnostic tools, and offer antimicrobial protection makes them versatile materials for modern medicine. With continued advancements in nanotechnology and polymer science, the scope of applications for PNCs in healthcare is expected to expand, offering innovative solutions for a variety of medical challenges [36 - 40].

APPLICATION IN THE SUSTAINABLE RENEWABLE ENERGY SECTOR

Polymeric Nanocomposites (PNCs), as a new category of material, have gained significant attention for use in the sustainable renewable energy sector [41, 42]. The fact that they are lightweight, possess high mechanical strength, and can assimilate functional nanomaterials to boost the efficiency of renewable energy technologies means that more advanced PNCs can be designed [43, 44]. This discussion focuses on several case studies illustrating how PNCs are now being applied to energy generation, energy storage, and energy conservation and efficiency, showing how they can significantly contribute to future sustainable energy systems.

Photovoltaics (Solar Energy)

Polymeric nanocomposites have been revealed to possess high adaptability in the creation of innovative materials for Photovoltaic (PV) components, specifically Organic Solar Cells (OSCs). In this application, PNCs allow for the optimization of solar power conversion technology by ameliorating the charge transfer characteristics, photovoltaic response, and mechanical flexibility of the devices.

a. **Organic Solar Cells (OSCs)**: In OSCs, conducting polymers are used to form the active layer. By incorporating nanoparticles such as metal oxides (*e.g.*, TiO_2, ZnO) or carbon-based materials (*e.g.*, graphene, carbon nanotubes), the electrical conductivity and stability of the device can be significantly enhanced (Kumar *et al.*, 2020). These nanocomposites not only improve the Power Conversion Efficiency (PCE) of solar cells but also provide better environmental stability, making them suitable for large-scale applications in renewable energy systems [45].
b. **Enhancement of Light Absorption**: Metal nanoparticles such as gold (Au) and silver (Ag) have been incorporated into polymer matrices to enhance the light absorption capabilities of photovoltaic devices. These materials improve the interaction between incident light and the active layer, resulting in higher energy harvesting efficiency [46]. Such PNCs are ideal candidates for developing next-generation, flexible, and efficient solar panels.

Energy Storage (Batteries and Supercapacitors)

Polymeric nanocomposites are also widely used in energy storage devices, such as batteries and supercapacitors, which are essential components in renewable energy systems. These materials offer a unique combination of high surface area, mechanical flexibility, and electrochemical stability, which are critical for increasing the performance of energy storage technologies.

a. **Batteries:** PNCs are utilized as electrode materials in lithium-ion and sodium-ion batteries to enhance cycle stability, energy density, and charge/discharge rates. The conductivity and capacity of the electrodes can be greatly increased by adding carbon-based nanofillers, such as graphene or Carbon Nanotubes (CNTs), to polymer matrices [45]. Furthermore, the application of metal oxide nanoparticles, such as Manganese Oxide (MnO_2), can enhance the battery's overall efficiency and rate capability, improving performance in applications involving the storage of renewable energy.
b. **Supercapacitors:** Using electrostatic charge instead of electrochemical reactions, supercapacitors also use PNCs to store energy. Increased capacitance and cycle life are the outcomes of incorporating conductive nanomaterials, like graphene or carbon nanotubes, into polymers [42].

Applications in electric vehicles and renewable energy grids require these composite materials' increased surface area for charge storage, enhanced electrical conductivity, and mechanical flexibility [44].

Hydrogen Storage and Production

Hydrogen is considered a key energy carrier for a sustainable future due to its high energy density and clean combustion properties. Polymeric nanocomposites are gaining attention in hydrogen storage and production due to their ability to enhance material properties such as surface area, stability, and hydrogen absorption capacity.

a. **Hydrogen Storage**: Nanocomposites that incorporate hydrogen-absorbing materials like Metal-Organic Frameworks (MOFs) or carbon nanotubes into polymers have been shown to improve the hydrogen storage capacity. These materials provide a high surface area for hydrogen molecules to adsorb, while the polymer matrix ensures structural stability during storage [29]. As a result, PNCs are being explored as alternatives to traditional hydrogen storage methods, offering improved storage densities and safety.
b. **Hydrogen Production (Electrolysis)**: Polymeric nanocomposites are also being developed for use in water electrolysis, a process that splits water into hydrogen and oxygen using electrical energy. The incorporation of conductive nanomaterials into polymer matrices can enhance the efficiency and stability of the electrolysis process, making it more viable for large-scale hydrogen production [28].

Energy Harvesting from Environmental Sources

Energy harvesting from ambient environmental sources such as mechanical vibrations, thermal gradients, and electromagnetic radiation is a growing area of interest in the renewable energy sector. Polymeric nanocomposites can be engineered to capture and convert these forms of energy into usable electrical power, offering new opportunities for sustainable energy generation.

a. **Thermoelectric Energy Harvesting**: In thermoelectric devices, PNCs can convert heat gradients into electrical energy. By incorporating thermoelectric materials such as Bismuth Telluride (Bi_2Te_3) or Lead Telluride (PbTe) into polymer matrices, the thermoelectric performance can be significantly enhanced [46]. This approach is useful for harvesting waste heat from industrial processes, automotive engines, or even wearable devices, providing a sustainable source of energy.
b. **Piezoelectric Energy Harvesting**: Polymeric nanocomposites are also used in piezoelectric devices to convert mechanical energy from vibrations or

movements into electrical energy. By incorporating piezoelectric nanoparticles, such as Zinc Oxide (ZnO) or Polyvinylidene Fluoride (PVDF), into polymer matrices, the energy conversion efficiency can be improved [41]. These devices are particularly useful for harvesting energy from mechanical vibrations in industrial equipment or human motion in wearable applications.

Polymeric nanocomposites are playing an increasingly important role in the sustainable renewable energy sector. Their versatility and ability to combine the benefits of polymers with functional nanomaterials make them ideal candidates for a variety of applications in energy generation, storage, and harvesting. Shortly, these materials are expected to further enhance the efficiency, performance, and cost-effectiveness of renewable energy technologies, contributing to the transition to a more sustainable and environmentally friendly energy landscape.

CHALLENGES IN POLYMERIC NANOCOMPOSITES FOR RENEWABLE ENERGY

Despite the promising potential of polymeric nanocomposites in renewable energy applications, there are several challenges that need to be overcome before these materials can be widely adopted in the industry.

i. **Scalability and Cost-Effectiveness** The scalability of production techniques and related expenses are two of the main obstacles to PNC commercialization. High-performance nanocomposites are frequently synthesized using costly and intricate procedures like sol-gel processing or chemical vapor deposition, which are not necessarily practical for large-scale production [17]. Furthermore, a lot of the high-performance nanomaterials utilized in PNCs, like carbon nanotubes and graphene, are costly and challenging to manufacture in large quantities. For these materials to be widely used in renewable energy systems, scalable and affordable fabrication techniques must be developed [44]. To reduce production costs, future developments in this field will concentrate on streamlining synthesis pathways and employing more affordable and abundant nanomaterials.

ii. **Environmental Impact and Sustainability** Even though polymeric nanocomposites have many performance benefits, it is important to carefully evaluate their environmental impact, especially when it comes to applications involving renewable energy. Petroleum-based resources are the source of many of the polymers used in nanocomposites, and hazardous chemicals and solvents may be utilized in the synthesis of some nanoparticles, which could be harmful to the environment. Future studies will concentrate on creating more ecologically friendly and safe nanomaterials as well as more sustainable polymers made from renewable resources, like bio-based polymers [33]. To make sure that PNCs do not contribute to environmental pollution, it is also necessary to handle their recyclability and disposal at the end of their lifecycle.

iii. **Compatibility and Dispersion of Nanomaterials in Polymers:** Another major issue that impacts PNC performance is the dispersion of nanoparticles within the polymer matrix. Since agglomeration of nanoparticles might decrease the composite's efficacy, achieving uniform dispersion is essential for maximizing the material's attributes. A careful selection of materials and surface modification procedures is necessary to achieve excellent compatibility between the nanoparticles and the polymer matrix. To guarantee even dispersion and improved performance, future studies will probably concentrate on creating more effective techniques for functionalizing nanoparticles and enhancing their interaction with the polymer matrix [29].

iv. **Long-Term Performance and Reliability:** Ensuring PNCs' long-term performance and dependability presents another difficulty, especially in demanding renewable energy applications. Materials used in solar cells, energy harvesting systems, and energy storage devices must be able to endure mechanical stress, repetitive cycling, and exposure to the environment without losing their functionality. Future studies must concentrate on creating PNCs with improved mechanical and chemical stability to solve this problem. To increase the lifespan of renewable energy systems that depend on these composites, it will be essential to incorporate nanomaterials that provide exceptional strength, durability, and resistance to deterioration [46].

From energy generation and storage to energy harvesting and management, polymeric nanocomposites have great potential to advance renewable energy systems. The creation of multifunctional nanocomposites and integration with smart grid systems are only two of the many fascinating potential future paths; however, several obstacles need to be overcome. To fully utilize these materials in the renewable energy industry, challenges related to scalability, cost-effectiveness, environmental impact, and long-term performance must be resolved. Polymeric nanocomposites are expected to become more significant in forming a more sustainable and energy-efficient future as research advances.

FUTURE DIRECTIONS IN POLYMERIC NANOCOMPOSITES FOR RENEWABLE ENERGY

Improvement in Material Performance and Stability

In PNCs, enhancing the total material performance—especially in terms of efficiency, stability, and lifetime—is one of the most important areas for future development. For example, despite their enormous promise, Organic Solar Cells (OSCs) continue to confront difficulties because of their poor stability in the face of environmental stressors such as moisture, temperature changes, and UV radiation [6]. The durability of the devices can be increased by including functional nanomaterials, such as metal oxide nanoparticles or carbon-based materials [44]. To guarantee increased stability and endurance, especially for large-scale applications in solar energy systems, future research will probably concentrate on improving the encapsulating processes and surface modification of these nanocomposites.

Designing High-Performance Nanocomposite Electrodes for Energy Storage

Designing high-performance electrodes for energy storage systems like lithium-ion and sodium-ion batteries and supercapacitors is another exciting avenue for the future. To improve energy density, power density, and cycle stability in energy storage devices, polymeric nanocomposites can offer improved mechanical qualities, increased conductivity, and a larger surface area. Future advancements will concentrate on creating new polymer matrices and hybrid nanofillers that minimize degradation during multiple cycles of charging and discharging, thereby providing better electrochemical performance and cycle life [43]. To make these materials feasible for widespread use in energy storage devices, more economical and environmentally friendly production techniques must be developed.

Advanced Functionalization of Polymers for Multifunctional Applications

The versatility of polymeric nanocomposites will be emphasized further in the future. PNCs are naturally adaptable and may be designed to carry out a variety of tasks, which is crucial in the field of renewable energy. For example, in energy harvesting applications, materials that can transform mechanical and thermal energy into electrical energy can be created by incorporating materials such as piezoelectric and thermoelectric nanoparticles into polymer matrices. To improve the overall efficiency and sustainability of energy conversion technologies, future research will concentrate on creating hybrid materials that combine these functions to produce systems that can gather energy from numerous sources concurrently [41].

Integration with Smart Grid and IoT Systems

The relevance of polymeric nanocomposites is anticipated to grow as smart grids and Internet of Things (IoT) technologies are progressively integrated with renewable energy systems. PNCs can be designed to integrate with wearable technology, energy management systems, and sophisticated sensors to deliver real-time information on energy generation, use, and storage. For example, PNC-based energy storage devices can be incorporated into smart grid systems to increase grid stability and facilitate more effective renewable energy distribution and storage. To optimize energy flow and storage, future research will concentrate on creating sensors and actuators based on nanocomposite technology that can communicate with energy systems and offer real-time feedback [46].

Waste-to-Energy Technologies

One of the areas with a lot of potential for future energy generation is the transformation of waste materials into electricity. In waste-to-energy technologies, polymeric nanocomposites can be extremely important, particularly when used as energy-harvesting materials. PNCs, for instance, can be used to increase the efficiency of thermoelectric generators, which produce electrical energy from waste heat from domestic appliances, automobile engines, and industrial processes. The creation of thermoelectric and piezoelectric nanocomposites that effectively transform mechanical vibrations or low-grade waste heat into usable electrical energy is probably the main emphasis of future research [29].

CONCLUSION

Investigations into PNCs' potential to enable new generations of energy technologies have been triggered by their fast advancement in renewable energy applications. The use of polymeric nanocomposites has been found to enhance the performance of energy generation, storage, and conversion systems, as they are constituted from basic polymers and functional nanomaterials that offer an innovative amalgamation of lightweightness, flexibility, and tunable characteristics. Despite the progress made so far, this raises other issues about obstacles and future orientations that must be considered to achieve PNCs' potential in renewable energy.

In general, in the renewable energy field, polymer nanocomposites demonstrate a high level of efficiency as materials used for various purposes. The characterization, classification, and processing strategies covered in this research shed light on how these materials might improve energy harvesting, conversion, and storage technologies. Optimizing PNC characteristics for particular applications requires careful consideration of processing and characterization procedures. The creation and use of polymer nanocomposites will be essential to the advancement of renewable energy technology as the need for efficient and sustainable energy solutions increases.

REFERENCES

[1] Razack, R.K. and Sadasivuni, K.K., 2025. Advancing nanogenerators: the role of 3d-printed nanocomposites in energy harvesting. Polymers, 17(10), p.1367.

[http://dx.doi.org/10.1016/j.est.2020.101829]

[2] Oladele, I.O., Adelani, S.O., Taiwo, A.S., Akinbamiyorin, I.M., Olanrewaju, O.F. and Orisawayi, A.O., 2025. Polymer-based nanocomposites for supercapacitor applications: a review on principles, production and products. RSC advances, 15(10), pp.7509-7534.
[http://dx.doi.org/10.1016/j.rser.2021.110798]

[3] Adeoti, M.O., Jamiru, T., Adegbola, T.A., Suleiman, I., Abdullahi, M. and Aramide, B.P., 2025. Selection criteria of polymer nanocomposites for electrical energy storage applications: A concise review. Express Polymer Letters, 19(2).
[http://dx.doi.org/10.1166/jnn.2021.19656]

[4] Somandi, K. and Choonara, Y.E., 2025. Exploring the 3D bioprinting landscape in the delivery of active pharmaceutical compounds for therapeutic and regenerative medicine applications. Journal of Biomedical Materials Research Part B: Applied Biomaterials, 113(10), p.e35654.
[http://dx.doi.org/10.1088/1757-899x/402/1/012113]

[5] Pereda, E.D.C., Rojas, H.A.G. and Egea, A.J.S., 2025. Dynamic modelling of needle-tissue interaction applied to soft tissue damage during needle extraction. IEEE access.
[http://dx.doi.org/10.1016/j.matpr.2022.02.498]

[6] R. Barua, N. Biswas, and D. Das, "Emergent applications of organ-on-a-chip (OOAC) technologies with artificial vascular networks in the 21st century", In: *Adv. Healthc. Inf. Syst. Admin.*, 2024, pp. 198-219.
[http://dx.doi.org/10.4018/979-8-3693-1214-8.ch010]

[7] Chikwendu, O.C., Emeka, U.C. and Onyekachi, E., 2025. The optimization of polymer-based nanocomposites for advanced engineering applications. World J. Adv. Res. Rev, 25, pp.755-763.
[http://dx.doi.org/10.1016/j.compositesa.2020.105807]

[8] S. Datta, Y. O. Waidi, and R. Barua, "3D printing of bioabsorbable polymeric composites in biomedical applications", In: *Apple Academic Press eBooks*, 2025, pp. 243-266.
[http://dx.doi.org/10.1201/9781003569817-12]

[9] Zizhe, L., Rauf, S., Xu, Z., Sagar, R.U.R., Faisal, F., Tayyab, Z., Rehman, R.U., Javed, R., Surulinathan, A., Zafar, Z. and Fu, X.Z., 2025. Advanced fabrication techniques for polymer-metal nanocomposite films: state-of-the-art innovations in energy and electronic applications. Chemical Science.
[http://dx.doi.org/10.1080/15583724.2021.1916974]

[10] Zhan, F. and Zhao, L., 2025. Intelligent biobased biodegradable materials. in research and applications of bio-based degradable materials (pp. 321-329). Singapore: Springer Nature Singapore.
[http://dx.doi.org/10.4018/978-1-6684-9224-6.ch003]

[11] Gokul Eswaran, S., Rashad, M., Santhana Krishna Kumar, A. and EL-Mahdy, A.F., 2025. A comprehensive review of mxene-based emerging materials for energy storage applications and future perspectives. Chemistry–An Asian Journal, 20(4), p.e202401181.
[http://dx.doi.org/10.1039/d1ta00834d]

[12] Tadi, S.R., Shenoy, A.G., Bharadwaj, A., CS, S., Mukhopadhyay, C., Sadani, K. and Nag, P., 2025. Recent advances in the design of SERS substrates and sensing systems for (bio) sensing applications: Systems from single cell to single molecule detection. F1000Research, 13, p.670.
[http://dx.doi.org/10.4018/979-8-3693-1306-0.ch010]

[13] Nazary, A., 2025. Functionalized nanomaterials for chemiresistive gas sensors. functionalized nanomaterials for electronic and optoelectronic devices: design, fabrications and applications, pp.365-403.

[14] Hegde, K.A., Talla, G. and Gangopadhyay, S., 2025. Tool condition monitoring and hole quality analysis during micro-drilling of NiTi shape memory alloy using artificial neural network. Measurement, 253, p.117487.
[http://dx.doi.org/10.4018/979-8-3693-7250-0.ch002]

[15] Bagheri, L., Jafari-Gharabaghlou, D., Dashti, M.R. and Zarghami, N., 2025. An update on implication of POSS-based nanocomposites in bone tissue engineering: a review. Journal of Biomaterials Science, Polymer Edition, pp.1-24.
[http://dx.doi.org/10.1002/jbm.a.36652]

[16] Shen, Y., Sun, Y., Liang, Y., Xu, X., Su, R., Wang, Y. and Qi, W., 2025. Full-color peptide-based fluorescent nanomaterials assembled under the control of amino acid doping. Nanoscale Horizons, 10(1), pp.158-164.
[http://dx.doi.org/10.4018/978-1-6684-8325-1.ch009]

[17] Liu, Z., Zhang, P., Pei, H., Wang, Z., Li, L., Ma, X., Wang, J. and Huang, D., 2025. Enhance electrohydrodynamic direct-writing potential in bone tissue engineering: design innovations, multidisciplinary insight, and future direction. Advanced Functional Materials, p.e19074.
[http://dx.doi.org/10.1177/0954411919891654]

[18] Castro-Rodríguez, B., Carranza-Chávez, F.J., Beltrán, A., Zavala-Guillén, I., Sauceda-Carvajal, D., Rodríguez-Hernández, A.G., Soto-Herrera, G. and Borbón-Núñez, H.A., 2025. Evaluation of the thermal properties of water and water/ethanol carbon-based nanofluids. Fullerenes, Nanotubes and Carbon Nanostructures, pp.1-14.
[http://dx.doi.org/10.4018/979-8-3693-3625-0.ch008]

[19] Harun-Ur-Rashid, M., Foyez, T., Krishna, S.B.N., Poda, S. and Imran, A.B., 2025. Recent advances of silver nanoparticle-based polymer nanocomposites for biomedical applications. RSC advances, 15(11), pp.8480-8505.
[http://dx.doi.org/10.1039/d1bm01835k]

[20] A. Kausar, I. Ahmad, M. Maaza, M.H. Eisa, and P. Bocchetta, "Cutting-edge green polymer/nanocarbon nanocomposite for supercapacitor—state-of-the-art", *Journal of Composites Science,* vol. 6, no. 12, p. 376, 2022.

[21] Narayanan, K.B., 2025. Nanotopographical features of polymeric nanocomposite scaffolds for tissue engineering and regenerative medicine: a review. Biomimetics, 10(5), p.317.
[http://dx.doi.org/10.3389/fbioe.2021.738573]

[22] Losetty, V., Lakkaboyana, S.K., Chappidi, H.Y., Venkateswarlu, K., Trilaksana, H., Koduru, J.R., Yuzir, A., Atanase, L.I., Seepana, P.K. and Knani, S., 2025. Transformative applications of Polymer-Based metal oxide nanocomposites in medicine, industry, and environmental remediation: A review. Journal of Inorganic and Organometallic Polymers and Materials, pp.1-33.
[http://dx.doi.org/10.1002/adfm.202005776]

[23] Ezzat, H.A., Elhaes, H., Ibrahim, M.A. and Shahat, M.A., 2025. Theoretical and experimental investigation of a CuO and graphene embedded polyethylene oxide counter electrode for efficient DSSCs. Scientific Reports, 15(1), p.25049

[24] M. Okamoto, "Polymer Nanocomposites", *Eng,* vol. 4, no. 1, pp. 457-479, 2023.
[http://dx.doi.org/10.3390/eng4010028]

[25] Das, A., Sengupta, P., Chatterjee, S., Khanam, J., Mondal, P.K., Romero, E.L., Manakhov, A.M., Thomas, S., Mahmood, S. and Ghosal, K., 2025. Development and evaluation of magnetite loaded alginate beads based nanocomposite for enhanced targeted analgesic drug delivery. Magnetochemistry, 11(2), p.14.
[http://dx.doi.org/10.1016/j.msec.2020.110863]

[26] Nayem, N.I., Hossain, M.S., Rashed, M.A., Anis-Ul-Haque, K.M., Ahmed, J., Faisal, M., Algethami, J.S. and Harraz, F.A., 2025. A sensitive and selective electrochemical detection and kinetic analysis of methyl parathion using Au nanoparticle-decorated rGO/CuO ternary nanocomposite. RSC advances, 15(19), pp.15348-15365.

[27] Tiwari, S., Jain, N., Bhosle, S., Ganure, K.A. and Chandola, M., 2025, February. Biodegradable Polymeric Nanocomposite for Wound Healing Application. In Macromolecular Symposia (Vol. 414, No. 1, p. 2400181).

[http://dx.doi.org/10.1002/app.48555]

[28] J. Feng, Y. Wang, and J. Liu, "Polymeric nanocomposites for hydrogen production via electrolysis", *Journal of Energy Chemistry,* vol. 52, pp. 286-299, 2021.
[http://dx.doi.org/10.1016/j.jechem.2020.09.023]

[29] Bannenberg, L.J., Krishnan, G., Boshuizen, B. and Schreuders, H., 2025. Palladium-PTFE metal–polymer nanocomposite film produced by cosputtering for hydrogen sensing applications. ACS Applied Energy Materials, 8(9), pp.5664-5674.
[http://dx.doi.org/10.1016/j.mtener.2020.100485]

[30] Omar, I.M., Emran, K.M. and Ali, S.M., 2025. Structural applications of polymer and two-dimensional nanocomposites: a critical review. Polymers and Two-Dimensional Nanocomposites, pp.567-584.
[http://dx.doi.org/10.1016/j.mser.2019.100514]

[31] Law, J., Stickley, E., Looi, T., Diller, E. and Podolsky, D., 2025. A novel articulating bone cutting tool for minimally invasive craniosynostosis surgery. Journal of Medical Devices, 19(4), p.041004.
[http://dx.doi.org/10.1177/09544119221143860]

[32] S. K. Verma, A. Prasad, and V. Katiyar, "State of art review on sustainable biodegradable polymers with a market overview for sustainability packaging", *Mater. Today Sustain.,* p. 100776.

[33] Xie, S., Liu, H., Yang, K., Wang, T., Zhang, H. and Li, Z., 2025. Versatile copper-chalcogenide-based nanoparticles for the treatment of brain diseases. Nano Biomedicine & Engineering, 17(1).
[http://dx.doi.org/10.1201/9781003569817-3]

[34] Hamzehlouy, A., Zarei, S., Salah Othman, R., Shahi, F., Afshar, H., Afshar Taromi, A. and Khonakdar, H.A., 2025. Recent advances in biomedical applications of mxene-integrated electrospun fibers: A Review. Polymers for Advanced Technologies, 36(2), p.e70112.

[35] Azad, M.A., 2025. Evaluating the role of lean manufacturing in reducing production costs and enhancing efficiency in textile mills. Authorea Preprints.

[36] A. Prasad, S.M. Bhasney, V. Prasannavenkadesan, M.R. Sankar, and V. Katiyar, "Polylactic acid reinforced with nano-hydroxyapatite bioabsorbable cortical screws for bone fracture treatment", *J. Polym. Res.,* vol. 30, no. 5, p. 177, 2023.

[37] Kaur, H., Garg, K., Athwal, S., Thakur, R., Singh, S., Singh, S., Bariki, R. and Mohan, C., 2026. Development of Biodegradable Materials and Advanced Coating Technologies for Biomedical Devices. In Advanced Manufacturing Technologies in Biomedical Science (pp. 92-115). CRC Press.

[38] Panjabrao, A.S., Dash, K.K., Kathuria, D., Shams, R., Chavan, P., Mukarram, S.A. and Kovács, B., 2025. Sustainable 3D-Printed food packaging from agricultural waste: A review of materials, properties, and applications. Journal of Agriculture and Food Research, 22, p.102061.

[39] A. Prasad, A. Kumar, K.K. Gajrani, Ed., *Biodegradable Composites for Packaging Applications.* CRC Press, 2022.

[40] Latibjonov, A., 2025. Biodegradable materials in orthopedics. International Journal of Artificial Intelligence, 1(4), pp.222-224.

[41] L. Liu, Z. Li, and J. Zhang, "Recent advances in polymer-based nanocomposite materials for solar energy applications", *Nano Energy,* vol. 80, p. 105561, 2021.
[http://dx.doi.org/10.1016/j.nanoen.2020.105561]

[42] Visan, A.I., Negut, I. and Hapenciuc, C., 2025. Recent Advances in thermoelectric materials for biomedical applications: energy harvesting and wearables.
[http://dx.doi.org/10.1016/j.nanoen.2019.104362]

[43] Q. Wang, Y. Zhang, and X. Li, "Polymeric nanocomposites in lithium-ion and sodium-ion batteries: Recent developments and future perspectives", *Energy Environ. Mater.,* vol. 3, no. 2, pp. 145-158, 2020.

[http://dx.doi.org/10.1002/eem2.11147]

[44] W. Zhang, L. Wang, and Y. Li, "Supercapacitors based on polymeric nanocomposites: A review of recent advances", *Nano Energy,* vol. 82, p. 105738, 2021.
[http://dx.doi.org/10.1016/j.nanoen.2020.105738]

[45] S.H. Kim, S.W. Lee, and H. Lee, "Polymeric nanocomposites for hydrogen storage", *Mater. Today Energy,* vol. 18, p. 100485, 2020.
[http://dx.doi.org/10.1016/j.mtener.2020.100485]

[46] S.D. Sharma, R. Kumar, and S. Singh, "Thermoelectric nanocomposites for energy harvesting applications", *Nano Energy,* vol. 68, p. 104362, 2020.
[http://dx.doi.org/10.1016/j.nanoen.2019.104362]

CHAPTER 3

Piezoelectric Polymer Composites: A Comprehensive Study on Energy-Harvesting Applications

Ratnamala Ganjir[1], Nitin Kumar[2,*], Nripesh Kumar[2,3], Pankaj Verma[4] and Sonika[5]

[1] Department of Physics, Government J. Yojanandam Chhatisgarh College, Raipur 492001, Chhattisgarh, India

[2] Department of Physics, National Institute of Technology Mizoram, Aizawl 796012, Mizoram, India

[3] Department of Physics, Bharat Institute of Engineering and Technology, Hyderabad 501510, India

[4] Department of Applied Sciences, Galgotias College of Engineering and Technology, Knowledge Park-II, Greater Noida, Uttar Pradesh, India

[5] Department of Physics, Rajiv Gandhi University, Rono Hills, Doimukh, Itanagar 791112, Arunachal Pradesh, India

Abstract: Piezoelectric energy harvesting is a promising technology; basically, this technology is used to convert ambient waste energy into usable electrical energy. This technique is typically employed to transform diffuse wasted energy into usable electrical energy. For energy harvesting applications, polymeric piezoelectric composites are regarded as a key study area because they provide the convenience of mechanical flexibility. The nature of the piezoelectric phenomena, the fundamental theory underlying Piezoelectric Energy Harvesting (PEH) devices, and the configuration used to fabricate PEH are explained at the outset.

Keywords: Energy harvesting, Piezoelectric Composites, Semiconductor.

INTRODUCTION

Over the past three decades, advancements in semiconductor manufacturing technologies have resulted in remarkable progress in tiny/small devices (electronic), including portable devices such as sensors, electronics, and transmitters. These advancements have significantly expanded functionality, improved energy efficiency, and drastically reduced device sizes. Furthermore,

---

[*] **Corresponding author Nitin Kumar:** Department of Physics, National Institute of Technology Mizoram, Aizawl-796012, India; E-mail: nitinphysicskushawaha@gmail.com

the continuous improvement in battery energy density has enabled numerous devices to operate for extended periods using battery power alone. However, certain applications, like sensors installed in remote areas or within the human body, pose challenges regarding individual power connections. Consequently, batteries have been widely used for their convenience, despite their power cost drawbacks. Nevertheless, it is impractical to repeatedly replace batteries on a large scale, such as in the Trillion Sensor Universe.

Consequently, researchers and engineers have been driven to explore alternative power solutions for small electronic devices. Several potential solutions have emerged, including the following.

i. Energy Harvesting (EH): This method involves capturing and converting ambient energy sources, such as light, heat, vibration, or radio waves, into electrical energy to power the devices. Energy harvesting techniques offer the potential for a continuous or intermittent power supply, reducing reliance on batteries.
ii. Wireless Power Transfer (WPT): Technologies like inductive charging and resonant inductive coupling enable devices to receive power wirelessly, eliminating the need for physical connections or frequent battery replacements.
iii. Micro-scale Power Generation (MPG): Researchers are investigating the use of micro-scale power generators, such as micro fuel cells or micro wind turbines, to generate electricity and provide self-sustainability for small devices.
iv. Power Management and Optimization: Enhancements in power management techniques, low-power design, efficient power conversion circuits, and intelligent power management algorithms can help extend battery life and reduce overall power consumption.

These alternative power solutions offer potential pathways to overcome the limitations of battery dependence and pave the way for long-term, sustainable power solutions for small electronic devices in various applications.

Currently, numerous wireless sensor nodes rely on battery power and operate with a limited energy budget. The sheer scale of networks, consisting of thousands of physically embedded nodes, makes it impractical to continuously replace batteries [1]. Notably, the WiseNET platform, created by the Swiss Centre for Electronics and Microtechnology (CSEM) [2, 3, 4], serves as an example of such wireless sensor networks. In order to ensure the longevity of sensor nodes, it is essential to focus on the low-power characteristics of the components and the system architecture design.

ENERGY HARVESTING

Power generation, also known as energy scavenging, encompasses the process of gathering and accumulating diverse amounts of energy from the immediate surroundings. This energy can be harnessed from sources like solar radiation, thermal gradients, wind, and electromagnetic waves. The objective of power generation or energy scavenging is to capture and utilize these available energy sources, enabling a sustainable and efficient power supply for various applications. This collected energy is then transformed into electrical power, which can be stored for future utilization. Mechanical Vibration (MV), mechanical strain and stress, and thermal energy are derived from the heat sources. In addition, the biological and chemical reactions are frequently employed as power sources in this process.

Energy harvesting is not only essential for sustaining self-powered systems, but it also offers a viable and economically applied alternative to batteries. Moreover, it plays a crucial role in reducing greenhouse gas emissions and preserving the environmental situation [5]. Characteristically, an Energy Harvesting System (EHS) comprises three key components [6].

i. Energy source: This refers to the origin of the energy from which the electrical power is harvested. The energy source can either be ambient, existing in the surrounding atmosphere (such as heat, sunlight, wind), or external, involving explicitly deployed energy sources (such as vibrations, lightning, human heat) [7].
ii. Harvesting mechanism: This component encompasses the structure or mechanism responsible for converting ambient energy into electrical energy. It facilitates the extraction and transformation of energy from the chosen source.
iii. Load: The load represents the destination or recipient of the electrical energy generated through harvesting. It either consumes the energy directly or stores it for future use.

By understanding and optimizing these three components, energy harvesting systems can effectively harness available energy, reduce dependence on batteries, mitigate greenhouse gas emissions, and contribute to the sustainability of our environment.

A burgeoning area of research focuses on energy harvesting methods from road infrastructure, which involve capturing and storing lost energy from pavements for future utilization. What makes these methods particularly attractive is that they leverage existing extensive paved surfaces. Recent examples of such energy harvesting technologies include the use of conductive pipes, piezoelectric sensors, nanomaterials, thermoelectric generators, phase change specimens, barriers, induction heating methods, and so on [8, 9, 10]. These innovative approaches hold great potential in harnessing and utilizing previously wasted energy from road infrastructure, contributing to a more sustainable and efficient energy ecosystem.

The increasing importance of energy harvesting technology in the context of environmental and bodily sensing cannot be overstated. In light of this, scientists and engineers have begun focusing on the development of cutting-edge EH (Energy Harvesting) technologies that harness wasted ES (Energy Sources). These technologies aim to address the problems mentioned earlier by converting waste energy into usable electricity, providing an efficient power supply to low-power electronic devices. Energy harvesting technology offers a simplified approach to power generation and holds great potential for enabling sustainable and efficient power.

The objective of energy harvesting is to cover the lifespan of further devices and field nodes. Consequently, it helps with on-site battery charging (rechargeable), as well as old-fashioned types of equipment and supercapacitors. Various applications Utilize Energy Harvesting (EH) systems, such as car tire pressure monitors, displays, implantable medical devices, wireless weather stations, Internet of Things (IoT), Wireless Sensor Network (WSN) nodes, traffic alert signs, and numerous other devices [11, 12, 13, 14]. The EH devices, when combined with the power accumulation, modulation, and storage unit's setup, create an integrated EH (Energy-Harvesting) system. This system can achieve a range of duties, including powering multiple circuits for intermittent task applications and enabling the operation of self-powered ED (Electronic Devices).

PIEZOELECTRIC MATERIALS

The application of mechanical stress on a piezoelectric material, or conversely, the induction of mechanical stress, can result in the generation of an electrical field across its boundaries. This process is referred to as the piezoelectric effect, which was initially observed by the Curie brothers in 1880 [15].

The mechanical and electrical features of the piezoelectric specimen can be effectively characterized using two equations. These equations describe the relationship between various parameters, including Electric Displacement (D), Stress (T), Electric Field Strength (E), and Strain (S). The Permittivity Matrix (εT) represents the response of the material under a constant mechanical stress, while the Flexibility Coefficients Matrix (sE) represents the material's behaviour under a constant electric field strength. These equations incorporate both mechanical and electrical variables and can be expressed in matrix form to capture the direct and inverse effects [16, 17].

For direct effect

$D = d.T + \varepsilon T.E \{1\}$ [16, 17]

For the converse effect

$S = sE.T + dt.E \{2\}$ [16, 17]

Here, the term s refers to the elastic compliance of the specimen. The term dt refers to the transport matrix form. The terms ε and d refer to the dielectric permittivity and piezoelectric coefficient, respectively.

Piezoelectric materials can be grouped into four main classes, namely single crystals, ceramics, polymers, and polymer composites/nano-composites. Each class has its common examples, as depicted in Fig. (**1**) [18]. When selecting a piezoelectric specimen for a specific application, it is important to consider its piezoelectric features in terms of their compatibility with the desired functionality. Therefore, a broader perspective is necessary to assess the versatility of different material classes.

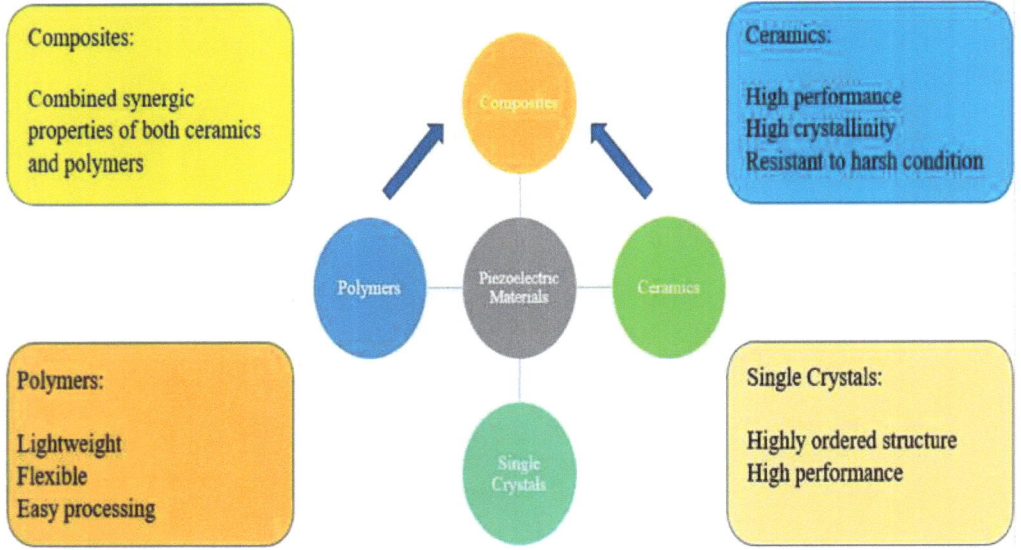

Fig. (1). Classification of piezoelectric material. Figure collected and reproduced from [18].

Ceramic piezoelectric materials, such as Lead Zirconium Titanate (PZT), can exhibit significant piezoelectric coefficients, reaching up to 500 pCN^{-1} [19]. Nevertheless, the mechanical characteristics related to the compositional behavior of ceramics may not be suitable for all applications. Fortunately, the presence of piezoelectricity is not limited to ceramics alone. Some polymeric materials also display piezoelectric properties. Although the piezoelectric coefficients of polymers are generally smaller, these materials offer greater compliance and are more easily processed.

PIEZOELECTRIC POLYMER

Piezoelectric polymer composites are materials that combine piezoelectric polymers with other substances to enhance their properties for energy harvesting applications. These composites are designed to efficiently convert mechanical energy, such as vibrations or deformations, into electrical energy through the piezoelectric effect.

The piezoelectric polymer serves as the active element that generates electrical charges when subjected to mechanical stress or strain. This polymer is typically embedded within a matrix material, such as a polymer or ceramic, which provides structural support and improves the mechanical properties of the composite.

Piezoelectric polymers, despite having a lower strain coefficient (d_{33}) compared to ceramics, exhibit excellent sensor characteristics due to their high voltage coefficients (g_{33}) resulting from their low permittivity value. These polymers are highly valuable for various energy harvesting applications due to their desirable features, including design flexibility, lightweight nature, low density, cost-effectiveness, and low refractive index [20]. Typically, these polymers are mainly semi-crystalline materials, associated with crystalline embedded sections in the amorphous matrix.

After the fabrication process involving cooling, the materials typically solidify into nonpolarizable phases. To acquire piezoelectric properties, it is necessary to stretch or draw these polymers uniaxially or biaxially, thereby converting nonpolar phases into polar ones. Alternatively, polarity can be induced through solution casting, and the alignment of dipoles can be achieved through electrode or corona poling treatment [9].

Polymers possess unique properties that differentiate them from inorganic materials, making them well-suited for niche applications in which ceramics/single crystals may not be as effective as claimed.

Piezoelectric polymers like PVDF (Polyvinylidene Fluoride) have a lower piezoelectric strain constant (d_{31}) compared to ceramics. However, they exhibit significantly higher piezoelectric stress constants (g_{31}), making them superior sensors. One of the advantages of piezoelectric polymeric sensors and actuators is their processing flexibility. Mostly, they are lightweight, durable, easily factory-made, and easily converted into complex forms. Polymers also possess notable features such as lower values of elastic stiffness, dielectric constant, and density, which contribute to their high voltage sensitivity, making them excellent sensor materials. Additionally, they exhibit low mechanical impedance and acoustic properties, which are crucial for numerous underwater applications and medical domains. Typically, polymers have high dielectric breakdown and field strength (high operating), enabling them to withstand much higher driving fields than ceramics. Furthermore, polymers have the capability to pattern electrodes on the film surface and selectively polarize specific zones. Such characteristics establish piezoelectric polymers as a distinct category for technical claims/features and facilitate the creation of useful device configurations [21].

The choice of piezoelectric polymer in the composite is crucial, as it determines the piezoelectric properties of the material. The scientific community's interest in piezoelectric polymers was sparked in 1969 with the discovery of piezoelectric effects in the synthetic polymer Poly (Vinylidene Fluoride) (PVDF) by Heiji Kawai [22] due to its high piezoelectric coefficients and excellent mechanical stability. However, other piezoelectric polymers like poly (Ethylene Terephthalate) (PET), Poly (Tetrafluoroethylene) (PTFE) [23], Polymethyldisiloxanes (PDMS) [24], Polyacrylonitrile (PAN) [25], Poly-Imides (PI) [26], Polyamides/nylons (PA) [27], and Poly (Vinyl Acetate) (PVA) [28] are also utilized. PVDF has been widely utilized in the development of piezoelectric devices for numerous applications, including piezo-speakers, musical instruments, sensors, and wearable energy harvesting devices [29, 30, 31, 32].

In semi-crystalline polymers, certain regions within the molecular chains show short- and long-term ordering, forming crystalline zones. By subjecting the material to a strong electric field at or above the glass transition temperature and then subsequently cooling it, the net dipole moment is achieved, resulting in a piezoelectric-like effect. PVDF (Polyvinylidene Fluoride) is a linear semi-crystalline polymer composed of repeating units (CH_2–CF_2), with a primarily head-to-tail structure, represented as CH_2–CF_2–(CH_2–CF_2)$_n$–CH_2–CF_2 [28]. The solid form of PVDF exhibits five crystalline phases: α phase and δ phase (TGTG') with a trans-gauche-trans-gauche configuration, β-phase (TTTT) with all trans configuration, and ϒ-phase and ε-phase (T3GT3G') [33, 34, 35, 36].

By subjecting PVDF to a durable electric field at or above the glass transition temperature, a net dipole moment can be induced, resulting in a polar phase. This polarization effect can be preserved by cooling the specimen, creating piezoelectric-type behaviour. PVDF has a Curie temperature of approximately 110°C, making it suitable for elevated temperature applications [37]. Incorporating fillers or additives into the piezoelectric polymer composite can significantly enhance its performance. These fillers can include inorganic materials such as piezoelectric ceramics like Lead Zirconate Titanate (PZT) or carbon-based materials like carbon nanotubes and graphene [38, 39, 40]. The addition of these fillers enhances the piezoelectric response, mechanical strength, and overall energy harvesting capabilities of the composite.

There is a significant interest in harnessing energy from numerous sources, like wind, rivers, or the human legs' kinetic energy during walking [41, 42]. The concept of utilizing piezoelectric specimens as energy harvesters arises from the desire to capture energy from the environment, which is normally wasted unnecessarily. Therefore, the storage and conversion of equivalent dissipated energy hold great importance for the economy and ecology. In contrast, the composites based on PVDF (Polyvinylidene Fluoride) exhibit fatigue resistance and better constancy in contrast to environmental features. Nevertheless, the energy attained from these composites is relatively low, typically in the microwatt range. As a result, ongoing intensive research aims to enhance the material properties and improve the performance of the existing devices.

The Piezoelectric Energy Harvesting (PEH) technology relies on the ability of materials to produce an electric field when exposed to mechanical force, known as the direct piezoelectric effect. Piezoelectric transducers are available in various forms and materials, making them appropriate for diverse applications. To maximize their effectiveness in different scenarios, it is essential to develop models that analyze the nature of piezoelectric devices in both temporal and frequency domains [43, 44]. Footstep power generation operates on the principle of utilizing piezoelectric sensors. By incorporating piezoelectric technology into flooring, the pressure exerted on the floor generates electrical energy, which is captured by floor sensors. These sensors are strategically positioned to optimize the output voltage. The electrical charge is then converted by a piezoelectric transducer. To monitor and utilize this generated power, a monitoring circuitry is employed, typically based on a microcontroller. This circuit allows users to screen the voltage and charge a battery, providing a versatile source (power) for various applications [9, 45, 46, 47].

In addition, while both ceramic and polymer materials can exhibit piezoelectric properties, ceramics are known for their stronger piezoelectric response and stiffness, whereas polymers offer flexibility and resilience. The comparison is summarized in the table below:

S. No	Property	Ceramic Materials	Polymer Materials
1	Piezoelectric Coefficient	High (*e.g.*, PZT)	Low to moderate (*e.g.*, PVDF)
2	Flexibility	Rigid and brittle	Flexible and ductile
3	Sensitivity	High electrical output	Lower electrical output
4	Durability	Prone to cracking under stress	Tough and impact-resistant
5	Temperature Stability	High thermal stability	Lower thermal stability
6	Application Suitability	Best for high-performance sensors/actuators	Ideal for wearable and flexible devices

PIEZOELECTRIC ENERGY HARVESTING (PEH)

Piezoelectric polymer composites offer several advantages when it comes to energy harvesting applications. These composites are lightweight, flexible, and can easily be shaped into various forms, making them highly adaptable for integration into different devices or structures. They also exhibit excellent durability and stability, ensuring reliable and long-term energy generation. The flexibility and conformability of piezoelectric polymer composites allow them to effectively harvest energy from diverse sources, like ambient vibrations, human motion, or even airflow. This versatility makes them suitable for a wide range of scenarios where mechanical energy needs to be converted into electrical energy [48, 49]. In short, piezoelectric polymer composites hold great promise for energy harvesting applications. They offer efficient and versatile solutions for converting mechanical energy into electrical energy, providing opportunities for sustainable power generation in various contexts.

The simplicity of their configuration, high conversion efficiency, and ability to integrate into complex schemes have led to increased interest in piezoelectric energy harvesting (PEH) systems. These systems are characterized by three essential phases (as depicted in Fig. **2**):

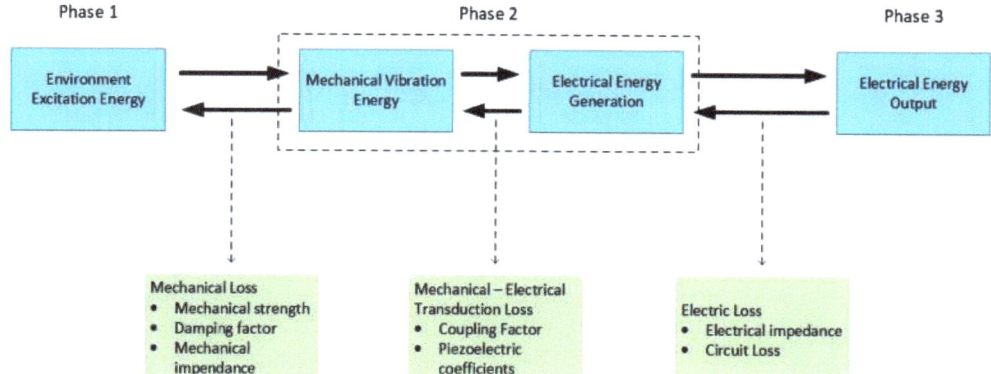

Fig. (2). Three essential stages/phases related to the piezoelectric energy harvesting. Figure collectedx and reproduced from [44].

a. The Mechanical-Mechanical Energy (MME) conversion: This stage/phase includes the mechanical strength of the piezoelectric energy harvester when subjected to significant stresses and the mixing of mechanical impedance.
b. The Mechanical-Electrical Energy (MEE) conversion: This stage/phase is closely related to the electromechanical coupling issues of the PEH structure and the piezoelectric coefficients. In this phase, the mechanical energy is transformed into electrical energy by leveraging the inherent characteristics of the piezoelectric specimen and the ability of the PEH structure to efficiently couple mechanical and electrical phenomena. The electromechanical coupling factor determines the effectiveness of this energy conversion process, while the piezoelectric coefficients play a vital role in quantifying the relationship between the applied mechanical strain or stress to generate further electrical voltage or current. Together, these factors enable the efficient conversion of mechanical energy into usable electrical energy within the PEH system.
c. The Electrical-Electrical Energy (EEE) conversion: This phase encompasses the process of transferring electrical energy within the PEH system through efficient electrical impedance matching. In this phase, the electrical impedance of the components involved in the energy transfer is carefully matched to ensure optimal transmission of electrical power. By matching the electrical impedance, maximum power transfer can be achieved, minimizing losses and enhancing the overall efficiency of the energy transfer process within the PEH system. This impedance matching plays a vital role in maintaining the integrity of the electrical energy and enabling its effective utilization within the system [49].

CONFIGURATIONS OF PIEZOELECTRIC ENERGY HARVESTERS

Piezoelectric energy harvesting often involves capturing vibrations or mechanical energy from sources with short acceleration or short motion frequencies. To effectively utilize this energy, piezoelectric elements with a flat and thin form factor are advantageous. This design enables the piezoelectric element to quickly respond to the motion of the host structure. Additionally, this form factor offers benefits such as reduced dimensions and lighter energy harvesting device weight. Consequently, most piezoelectric energy harvesting systems employ piezoelectric materials that are well-suited for these characteristics.

PEH devices offer various structural configurations, with the cantilever beam type being the most commonly utilized. This configuration is favored because of the gentle geometric design and the ability to produce a significant quantity of strain from mechanical vibrations. It involves bonding a thin layer of piezoelectric material to a non-piezoelectric layer, typically a conductive metallic layer, and securing one end to enable flexural mode operation.

The cantilever geometry is widely employed in PEHs, particularly for harvesting the mechanical energy from vibrations. This design is advantageous as it allows for the generation of large mechanical strain inside the piezoelectric material through vibration. Moreover, the construction of piezoelectric-based cantilevers is quite straightforward. An important factor is that the resonance frequency of the cantilever's fundamental flexural modes is significantly lower compared to further piezoelectric element's vibration modes. Consequently, the features of popular Piezoelectric Energy Harvesting (PEH) devices, to date, mostly belong to a Bimorph or Unimorph cantilever design.

If the mechanical vibrations are tested by means of a cantilever mass system, the structure undergoes oscillation, resulting in the generation of an AC voltage by the piezoelectric layer. The cantilever beam structure is categorized into two configurations depending on the cycle of piezoelectric layers bonded to the metallic layer. The first configuration is known as "bimorph," which utilizes two thin active layers of piezoelectric specimen (Fig. **3a**). The bimorph symmetry/structure offers twice the improvement in electrical energy output with no significant variations in device volume, in contrast with the Unimorph symmetry. Consequently, it has gained considerable recognition in the field of piezoelectric energy harvesting devices.

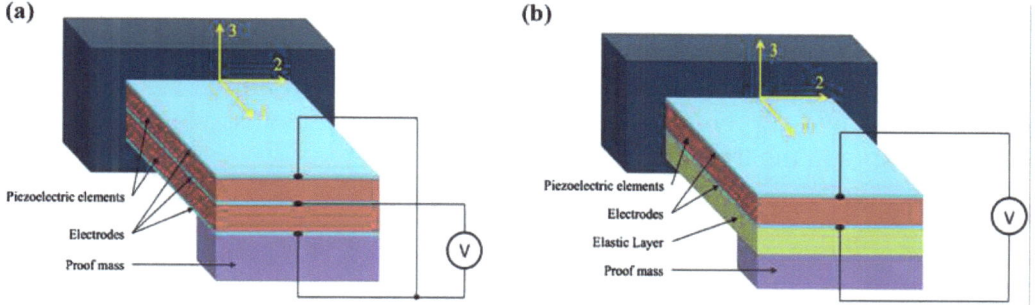

Fig. (3). (a) Bimorph and (b) Unimorph structure of PEH. Figure collected and reproduced from [50, 51].

The second configuration is referred to as "Unimorph," where only one active piezoelectric layer is present (Fig. **3b**). Piezoelectric cantilevers are typically polarized in a direction perpendicular to the planar direction of the cantilever. This arrangement is commonly used during fabrication due to its convenience. Apart from cantilevers, researchers have also explored energy harvesters by means of circular profiles, like cymbal transducers and piezoelectric diaphragms. These alternative designs offer additional options for harvesting energy from mechanical vibrations.

Cymbal transducers were specifically developed to handle high-impact forces in various applications. This transducer configuration involves a piezoelectric ceramic disc sandwiched between metal (steel) end caps on each section (Fig. **4**).

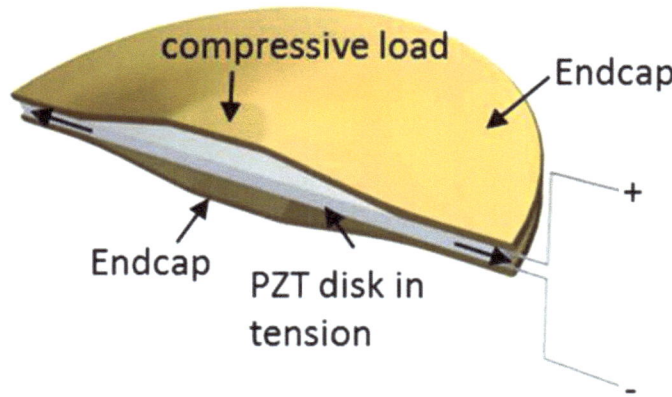

Fig. (4). The representation of a piezoelectric "cymbal" transducer scheme. Figure collected and reproduced from [52].

Steel is commonly used due to its higher yield strength compared to brass or aluminium, enabling the transducer to withstand greater force loading [53]. In this contest, if the axial stress is applied to the cymbal transducer, the steel caps (outer side) convert and magnify this stress into radial stress within the PZT disc. As a result, the d_{33} and d_{31}, both piezoelectric charge coefficients, work together to generate charge in the transducer.

The stack configuration used for Piezoelectric Energy Harvesters (PEHs) involves stacking multiple piezoelectric layers on top of each other (see Fig. **5a**). In this configuration, the poling direction of the layers is aligned with the existence of applied pressure. The d_{33} mode is utilized in this configuration as it has a larger coefficient compared to the d_{31} mode, making it appropriate for further applications requiring high-pressure sensitivity. Nevertheless, due to the higher stiffness in their symmetry configuration, the mechanical energy generated from the utilized pressure is reduced. To overcome this limitation and improve the output, mechanical amplifiers are often employed in conjunction with the stack configuration [54].

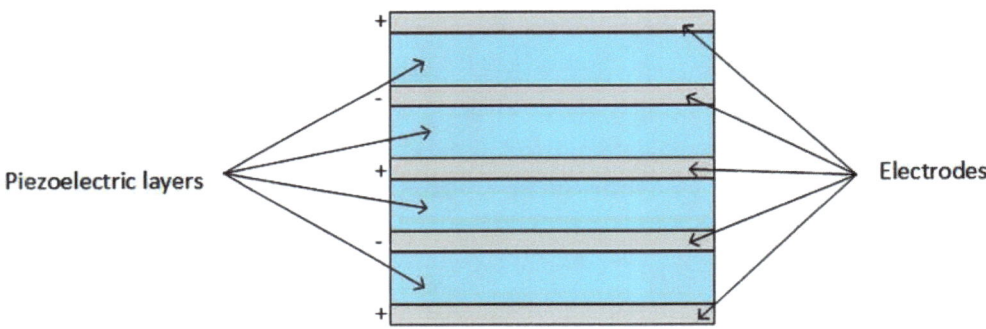

Fig. (5a). Stack-type transducer.

The circular diaphragm (CM) is another symmetry commonly utilized for Piezoelectric Energy Harvesting (PEH) devices. In this association, a disk-type thin layer of piezoelectric specimen is attached to a metal shim, and the entire procedure is established on the clamping ring edges (Fig. **5b**). To improve its presentation at low frequencies and enhance power output, pre-stress is applied to the piezoelectric specimen by fixing a proof mass at the centre of the circular diaphragm. The pre-stress can be introduced through the fabrication process of the piezoelectric-metal composite.

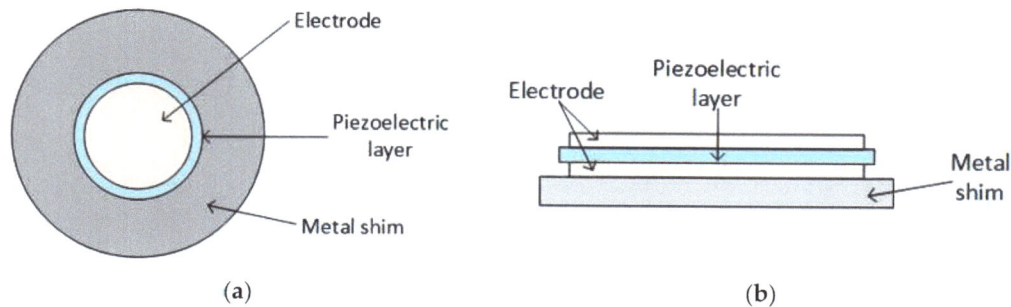

Fig. (5b). *Circular diaphragm transducer. Figure 5 (a-b) collected and reproduced from [55]*

The diaphragm structure offers many advantages associated with the cantilever configuration, particularly in its ability to activate in pressure mode. It can convert various types of pressure fields into an AC signal through a piezoelectric transformer. The diaphragm configuration facilitates energy harvesting from slowly varying, periodic pressure fields, periodic acoustic waves, and mechanical vibrations. Compared to a cantilever, this symmetry is considerably stiffer, resulting in a higher resonance frequency during vibration mode operation [8, 54].

CONCLUSION

Energy harvesting systems can effectively harness available energy, reduce dependence on batteries, mitigate greenhouse gas emissions, and contribute to the sustainability of our environment. The objective of energy harvesting is to cover the lifespan of further devices and field nodes. The concept of utilizing piezoelectric specimens as energy harvesters arises from the desire to capture energy from the environment, which is normally wasted unnecessarily. Therefore, the storage and conversion of equivalent dissipated energy hold great importance for the economy and ecology. Piezoelectric polymer composites offer several advantages when it comes to energy harvesting applications. These composites are lightweight and flexible and can be easily shaped into various forms, making them highly adaptable for integration into different devices or structures. They offer efficient and versatile solutions for converting mechanical energy into electrical energy, providing opportunities for sustainable power generation in various contexts.

REFERENCES

[1] V. Raghunathan, C. Schurgers, S. Park, and M.B. Srivastava, "Energy-aware wireless microsensor networks", *IEEE Signal Process. Mag.,* vol. 19, no. 2, pp. 40-50, 2002.
[http://dx.doi.org/10.1109/79.985679]

[2]   C.C. Enz, and A. El-Hoiydi, "J. -D. Decotignie, and V. Peiris, "WiseNET: an ultralow-power wireless sensor network solution,"", *Computer (Long Beach Calif)*, vol. 37, no. 8, pp. 62-70, 2004.
[http://dx.doi.org/10.1109/MC.2004.109]

[3]   B. Warneke, M. Last, B. Liebowitz, and K.S.J. Pister, "Smart Dust: communicating with a cubic-millimeter computer", *Computer (Long Beach Calif)*, vol. 34, no. 1, pp. 44-51, 2001.
[http://dx.doi.org/10.1109/2.895117]

[4]   S. Rajput, E. Farber, and M. Averbukh, "Optimal selection of asynchronous motor-gearhead couple fed by VFD for electrified vehicle propulsion", *Energies,* vol. 14, no. 14, 2021.
[http://dx.doi.org/10.3390/en14144346]

[5]   E. Maghsoudi Nia, N.A. Wan Abdullah Zawawi, and B.S. Mahinder Singh, "Design of a pavement using piezoelectric materials", *Materialwiss. Werkstofftech.,* vol. 50, no. 3, pp. 320-328, 2019.
[http://dx.doi.org/10.1002/mawe.201900002]

[6]   S. Rajput, M. Averbukh, and N. Rodriguez, *Energy Harvesting and Energy Storage Systems.,* 2022.
[http://dx.doi.org/10.3390/books978-3-0365-3709-2]

[7]   F.K. Shaikh, and S. Zeadally, "Energy harvesting in wireless sensor networks: A comprehensive review", *Renew. Sustain. Energy Rev.,* vol. 55, pp. 1041-1054, 2016.
[http://dx.doi.org/10.1016/j.rser.2015.11.010]

[8]   S. Rajput, M. Averbukh, and A. Yahalom, "Electric power generation using a parallel-plate capacitor", *Int. J. Energy Res.,* vol. 43, no. 8, pp. 3905-3913, 2019.
[http://dx.doi.org/10.1002/er.4492]

[9]   Y. Motey, P. Dekate, M. Kewate, and J. Aswale, "Footstep power generation system", *International Journal of Innovations in Engineering and Science,* vol. 2, no. 6, 2017.

[10]   S.P. Muduli, L. Lipsa, A. Choudhary, S. Rajput, and S. Parida, "Modulation of electrical characteristics of polymer–ceramic–graphene hybrid composite for piezoelectric energy harvesting", *ACS Appl. Electron. Mater.,* vol. 5, no. 6, pp. 3023-3037, 2023.
[http://dx.doi.org/10.1021/acsaelm.3c00078]

[11]   S Round, "Power MEMS 2008 and micro EMS", *PowerMEMS 2008 + µEMS 2008 Conf,* 2008

[12]   A.E. Kubba, and K. Jiang, "A comprehensive study on technologies of tyre monitoring systems and possible energy solutions", *Sensors (Basel),* vol. 14, no. 6, pp. 10306-10345, 2014.
[http://dx.doi.org/10.3390/s140610306]

[13]   W.W. Clark, and M.J. Ramsay, "Smart material transducers as power sources for MEMS devices", *International Symposium on Smart Structures and Microsystems,* 2000 Hong Kong

[14]   H. Landaluce, L. Arjona, A. Perallos, F. Falcone, I. Angulo, and F. Muralter, "A review of IoT sensing applications and challenges using RFID and wireless sensor networks", *Sensors (Basel),* vol. 20, no. 9, 2020.
[http://dx.doi.org/10.3390/s20092495]

[15]   J. Curie, and P. Curie, "Development by pressure of polar electricity in hemihedral crystals with inclined faces", In: *Bull Soc Minéral Fr* vol. 3. , 1880, pp. 90-93.

[16]   X. Du, J. Zheng, U. Belegundu, and K. Uchino, "Crystal orientation dependence of piezoelectric properties of lead zirconate titanate near the morphotropic phase boundary", *Appl. Phys. Lett.,* vol. 72, no. 19, pp. 2421-2423, 1998.
[http://dx.doi.org/10.1063/1.121373]

[17]   M. D. Donato, Ph.D. Thesis, "Politecnico di Torino", Turin, Italy 2015.

[18]   S. Sapkal, B. Kandasubramanian, and H.S. Panda, "A review of piezoelectric materials for nanogenerator applications", *J. Mater. Sci. Mater. Electron.,* vol. 33, no. 36, pp. 26633-26677, 2022.
[http://dx.doi.org/10.1007/s10854-022-09339-7]

[19]   P. Martins, A.C. Lopes, and S. Lanceros-Mendez, "Electroactive phases of poly(vinylidene fluoride): Determination, processing and applications", *Prog. Polym. Sci.,* vol. 39, no. 4, pp. 683-706, 2014.
[http://dx.doi.org/10.1016/j.progpolymsci.2013.07.006]

[20]   B. Maamer, A. Boughamoura, A.M.R. Fath El-Bab, L.A. Francis, and F. Tounsi, "A review on design improvements and techniques for mechanical energy harvesting using piezoelectric and electromagnetic schemes", *Energy Convers. Manage.,* vol. 199, p. 111973, 2019.
[http://dx.doi.org/10.1016/j.enconman.2019.111973]

[21]   J. Harrison, and Z. Ounaies, *Piezoelectric Polymers,* ICASE Report No. 2001-43, 2001.

[22]   H. Kawai, "The Piezoelectricity of Poly (vinylidene Fluoride)", *Jpn. J. Appl. Phys.,* vol. 8, no. 7, p. 975, 1969.
[http://dx.doi.org/10.1143/JJAP.8.975]

[23]   M. Wegener, W. Wirges, and R. Gerhard-Multhaupt, "Piezoelectric polyethylene terephthalate (PETP) foams – specifically designed and prepared ferroelectret films", *Adv. Eng. Mater.,* vol. 7, no. 12, pp. 1128-1131, 2005.
[http://dx.doi.org/10.1002/adem.200500177]

[24]   S. Schneegaß, and O. Amft, *Smart Textiles: Fundamentals.* Design, and Interaction, 2017.

[25]   G.K. Novikov, and A.I. Smirnov, Electret effect in polyolefins joined by x-ray radiation of an electric gas barrier discharge. Vol. 53. Russian Physics Journal, 2011, p. 1113. https://link.gale.com/apps/doc/A359411494/AONE?u=anon~7059874&sid=googleScholar&xid=a6fb740c

[26]   Z.O.J.S. Harrison, *Piezoelectric Polymers 2001; NASA/CR-2001-211 422 ICASE Report No. 2001-43.*

[27]   Z. Zhang, M.H. Litt, and L. Zhu, "Unified understanding of ferroelectricity in n-nylons: is the polar crystalline structure a prerequisite?", *Macromolecules,* vol. 49, no. 8, pp. 3070-3082, 2016.
[http://dx.doi.org/10.1021/acs.macromol.5b02739]

[28]   D. Koyama, and K. Nakamura, "Array configurations for higher power generation in piezoelectric energy harvesting", *Jpn. J. Appl. Phys.,* vol. 49, no. 7S, p. 07HD04, 2010.
[http://dx.doi.org/10.1143/JJAP.49.07HD04]

[29]   D.V. Bayramol, N. Soin, T. Shah, E. Siores, D. Matsouka, and S. Vassiliadis, Energy Harvesting Smart Textiles.*Smart Textiles: Fundamentals, Design, and Interaction.,* S. Schneegass, O. Amft, Eds., Springer International Publishing: Cham, 2017, pp. 199-231.

[30]   M. Hosseini, A.S.H. Makhlouf, Ed., *Industrial Applications for Intelligent Polymers and Coatings.* Springer International Publishing: Cham, 2016.

[31]   J.N. Aravind Dasari, Ed., *Functional and Physical Properties of Polymer Nanocomposites.* Wiley Online: New York, 2016.

[32]   Y-T. Wang, Y-C. Hu, and K-R. Chen, "A flexible polyvinylidene fluoride film-loudspeaker", *Journal of the Chinese Society of Mechanical Engineers, Transactions of the Chinese Institute of Engineers - Series C,* vol. 36, no. 1, pp. 59-66, 2015.

[33]   C-T. Pan, "Significant piezoelectric and energy harvesting enhancement of poly(vinylidene fluoride)/polypeptide fiber composites prepared through near-field electrospinning", *J. Mater. Chem. A Mater. Energy Sustain.,* vol. 3, no. 13, pp. 6835-6843, 2015.
[http://dx.doi.org/10.1039/C5TA00147A]

[34]   V.F. Cardoso, D M. Correia, C. Ribeiro, M.M. Fernandes, and S. Lanceros-Méndez, "Fluorinated polymers as smart materials for advanced biomedical applications", *Polymers (Basel),* vol. 10, no. 2, 2018.
[http://dx.doi.org/10.3390/polym10020161]

[35]   Q. Jing, and S. Kar-Narayan, "Nanostructured polymer-based piezoelectric and triboelectric materials and devices for energy harvesting applications", *J. Phys. D Appl. Phys.,* vol. 51, 2018.
[http://dx.doi.org/10.1088/1361-6463/aac827]

[36]   L. Ruan, X. Yao, Y. Chang, L. Zhou, G. Qin, and X. Zhang, "Properties and applications of the β phase poly(vinylidene fluoride)", *Polymers (Basel)*, vol. 10, no. 3, 2018.
[http://dx.doi.org/10.3390/polym10030228]

[37]   T. Yang, D. Xie, Z. Li, and H. Zhu, "Recent advances in wearable tactile sensors: Materials, sensing mechanisms, and device performance", *Mater. Sci. Eng. Rep.*, vol. 115, pp. 1-37, 2017.
[http://dx.doi.org/10.1016/j.mser.2017.02.001]

[38]   V.R. Mudinepalli, and F. Leng, "Dielectric and ferroelectric studies on high dense Pb(Zr0.52Ti0.48)O3 nanocrystalline ceramics by high energy ball milling and spark plasma sintering", *Ceramics*, vol. 2, no. 1, pp. 13-24, 2019.
[http://dx.doi.org/10.3390/ceramics2010002]

[39]   C. Zhao, "Deconvolved intrinsic and extrinsic contributions to electrostrain in high performance, Nb-doped Pb(ZrxTi1-x)O3 piezoceramics ($0.50 \leq x \leq 0.56$)", *Acta Mater.*, vol. 158, pp. 369-380, 2018.
[http://dx.doi.org/10.1016/j.actamat.2018.08.006]

[40]   L. Fan, J. Chen, Y. Ren, and X. Xing, "Structural correlation to piezoelectric and ferroelectric mechanisms in rhombohedral Pb(Zr,Ti)O3 ceramics by in-situ synchrotron diffraction", *Inorg. Chem.*, vol. 57, no. 6, pp. 3002-3007, 2018.
[http://dx.doi.org/10.1021/acs.inorgchem.7b02329]

[41]   C. Baur, D.J. Apo, D. Maurya, S. Priya, and W. Voit, *Advances in Piezoelectric Polymer Composites for Vibrational Energy Harvesting.*, 2014, pp. 1-27.
[http://dx.doi.org/10.1021/bk-2014-1161.ch001]

[42]   H. Li, C. Tian, and Z. Deng, "Energy harvesting from low frequency applications using piezoelectric materials", *Appl. Phys. Rev.*, vol. 1, p. 41301, 2014.
[http://dx.doi.org/10.1063/1.4900845]

[43]   C. Covaci, and A. Gontean, "Piezoelectric energy harvesting solutions: a review", *Sensors (Basel)*, vol. 20, no. 12, 2020.
[http://dx.doi.org/10.3390/s20123512]

[44]   B. Maamer, A. Boughamoura, A.M.R. Fath El-Bab, L.A. Francis, and F. Tounsi, "A review on design improvements and techniques for mechanical energy harvesting using piezoelectric and electromagnetic schemes", *Energy Convers. Manage.*, vol. 199, p. 111973, 2019.
[http://dx.doi.org/10.1016/j.enconman.2019.111973]

[45]   E. Maghsoudi Nia, N.A. Wan Abdullah Zawawi, and B.S. Mahinder Singh, "Design of a pavement using piezoelectric materials", *Materialwiss. Werkstofftech.*, vol. 50, no. 3, pp. 320-328, 2019.
[http://dx.doi.org/10.1002/mawe.201900002]

[46]   M. V Gudur, and A. Professor, *International Journal of Advanced Research in Electrical, Electronics and Instrumentation Engineering Piezoelectric Energy Harvesting using PZT in Floor Tile Design*, 2017.

[47]   S. Poddar, M. Dutta, D. Chowdhury, A. Dey, and D. Maji, "Footstep voltage generator using piezo-electric transducers", *Int J Sci Eng Res*, vol. 8, 2017no. 3, . Available from: http://www.ijser.org

[48]   S. Roundy, and P.K. Wright, "A piezoelectric vibration based generator for wireless electronics", *Smart Mater. Struct.*, vol. 13, no. 5, p. 1131, 2004.
[http://dx.doi.org/10.1088/0964-1726/13/5/018]

[49]   S.R. Anton, and H.A. Sodano, "A review of power harvesting using piezoelectric materials (2003–2006)", *Smart Mater. Struct.*, vol. 16, no. 3, p. R1, 2007.
[http://dx.doi.org/10.1088/0964-1726/16/3/R01]

[50]   K. Uchino, and T. Ishii, "Energy flow analysis in piezoelectric energy harvesting systems", *Ferroelectrics*, vol. 400, no. 1, pp. 305-320, 2010.
[http://dx.doi.org/10.1080/00150193.2010.505852]

[51] S. Priya, "A review on piezoelectric energy harvesting: materials, methods, and circuits", *Energy Harvesting and Systems,* vol. 4, 2017.
[http://dx.doi.org/10.1515/ehs-2016-0028]

[52] Y. Kuang, A. Daniels, and M. Zhu, "A sandwiched piezoelectric transducer with flex end-caps for energy harvesting in large force environments", *J. Phys. D Appl. Phys.,* vol. 50, no. 34, p. 345501, 2017.
[http://dx.doi.org/10.1088/1361-6463/aa7b28]

[53] H. Kim, "Energy harvesting using a piezoelectric 'cymbal' transducer in dynamic environment", *Jpn. J. Appl. Phys.,* vol. 43, p. 6178, 2004.
[http://dx.doi.org/10.1143/JJAP.43.6178]

[54] A. Toprak, and O. Tigli, "Piezoelectric energy harvesting: state-of-the-art and challenges", *Appl. Phys. Rev.,* vol. 1, 2014.
[http://dx.doi.org/10.1063/1.4896166]

[55] C. Covaci, and A. Gontean, "Piezoelectric energy harvesting solutions: a review", *Sensors (Basel),* vol. 20, no. 12, 2020.
[http://dx.doi.org/10.3390/s20123512]

CHAPTER 4

Highly Stable Nanocomposite PVDF-Based Energy Harvesting Piezoelectric Devices

Rahutosh Ranjan[1], Sushama Yadav[2], Sujeet Kumar Chaurasia[2], Nitin Srivastava[3,*] and Neelabh Srivastava[1,*]

[1] *Department of Physics, School of Physical Sciences, Mahatma Gandhi Central University, Motihari 845401, East Champaran, Bihar, India*

[2] *Centre for Nanoscience & Technology, Veer Bahadur Purvanchal University, Jaunpur 222003, Uttar Pradesh, India*

[3] *Department of Physics, Sri Chitragupt P. G. College, Mainpuri 205001, Uttar Pradesh, India*

Abstract: The topic of energy harvesting is very important because of the rising power consumption and environmental concerns. Extensive research is being done on Poly (Vinylidene Fluoride) PVDF and its copolymers for the purpose of developing energy harvesting devices. This chapter deals with the various types of piezoelectric devices and their working principle, along with their applications. The use of PVDF and its nanocomposites in piezoelectric devices makes them promising due to their outstanding properties, such as ease of processing, good flexibility, good biocompatibility, high stability, *etc*. Also, the applications of PVDF-based piezoelectric materials in energy harvesting devices such as sensors, biomedical science, flexible and hybrid nanogenerators, *etc.*, are discussed.

Keywords: Energy harvesting, Nanocomposite, Piezoelectric applications, Piezoelectric effect, Polymer composite.

INTRODUCTION

Energy sources (solar, thermal, wind, and mechanical vibrations) can be harnessed using specific materials that are capable of generating electricity or collecting charges [1]. Thus, to harness mechanical energy, piezoelectric materials are prominent candidates, manifesting greater potential, higher power density, and increased life span [1, 2]. In 1880, the Curie brothers first discovered the piezoelectric effect. *Piezoelectric* is a Greek word meaning "pressure electricity"

---

* **Corresponding authors Nitin Srivastava:** Department of Physics, Sri Chitragupt P. G. College, Mainpuri 205001, Uttar Pradesh, India; E-mail: srivastavanitin93@gmail.com;
Neelabh Srivastava: Department of Physics, School of Physical Sciences, Mahatma Gandhi Central University, Motihari 845401, East Champaran, Bihar, India;
E-mail: neelabh@mgcub.ac.in

Sushil Kumar Verma, Sonika & Arbind Prasad (Eds.)
All rights reserved-© 2025 Bentham Science Publishers

[3]. Piezoelectricity is observed in materials that are non-centrosymmetric (their crystalline state lacks a centre of symmetry), and this phenomenon is generally attributed to the electric dipole [3, 4]. Dipoles provide a macroscopic polarization vector, P, when they are mutually aligned in specific spatial areas (Weiss domains). Such alignment can occur spontaneously, or it can be induced in some materials (ferroelectrics) by either a mechanical drawing technique or an externally provided, strong electric field (poling method). Following a mechanical stress (σ), both the strain (ε), which is induced in the material, and the direction of P and/or intensity may change. The electro-mechanical behavior of the material is then described as a piezoelectric coefficient, d_{ij} = Di/j, where Di denotes the electric displacement's components and the direction denoted by 3 is recognized as the direction of pristine polarization (poling process). In d_{ij}, the j subscript signifies the direction of applied stress (for j = 1-3), as well as whether shear stress is present (for j = 4-6). The subscript i in d_{ij} denotes the direction (*i.e.*, x, y, or z) of the electric displacement resulting in the piezoelectric material, *i.e.*, the direction along which a voltage bias is produced [5 - 7]. The variable design of a system can benefit greatly from shear piezoelectric response, which has been demonstrated in numerous uniaxially oriented systems.

Piezoelectric materials are desirable for their use in Energy Harvesting (EH), actuators, sensor devices, and a variety of other fields of technology because they can directly generate electrical and mechanical energy [8, 9]. Due to its high energy conversion efficiency, simplicity of use, and miniaturization, the piezoelectric effect has been widely utilized to convert mechanical energy into electricity [10]. Due to their piezoelectric behavior, several materials can be used in harvesting energy, as well as in the fabrication of piezoelectric sensors and actuators. Barium Titanate ($BaTiO_3$) and Lead Zirconate Titanate (PZT) are traditionally popular Piezoelectric Materials (PEM) [11]. However, these are brittle and toxic. PEMs that are non-toxic, flexible, and simple to fabricate are therefore in demand for commercial applications [12, 13]. Fig. (**1**) depicts the various piezoelectric materials and their atomic structure. Although the piezoelectric effect is frequently linked to ceramic materials, numerous polymers also exhibit piezoelectric behavior. Despite frequently being less piezoelectric than their ceramic counterparts, piezoelectric polymers are frequently preferred for particular applications because of their flexibility, simplicity of production, and biocompatibility [14, 16]. Many different polymer families exhibit piezoelectric effects. In addition to the well-known ferroelectric polymer Polyvinylidene Fluoride (PVDF), various polyureas, polyamides, polypeptides, and polyesters also exhibit piezoelectric behaviour [17]. Piezoelectric technology is often used to generate electricity. Due to its high electro-mechanical coupling factor and piezoelectric constant, PZT is the most widely utilized piezoelectric material. However, due to its fragility, high cost, high density, and environmental risks, the development is incomplete at this point [11, 18].

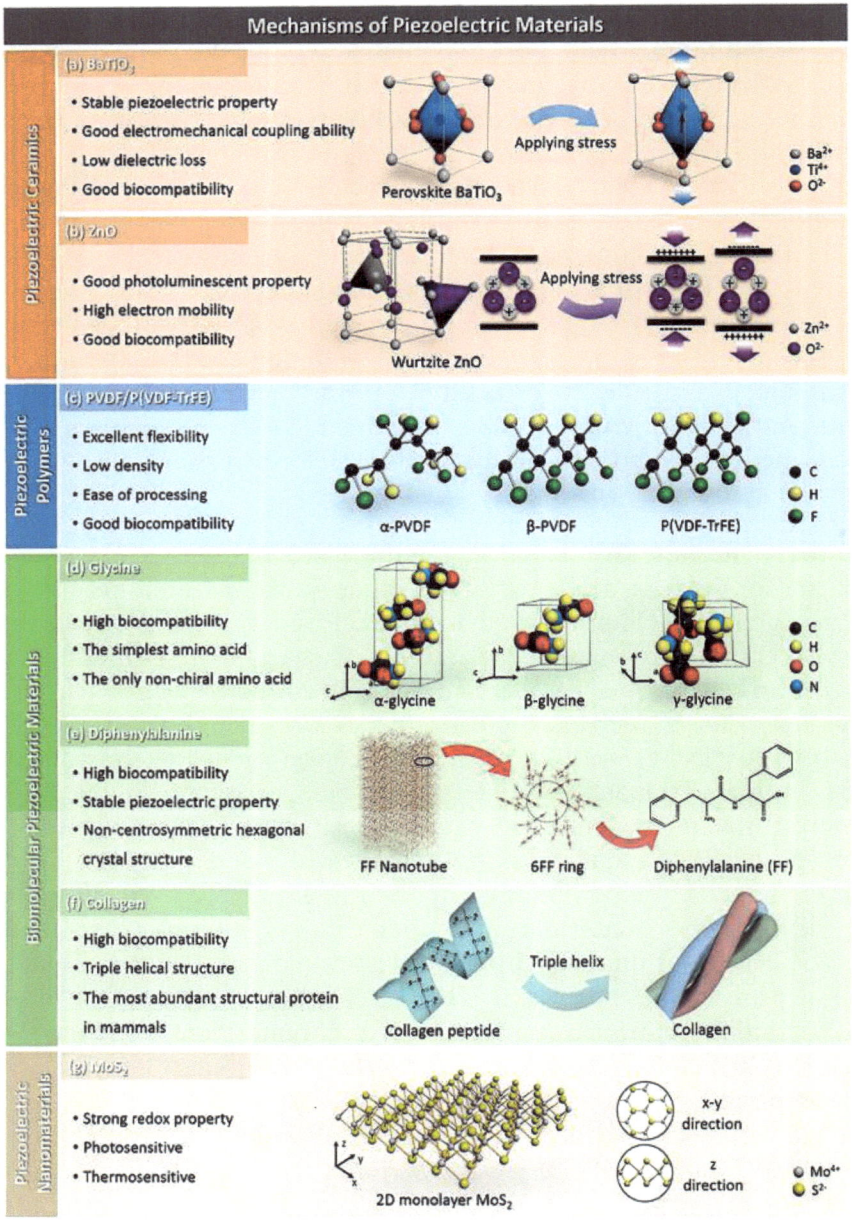

Fig. (1). Various piezoelectric materials and their atomic structure [3].

On the other hand, PVDF and its copolymers have a variety of advantages, including exceptionalresistance to halogen compounds and acids, outstanding biological compatibility, improved elasticity, and environmental friendliness [2, 9, 11, 19, 20]. In particular, these materials can be formed into curvilinear structures [21]. They are therefore good candidates to partially substitute PZT. PVDF's applications are significantly hampered by its relatively poor piezoelectricity as compared to PZT [14]. The piezoelectric effect serves as the foundation for piezoelectric EH devices. In contrast, the reverse piezoelectric effect is the basis for piezoelectric actuators. A semi-crystalline polymer called PVDF exhibits electro-active phenomena like ferroelectricity, pyroelectricity, and piezoelectricity [22]. PVDF and its copolymers have coefficients of piezoelectricity, typically of the equivalent order of degrees, or 20–40 pC/N, which is a step lower compared to that of PZT [23 - 25]. Therefore, for the advancement of piezoelectric devices involving PVDF, the piezoelectricity must be improved. A high concentration of the β-phase in PVDF material is preferred in order to have good performance of piezoelectric devices [19, 20, 26, 27]. The piezoelectric coefficient of Poly(vinylidene fluoride-co-Trifluoroethylene) PVDF-TrFE copolymer is high. TrFE and other copolymers can introduce artificial flaws to lower the energy required for the all-trans form to crystallize. The all-trans polar crystalline phase (β-phase) PVDF-TrFE is the only one that is stable at ambient temperature and generally exhibits good piezoelectric activity [19, 20, 28]. A large surface area to volume ratio can be further attained by using micro/nano structures to increase piezoelectric efficiency. The majority of nanofabrication techniques, however, are only available for inorganic materials like ZnO and PZT.

HISTORY OF PIEZOELECTRIC POLYMERS

This class of material is distinguished from single crystals and ceramics by having lengthy polymer chains and a high degree of flexibility [29]. They are more suited for applications requiring a lot of bending or twisting because of their natural flexibility, which allows them to endure much higher strain [2]. Some well-known piezoelectric polymers include PVDF, PVDF-TrFE, cellulose and its derivatives, Polyamides (PA), Polylactic Acids (PLA), *etc.*, out of which PVDF has received considerable attention [2]. Jacques and Pierre Curie discovered piezoelectricity for the first time in inorganic crystals in 1880 [3]. The majority of the piezoelectric materials employed in practical applications nowadays are inorganic, despite the fact that the phenomenon is still linked with ceramic materials [30]. In the instance of Lead Zirconium Titanate (PZT), the piezoelectric coefficient of ceramic materials can reach 500 pC/N [31]. Ceramics' mechanical characteristics and composition, however, are not appropriate for all applications. Fortunately, nothing fundamental prevents piezoelectricity from occurring in materials other than ceramics; in fact, piezoelectricity has been reported in a few polymeric materials. Piezoelectric coefficients are typically lower, but the materials are more flexible and simpler to work with. As a supplement to piezoelectric ceramics, piezoelectric polymers are not a replacement for them [30]. Heiji Kawai's discovery of piezoelectric properties in the synthetic polymer Poly(Vinylidene Fluoride) (PVDF) in 1969 sparked interest in piezoelectric polymers among a wider scientific audience [32]. The piezoelectric coefficients were significantly higher than anything that has previously been seen in polymers by at least an order of magnitude. The piezoelectric response's symmetry was also considerably different from what had previously been observed. Soon after, pyroelectric and ferroelectric behaviour in PVDF was confirmed, igniting interest in piezoelectric polymers for use in a variety of fields [33 - 35]. Following the discovery of piezoelectric behaviour in several more polymer families, piezoelectric-based polymers are now the focus of significant scientific study for applications in biomedical devices, EH, and wearable technology [36].

Classification of Piezoelectric Polymers

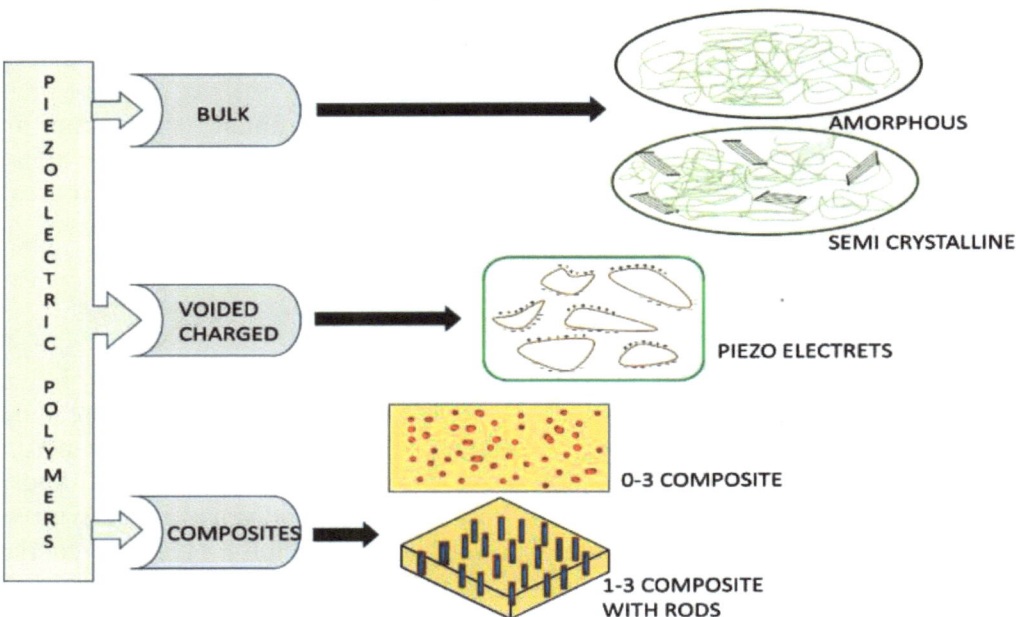

Fig. (2). Classification of piezoelectric polymers [37].

Bulk Piezoelectric Polymer

Because of their orientation and chemical makeup, bulk piezoelectric polymers exhibit the piezoelectric effect. Semi-crystalline and amorphous polymers are two further subclasses of bulk polymers [37]. Four important things should be kept in mind for any polymer to exhibit piezoelectric properties:

i. A permanent molecular dipole must be present in the material.
ii. It should have the ability to align the molecular dipoles.
iii. Once the alignment is achieved, it should be retained.
iv. The material should be able to withstand a higher strain when put under a lot of stress.

Voided Charged Polymers (VCP)

Another intriguing use of polymer thin films is in Cellular Polymers (VCPs), also known as piezoelectrets or polymers, which exhibit a greater piezoelectric constant than piezoelectric ceramics. The electric field ionizes the gas molecules in the thin sheets. Internal dipoles are created as a result of the opposite charges accelerating and implanting on each side of the voids [37]. A piezoelectric action can be produced in the substance by any deformation of the void. The piezoelectric coefficient is determined by the gas type, pressure, void form, and density [17]. All the necessary components for piezo- and pyroelectricity in VCPs, including non-uniform charge distribution, non-uniform strain, and adequate coupling between the two layers, were first recognized in early research by G. Dreyfus and J. Lewiner. These factors have a direct effect on the charge distribution on each side of the polymer's voids and the net dipole moment generated. To trap and store the electrical charges, the polymers used for VCPs must have voids, much like foam, and they must also be electrically stable. Charge trapping can take place at the interfaces of crystallites and their amorphous surroundings, in cages between adjacent molecules, or between electronegative groups on the chain. We have two separate piezoelectric coefficients, namely quasistatic and dynamic, in the VCPs because the piezoelectric coefficient can be changed in accordance with the frequency of the electrical field [38]. For zero or low frequencies, the quasistatic piezoelectric coefficient is used, but the dynamic piezoelectric coefficient is used for frequencies greater than 100 Hz. According to a thermodynamic layer model of a piezoelectret, D_{33} varies linearly with the compressibility and inversely to the foam's Young's modulus [39].

Piezoelectrics Composites

Inorganic piezoelectric materials incorporated within a polymer material form a piezo-composite. The polymer used in these materials is non-electroactive. By mixing polymers with piezoelectric polymers, it is possible to combine the ceramic's high dielectric constant and huge piezoelectric coefficient with the polymer's superior mechanical flexibility [37]. The microstructural organization of the component phases in composites (connectedness) was described by Newnham et al. [40] and later modified by Pilgrim et al. [41]. For a composite with two phases, there are 16 potential connectivity patterns. The phases range from (0-0), which denotes that neither is self-connected, to (3), which denotes that each phase is self-connected in 3-D. Commercial (1-3) composites are produced by businesses like Smart Materials with ceramic rods.

INTRODUCTION TO PVDF

Due to the toxicity and non-biocompatibility of inorganic ceramic piezoelectric materials, the need for flexible and biocompatible PEMs can be fulfilled by polymer-based materials. Various organic polymers such as diisopropylammonium bromide, biomolecular materials, nylon, PVDF, and P(VDF–TrFE) have been developed to counter the inorganic PEMs [10]. Polyvinylidene fluoride (PVDF) has demonstrated its potential as a favorable candidate for replacing the ceramic PEMs [20]. PVDF is a semi-crystalline polymer that possesses easy fabrication, high resistance to halogens and acids, transparency, and great mechanical and piezoelectric properties [11]. The monomers "CH_2CF_2" make up the PVDF bio-piezoelectric polymers. The difference in electronegativity between the fluorine and hydrogen atoms causes their molecular dipoles to appear.

How is Piezoelectricity Generated?

Piezoelectricity is a property of a substance that produces an electrical current when subjected to mechanical stress or pressure. When piezoelectric materials are stressed from the outside, the distance between the positive and negative charge centres changes, causing the material to polarize in the direction of the stress [[42]]. The direct piezoelectric effect occurs in this scenario when the surface-free charges are partially released to produce piezoelectricity. Similar to how piezoelectric materials can bend geometrically when put under an external electric field, this phenomenon is known as the inverse piezoelectric effect [42]. The term "piezoelectric material" refers to substances that exhibit this type of behavior. Another piezoelectric polymer is PVDF. Since the PVDF α- and ε-phases are non-polar, as already mentioned, piezoelectric action is not present. Piezoelectricity is produced in it by the β-, δ-, and γ-phases. Even though they are not placed symmetrically, piezoelectric crystal charges are typically balanced. This causes the charges to become unbalanced. Because of the way that charges interact, net charges show up on both sides of crystals. By pressing it in this manner, a voltage can be generated in it. This unique property can be used to generate electricity from environmental vibrations, for example, human motion, for smart textiles and medical applications [43]. The production of piezoelectricity now depends on the electroactivity of the PVDF material and its crystallinity. Since PVDF material is typically in a non-polar phase, it is not piezoelectric. It is made of thick, wide sheets of translucent plastic that are stretched and poled to give them their piezoelectric capabilities. When forces are applied to an electrospun piezoelectric web, the piezoelectric effect causes voltage to build up on both sides of the web.

Phases of PVDF

There are five possible ways that the crystal structures of PVDF can be arranged (α, β, γ, δ, and ε), with the α- and β-phases being the most prevalent [26]. There is little piezoelectricity seen in the α-phase of PVDF because it has dipoles arranged in reverse parallel order. The parallel arrangement of the dipoles in the β-phase results in a higher dipole moment per unit cell and better piezoelectric characteristics [3]. The five different crystalline phases of PVDF (α, β, γ, δ, and ε) are obtained due to differences in processing techniques [11]. Crystalline α-phase is the most common, but the β-phase, which has exceptional piezoelectricity, is also of the highest importance. The unit cell (monoclinic) of the α-polymorph is inactive in piezoelectric terms and has the TGTG configuration, with F and H atoms placed alternative in the chain. In addition, the α-phase is the most thermodynamically stable phase [20, 42, 44]. As opposed to this, the β-polymorph has a piezoelectrically active orthorhombic unit cell with the all trans(TTTT) configuration, which shows the highest dipole moment and piezoelectric coefficient [2]. Orthorhombic unit cells with the T3GT3G conformation form the γ-polymorph's unit cell. The δ-polymorph is the polar counterpart to the α-form, whereas the "ε-polymorph" is the anti-polar counterpart to the "γ-form". [14, 27]. Thus, the phases α and ε are antipolar, while β, γ, and δ are polar. The structural configuration of the three most important phases of PVDF is shown in Fig. (**3**).

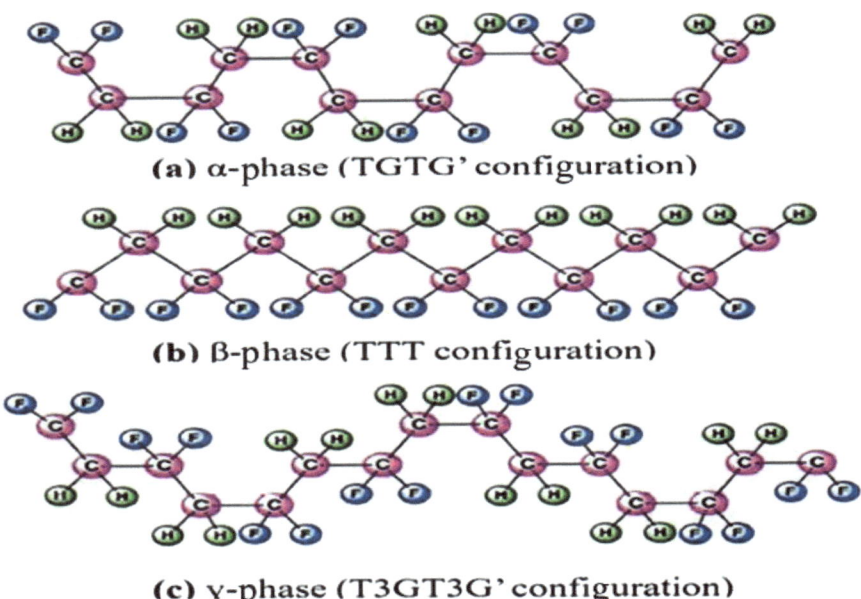

Fig. (3). Structural configuration of α, β, and γ-phases [45].

β-phase Enhancement

The α-phase predominates in the majority of newly produced PVDF films because of its steady thermal characteristics. In contrast, the all-trans conformation has large energy barriers, making the development of the β-phase challenging. In contrast, PVDF or copolymers' piezoelectricity is dependent on the electroactive phase, particularly the β-phase. As discussed above, the β-phase of PVDF is mostly desired for the piezoelectric behavior. The β-phase PVDF shows piezoelectric behavior, but the β-phase is difficult to synthesize, which prohibits its large-scale use [46]. The enhancement of the β-phase in PVDF has opened new avenues for the advancement of flexible piezoelectric devices utilizing PVDF as a base material. In addition, the β-phase enhancement in PVDF also improves its piezoresponse. In contrast, PVDF or copolymers' piezoelectricity is dependent on the electroactive β- and γ-phases, particularly the β-phase [43]. In order to enhance the production of β-phases or the transition from α- to β-phases, various approaches have been made. The β-phase has previously been reported to be created by melting, solvent casting, copolymerizing, or adding fillers to the existing PVDF α-phase [9, 19], as shown in Fig. (4). To convert PVDF from α-phase to β-phase, it has been investigated to add organic, inorganic, or hybrid organic-inorganic additives such as $BaTiO_3$, cellulose, montmorillonite clays, TiO_2, carbon nanotubes (CNT), gold, palladium, PMMA, cerium(III), and a complex of fillers into a PVDF polymer matrix [19]. In most of the reported literatures, it is found that β-phase content of PVDF can be augmented through the incorporation of certain fillers, including but not limited to clays, carbon nanotubes, graphene, and metal oxides, as well as through polymer blending techniques with other polymers such as PVDF-TrFE, polymethyl methacrylate (PMMA), or ionic liquids. Furthermore, it is also observed that subjecting the sample to uniaxial or biaxial stretching in particular orientations can augment the β-phase composition of PVDF [2, 43]. Depending on the material of origin, various fillers are categorized into multiple groups, such as ceramic fillers, carbon-based fillers, hybrid fillers, and bio-based fillers [2]. These play a key part in increasing the piezo-responsiveness of polymers.

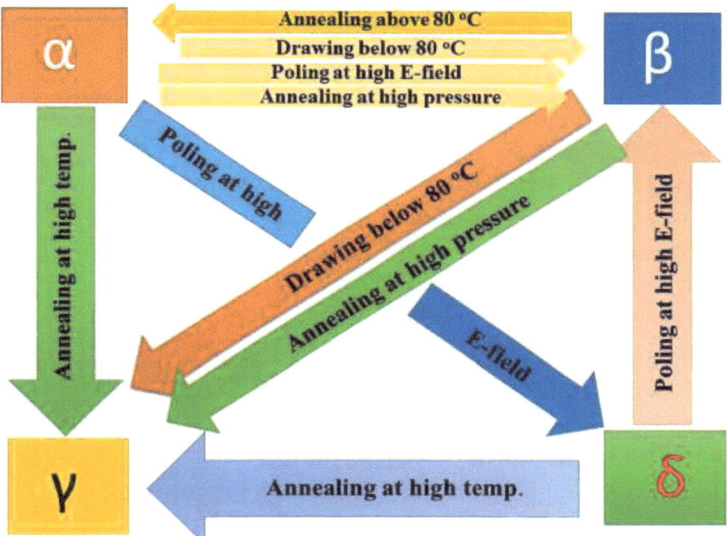

Fig. (4). Transformation occurring among different phases of PVDF [47].

Piezoelectric PVDF Fabrication Method

Several fabrication techniques have been developed for the synthesis of materials based on piezoelectric PVDF. Some of these techniques include melt blending, Microelectromechanical Systems (MEMS) technologies, 3D printing, self-poling, electrospinning, soft lithography, drop-casting, solution casting, and hot pressing.

Melt Blending

Melting dielectric polymers with other polymers is one fabrication technique. The process of melt-blowing involves the extrusion of a polymer melt through orifices, which is subsequently subjected to high-velocity hot air to produce ultrafine fibers. This particular technique is a single-step process. A nonwoven web can be created by blowing nanofibers made of molten dielectric polymer into a collection on a rotating drum or a forming belt with a vacuum below the surface. It has been observed that polymers' increased micromorphology and crystallisation tendencies boost their dielectric properties [20].

Self-poling

For very sensitive piezoelectric devices, particularly ferroelectric nanogenerators, self-poling provides a straightforward, affordable, and expandable production procedure. By introducing outside agents, CH_2/CF_2 dipoles are oriented using this technique. A porous electret is utilized to create an arrangement possessing a d_{33} value comparable to that of PZT. In a study, the regionally directed CH_2/CF_2 dipoles of the porous film are produced by combining ytterbium salt with PVDF [48].

Electrospinning

The most common and efficient method for creating piezoelectric PVDF with piezoelectric characteristics is probably electrospinning [1, 14]. An instantaneous piezoelectric structure can be created *via* electrospinning [20]. The fundamental setup consists of a syringe, collector, drum, high-voltage power supply, needle, and pump. Upon application of a high voltage, the viscous polymer undergoes uniaxial elongation and subsequently emerges through a small nozzle in the form of electrically charged droplets. Charged droplets then condense into a jet (a Taylor cone) and are propelled in the direction of the collector (Fig. **5**) [44]. Collectors might be in the form of discs, plates, or drums. Then, the collector is used to gather nonwoven fibres. In the electrospinning process, viscoelastic polymer solutions are stretched in an electric field with a high voltage and solidified as the solvent is volatilized to create nonwoven nanofiber mats. A macroscale substance is electrospun into a series of uninterrupted, continuous nanoscale fibers. Electrospinning has the potential to create structures without a substrate, with easy procedures and large inputs. The technique employed in the production of nano-fibers typically yields circular cross-sectional shapes. However, it is noteworthy that alternative geometries, including rectangular and coaxial shapes, are also viable options. With a high fraction of β-phase and crystallinity, electrospinning is an effective technique for producing PVDF nano-fibers [14]. This technique works by reorienting molecular dipoles in response to an applied voltage. By adjusting the necessary factors of electrospinning, including temperature, solvent, rate of feeding, and distance between the tip and the collector, PVDF fibres with an overwhelming majority of the β-phase may be produced [49]. Another benefit is that due to polymer jet elongation and whipping, poling may be done during the electrospinning process, enabling electrospun PVDF films to display piezoelectricity without any need for post-poling. The electric forces used in electrospinning can complete the poling and stretching process.

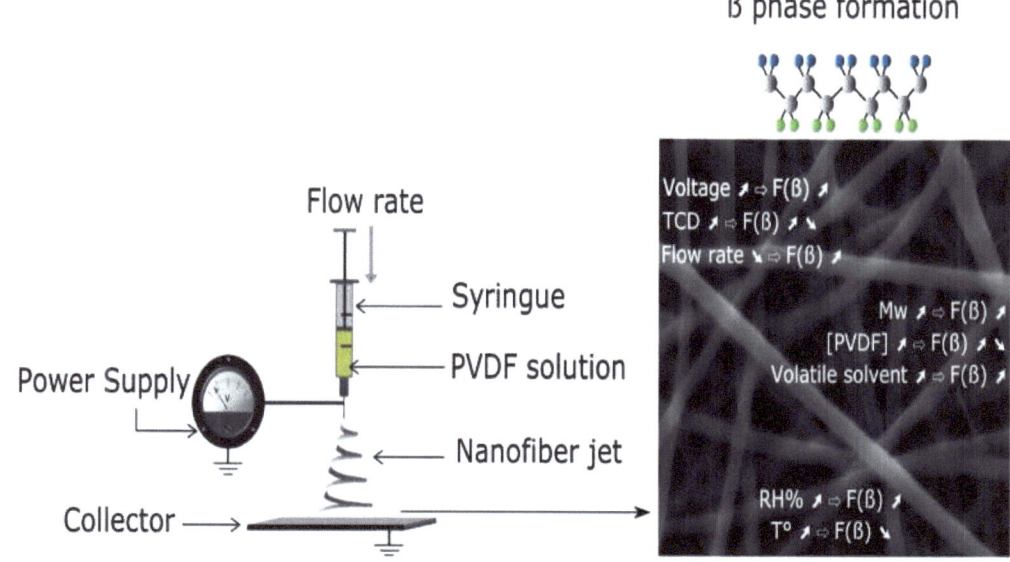

Fig. (5). Schematic diagram of the electrospinning method [44].

Table 1 shows the phase evolution effects of various techniques on piezoelectric films made of PVDF or its copolymers.

Table 1. Phase evolution effects of various techniques on piezoelectric films made of PVDF or its copolymers [14].

Methods	Positive effects	Negative effects	Typical parameters
Adding Filler	(1) formation of β-phase (nucleate agent and charge accumulation in poling). (2) lower electric field for poling. (3) increase in crystallinity degree (uniform distribution).	1) crystal defects. (2) failure in poling process.	(Content: less than 1 wt% (large specific surface area)

(Table 1) cont.....

Methods	Positive effects	Negative effects	Typical parameters
Stretching	1) transformation from α- to β-phase. (2) alignment of dipoles. (3) crystallinity increment (realignment of molecular chains in the amorphous area).	(1) decrease in crystallinity. (2) crystal defects (Higher elongation ratio).	Ts < 100°C, R > 4
Thermal treatment	(1) formation of β-phase. (2) increment in degree of crystallinity (realignment of molecular chains' orientation and position by thermal energy).	(1) γ-phase formation (Easier motion of conformers). (2) increment in degree of crystallinity (amorphous region expansion accompanied by defect formation).	T < 140°C
Poling	1) alignment of dipole moments. (2) transformation from α- to β-phase.	electric breakdown, which occurs under conditions of high electric field and/or high temperature.	E_{max} < 160 MV/m
Filler alignment	(1) alignment of molecular chains as a result of induced stress by magnetic or electric field. (2) formation of β-phase and transformation from α- to β-phase. (3) increase in degree of crystallinity.	Degree of crystallinity (the suppression of the crystallization process at high magnetic fields).	Lack of enough experimental data

When fabricating piezoelectric PVdF materials, a quantitative description of process parameters is crucial for reproducibility and optimization. Here are some key process parameters to consider:

1. Poling voltage: The electric field strength used to align the dipoles in the PVdF material.
2. Poling temperature: The temperature at which the poling process is performed, which can affect the material's piezoelectric properties.
3. Poling Time: The duration of the poling process, which can impact the degree of dipole moment.
4. Film thickness: The thickness of the PVdF film, which can influence its piezoelectric properties and performance.
5. Stretching ratio: The ratio of the stretched to upstretched length of the PVdF film, which can affect its crystallinity and piezoelectric properties.

PVDF-based cOpolymers, Composites, and Nano-composites

A composite material consisting of a polymer matrix with a dispersed ceramic phase is commonly referred to as a ceramic/polymer composite. A "composite" is a combination of two or more components, one of which is a stiff and durable substance known as "filler" and another is a "binder" or "matrix" (polymer) that preserves the filler's position in the mixture [50]. Polymers' mechanical qualities enable them to be used in applications like flexible and wearable electronics, where conventional piezoelectric ceramic crystals are inefficient. They have substantially greater piezoelectric stress constants than ceramics, which makes them superior sensors [20]. Ceramic particles that are piezoelectric, such as micro- and nano-sized $BaTiO_3$, PZT, PMN-PT, and ZnO, are used in Polymer Nanocomposites (NPCs) [37]. Piezoelectric ceramics can be easily blended to create piezocomposites along with polymers. Piezoelectric ceramics are being employed by the researchers as a filler material for specialized applications, including sensors and actuators, owing to the low stiffness and high flexibility of polymers [51]. The majority of these filler materials are made of piezoelectric ceramics like $BaTiO_3$, PZT, PMN-PT, and ZnO [37]. PZT and its nanoparticles are frequently tested in combinations with polymers like PDMS, PVDF, epoxies, and resins due to their high piezoelectric capability. PVDF composites with nano-structured components can perform better. They have a substantial capacity to combine mechanical and electrical characteristics, are structurally flexible, have a high piezoelectric coefficient, have high elastic adherence, are naturally biocompatible, have appropriate properties, are mechanically strong, and have a good corrosive resistance to acidic substances, solvents, and bases.

PVDF and its copolymers have received considerable research attention. Additionally, because of their relatively low density, extremely lightweight, low Refractive Index (RI), and low price index, they are appropriate for a variety of energy harvesting applications [20]. Furthermore, a multitude of design configurations can be achieved through the patterning of electrodes on the surface. For wearable electronics, it is necessary to have lightweight, flexible power supply modules with great energy storage performance. To boost capacitive energy storage, structural flaws were introduced into chains to produce copolymers and terpolymers of PVDF. The science and technology of piezoelectric polymers have been dominated by the fabricated copolymers, such as PVDF-TrFE, PVDF-HFP, and Polyvinylidene Fluoride with Chlorotrifluoroethylene (PVDF-CTFE), which have higher energy densities than mono polymer PVDF [45]. The main properties of PVDF composites and their uses are given in Table **2**. The formation of the phase through copolymerization can be improved without the need for further post-treatments by copolymerizing PVDF with TrFE, also known as $-(CHF-CF_2)-$, and tetrafluoroethylene (TeFE; $-(CF_2-CF_2)-$), in a random order [44]. The development of the phase is constrained by the additional steric barrier caused by fluorine atoms. Additionally, annealing, mechanical stretching, or electrical poling treatments increase the CF_2 dipoles' degree of crystallinity and alignment, which results in greater pyroelectric and piezoelectric effects than those of PVDF. TrFE content has a significant impact on the structural and piezoelectric characteristics of P(VDF-TrFE) [44]. Some of the PVDF copolymers are described below:

Table 2. Main features along with applications of PVDF composites [58].

S. No.	Composite	Improved properties	Applications
1.	PVDF-HFP	Piezoelectric effect, biocompatibility, and chemical stability.	Development of micro energy in flexible, wearable electronics and wireless sensor networks.
2.	P(VDF-TrFE)	Flexibility, chemical stability, and biocompatibility Type of ferroelectric material with a significant piezoelectric effect.	Electrodes were screen-printed on a multi-layered P(VDF-TrFE) structure.
3	Poly(vinylidene cyanide-vinyl acetate) (P[VDCN-VAC]), and(vinylidene fluoridetetra fluoroethylene) (P[VDFTeFE]), and poly(vinylidene fluoridehexafluoropropylene) (P[VDF-HFP])	Lightweight, flexible, and inexpensive with a piezoelectric effect.	Flexible tactile sensors.
4	Lanthanum-doped lead zirconate titanate (PLZT) – PVDF composite with PLZT	In a 50 vol% CNT film, the polar phase was obtained. Piezoelectric constant augmented to 30 pC/N for 50 vol% PLZT-PVDF. Energy density raised to 29.1 mg-cm^{-3} Piezoelectric output.	Energy-harvesting devices.
5	PVDF fibers were prepared using electrospinning – graphene oxide (GO) and reduced graphene oxide (rGO)	The piezoelectric capabilities are aided by RGO and GO nanofillers. The PVDF chain interacts strongly with the filler substance at the interface.	Energy-harvesting devices, portable electronics, wearable devices, and pressure sensors.
6	$Ca_3(PO_4)_2$ nanorod-incorporated PVDF films have been prepared via an in situ flexible piezoelectric nanogenerator	The formation of electroactive crystals was promoted by adding Ca3(PO4)2 at various molar concentrations. Additionally, by using an in situ technique (interfacial polarization), it produces a high dielectric value.	Charge cell phones, small portable electronic gadget.

S. No.	Composite	Improved properties	Applications
7	PVDF-ZnO nanofiber mats prepared with electrospinning	Modification in the crystal structure and improvement in the electrical output.	As an increased energy-scavenging interface, it can offer a simple yet effective, adaptable, and affordable method for self-powering microelectronics for a variety of applications.
8	BaTiO$_3$–PVDF composite nanofibers	Improvement in the polymer phase crystallization and an increase in piezoelectrically generated voltage.	Provides an easy and flexible way for self-powering nanogenerators and microelectronics for a variety of applications in industry, medicine, and other fields.

PVDF-HFP

One of the most popular PVDF copolymers is PVDF-HFP, which can be created by polymerizing VDF (Vinylidene Fluoride) and HFP (Hexafluoropropylene), as shown in Fig. (**6**). Due to the large CF$_3$ groups, copolymerization with HFP significantly lowers the degree of crystallinity, which has an adverse impact on achieving high piezoelectricity [14]. Contrarily, flexibility significantly improves compared to pure PVDF [52]. The increase in fluorine content brought on by the addition of HFP also increases hydrophobicity [53]. According to reports, while its pyroelectric coefficient is larger, its piezoelectric coefficient (d$_{31}$) is comparable to that of PVDF [54].

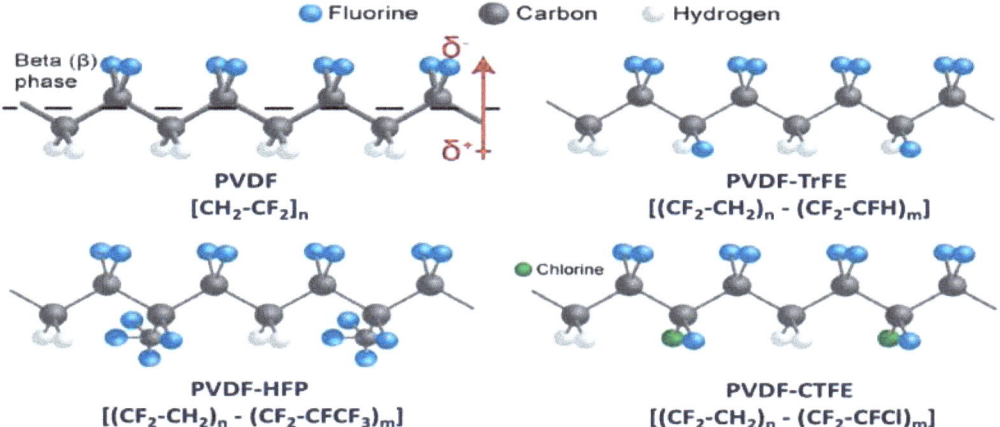

Fig. (6). Structural configuration of PVDF and its copolymers: PVDF–TrFE, PVDF–HFP, and PVDF–CTFE [57].

PVDF-TrFE

PVDF-TrFE is another copolymer of PVDF that can be synthesized through the polymerization of VDF and TrFE, as illustrated in Fig. (**6**). The incorporation of TrFE into the PVDF crystal structure results in a modification of the ferroelectric phase due to an increase in the unit cell size and interplanar distance [55]. The electromechanical coupling factor is enhanced by the superior crystallinity and preferred orientation of well-grown PVDF-TrFE crystals in comparison to PVDF [13]. The synthesis conditions and mass ratio of VDF/TrFE have a significant impact on the piezoelectricity of copolymers. TrFE is added, which makes it simple to form the β-phase. This allows for its polarization without stretching [56].

PVDF NANOFIBERS

The phrase "nanofiber" has several meanings. Nanometer-scale fibres, nanocomposite fibres, and nanostructured surface fibres fall under this category. Nanomaterials, which include nanofibers, have at least one dimension that is equal to or smaller than 100 nanometers. Additionally, they have a length-to-diameter ratio (L/D) that is greater than 100 [44]. The International Standards Organisation (ISO) defines "nanofiber" as a fibre having a mean diameter less than 100 nm; however, the engineered fibres industry also defines nanofibres as fibres with a diameter less than 1000 nm. The ability to create materials with distinctive and better properties or functionalities in comparison to the same materials at a larger scale is made possible by the reduction in the size of materials and their entry into the sphere of nanomaterials. When it comes to fibres, a decrease in diameter primarily causes an increase in specific surface area. High nanofiber surface area changes bioactivity (for example, by increasing cell proliferation), electroactivity, strength, *etc* [44]. Nanofibers are used in many different contexts, including filter materials, electronic technology (including nanoelectronics, optical sensors, battery electrode materials, and capacitors), biomedical science (including tissue engineering and medication delivery), and catalyst supports [59]– [64]. For producing nanofibers, a variety of top-down or bottom-up techniques are available.

Recently, there has been a lot of interest in studying and using PVDF nanofibers with various nanoparticles. Using different structures, such as tiny molecules, polymers, and nanostructures, leads to better and multifunctional features. By adjusting production settings, nanofiber morphology can be altered, which is essential for improving material qualities [65]. An example of the advantageous impact of nanofiber morphology on piezo/pyro/ferroelectric properties is observed through the enhancement of phase due to the reduction in diameter resulting from increased thinning [66]. The hydrophobic, controllable structural characteristics of PVDF nanofibers with nanostructures—length to diameter ratio, high surface to weight ratio, and small diameter (tens to hundreds of nanometers)—as well as piezo/pyro/ferroelectric properties (easily controlled by adjusting production parameters)—make them more noticeable [66]. Nanofibers can exhibit a variety of morphologies, including mesoporous fibres, depending on the synthesis conditions. These topologies give nanofibers a greater specific surface area, which can improve their characteristics, particularly their capacity for sensing [65].

The following discussion covers the procedures for making PVDF nanofibers, as well as some of their main traits and prospective applications in several economic fields. Numerous methods, including centrifugal spinning, blast spinning, and electrospinning, can be used to create nanofibers. Over the last ten years, the majority of research has focused on electrospinning with polymer solutions. Nanofibers can be produced using a variety of spinning techniques, including centrifugal, blast, and electrospinning. The majority of recent research has concentrated on electrospinning with polymer solutions [66].

APPLICATION OF PIEZOELECTRIC MATERIALS

Polyvinylidene Fluoride (PVDF) is a semi-crystalline polymer that exhibits piezoelectricity in response to the application of pressure or mechanical force. When pressure or mechanical force is applied to PVDF, voltage is produced. It vibrates when electricity is added, which is also true in reverse [58]. In sensor applications, the first property is used, while in actuator applications, the reverse process is used. Force is applied to it. Depending on the chain conformation structure, it has five crystal phases: α, β, γ, δ, and ε. Due to its flexible, biocompatible, nontoxic nature and excellent ferroelectric, piezoelectric, and pyroelectric capabilities, it is employed in nanogenerator applications [2, 9, 26]. Barium titanate, lead zirconate titanate, and barium zirconate titanate are examples of piezoelectric ceramics. Some polymers, including PVDF and PVDF-TrFE, also exhibit the piezoelectric effect. A ferroelectric polymer with effective pyroelectric and piezoelectric characteristics is PVDF. It is advantageous in sensor and battery applications because of these qualities. PVDF is typically between 50 and 60 percent crystalline and has a glass transition temperature (T_g) of about 35 °C [67]. Owing to their potential for sensor, EH, and actuation purposes, mechanical vibrations are converted into an electric charge or electric field. The most effective and practical techniques for synthesizing piezoelectric Polyvinylidene Fluoride (PVDF) are being sought after due to the growing interest in its use in Nanogenerators (NGs), sensors, and microdevices [10, 46]. For making PVDF nanofiber films with a higher proportion of the β-phase, electrospinning works well. The flexible nanogenerator was created from the

Polymer Nanocomposite (PVDF-HFP/CoZnO). PVDF and its copolymers are used in a variety of products, such as MEMS, wearable electronics, nanogenerators, and sensors [11, 46, 58]. For applications such as wastewater treatment, purification and filtration technologies, electrolytes, catalysis, biological reactors, the distillation process, separating gases, separators for battery packs with lithium ion, *etc.*, PVDF has a high chemical stability [68 - 70]. It finds use in flexible actuators, biomaterials, health monitoring systems, and wearable functional devices due to its high piezoelectricity, pyroelectricity, and ferroelectricity, as well as lower density, rigidity, and flexibility [42, 71, 72]. Piezoelectric ignition technologies assess the mechanical and physical changes under stress loading in electronic devices [73].

A carbon chain serves as the foundation of polymer piezoelectric materials, which are more flexible than single crystals and ceramics. Due to their considerable flexibility, they can tolerate higher strain, making them better suited for applications requiring significant bending and twisting [42]. For example, PVDF, P(VDF-TrFE), cellulose and its derivatives, polyamides, and polylactic acids are among the piezoelectric polymers that are extensively employed in various applications such as pressure sensors, vibrometers, audio sensors, ultrasonic sensors, impact sensors, tactile sensors, display equipment, piezoelectric catalysis for wastewater treatment, chemical sensors, and medical sensors [2, 14, 58, 74, 75], as shown in Fig. (7).

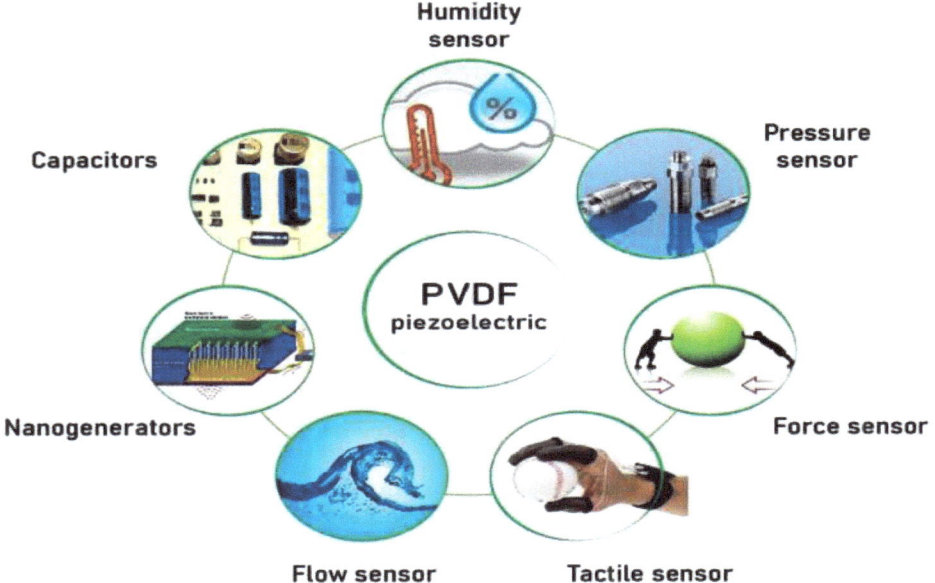

Fig. (7). Applications of PVDF-based piezoelectric polymers [1].

Application in Energy Generation

The utilization of the piezoelectric conversion effect exhibited by piezoelectric materials serves as the underlying principle for the creation of energy harvesting devices [76], as illustrated in Fig. (**8**). Under pressure from external factors, the piezoelectric material deforms and produces a voltage, transforming mechanical energy into electrical energy. Piezoelectric Nanogenerators (PENGs) are among the simplest examples of energy harvesters [46, 77]. Additionally, recent years have concentrated on the development of Piezoelectric Energy Harvesters (PEH) from human body movement. This technology has led to the development of products like the piezoelectric dancing floor, electronic skin and health monitoring patches, e-textiles, and piezoelectric shoes, which use vibrations produced by movements like walking, running, breathing, and dancing to power up small electronic devices.

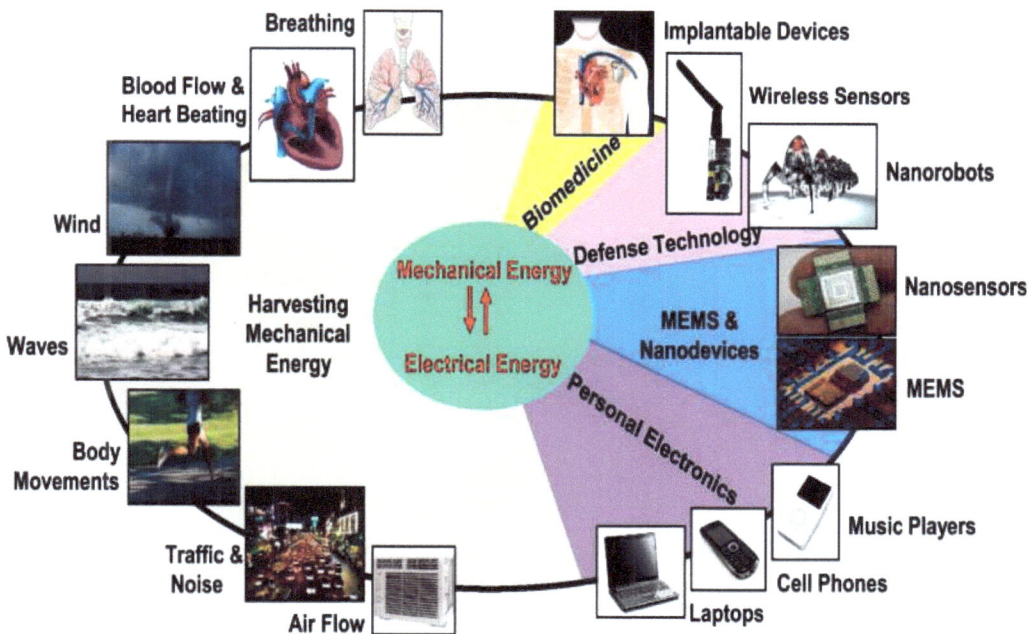

Fig. (8). Various uses of PVDF-based PEMs as an energy harvester [78].

Piezoelectric Nanogenerator (PENG)

A nanogenerator is a technological device that generates electrical energy through the conversion of mechanical or thermal energy, which is triggered by minute physical alterations. Three prevalent techniques are used in it: pyroelectric, triboelectric, and piezoelectric nanogenerators [79]. Electrical energy may be produced using both triboelectric and piezoelectric nanogenerators. However, a time-dependent temperature variation can be exploited to generate thermal energy with pyroelectric nanogenerators. PENGs have the capacity/ability to harvest random electrical energy into electrical energy using nano-scaled PEMs. The potential to convert weak and irregular mechanical energy of the environment into electric current makes PENGs a potential candidate for attention. Harvesting mechanical energy using PENGs can be beneficial, as mechanical energy is omnipresent in the surroundings [46]. Thus, PENGs are a viable alternative to conventional chemical batteries and provide energy sustainably. The first PENG was developed by Z.Wang in 2006 using ZnO nanowires [80]. After that, many other PEMs have been used to develop PENGs. The development history of piezoelectric nanogenerators in recent years is shown in Fig. (9). The nanogenerator is anticipated to be used in a variety of situations where periodic kinetic energy is present, including large-scale processes like wind and ocean waves, as well as smaller ones like heartbeat-induced muscle contractions and lung inhalation [1, 58]. The utilization of implantable telemetric energy receivers, transparent and flexible devices, self-powered nano/micro devices, and smart wearable systems is common.

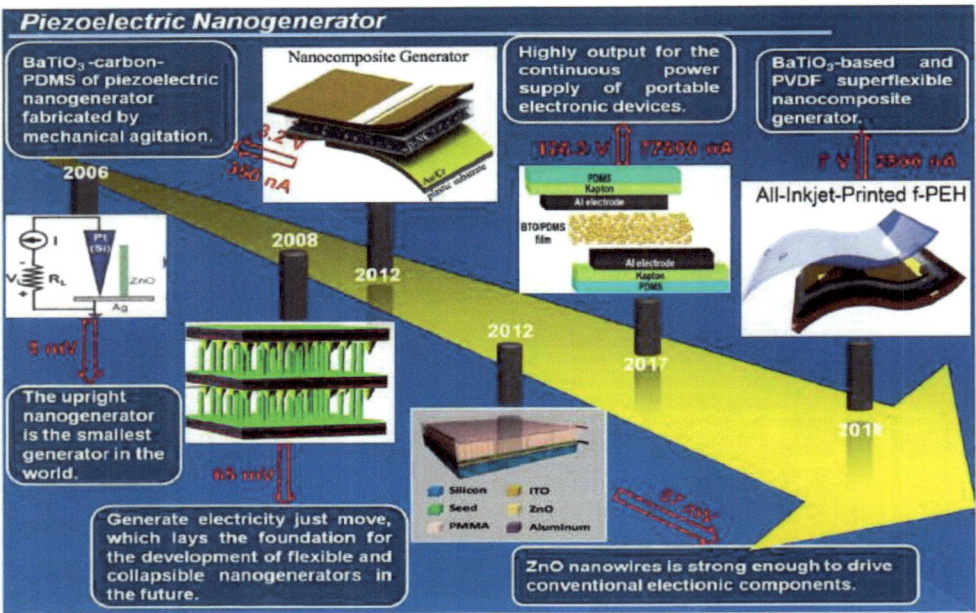

Fig. (9). The development of high-output piezoelectric nanogenerators since their commencement in 2006 [46].

- **PENGs working:** The basic structure of PENG is described as a PEM sandwiched between two electrodes, as depicted in Fig. (**10**). The piezoelectric effect is observed when external mechanical stress is applied to the sandwiched PEM. This changes the dipole moment, causing polarization in the structure. Further, the polarization leads to the generation of a potential difference between the electrodes, causing the flow of electrons through the respective electrodes. Thus, the electrons return to the same state when the stress is released [10, 81]. Various organic materials such as di-isopropylammonium bromide, bimolecular material, nylon, PVDF, and PVDF composites have been developed to counter the inorganic PEMs in PENGs.

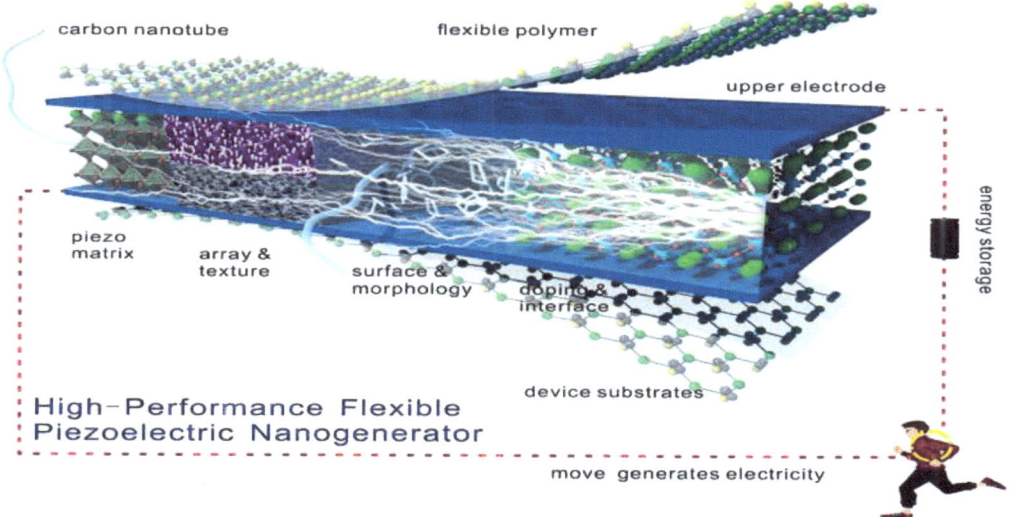

Fig. (10). Structure of flexible PENGs [46].

- **PVDF-based PENGs:** A variety of materials, including ZnO, CdS, GaN, InN, $NaNbO_3$, PZT, $BaTiO_3$, and PVDF, are commonly being employed in the construction of PENGs. However, PZT is not commonly used in PENGs due to the toxicity of lead [46]. In addition, $BaTiO_3$ lacks flexibility, and while ZnO is easy to process, it possesses poor piezoelectric properties. Therefore, the PEMs based on organic polymers are used to develop flexible and stable PENGs [46]. However, polymer-based PENGs are limited to low output efficiency. PVDF can be considered to develop PENGs because of its flexibility, easy processing, chemical resistance, and biocompatibility. Apart from material selection, the performance of PENGs is also dependent on several other factors, as shown in Fig. (**11a** and **11b**).

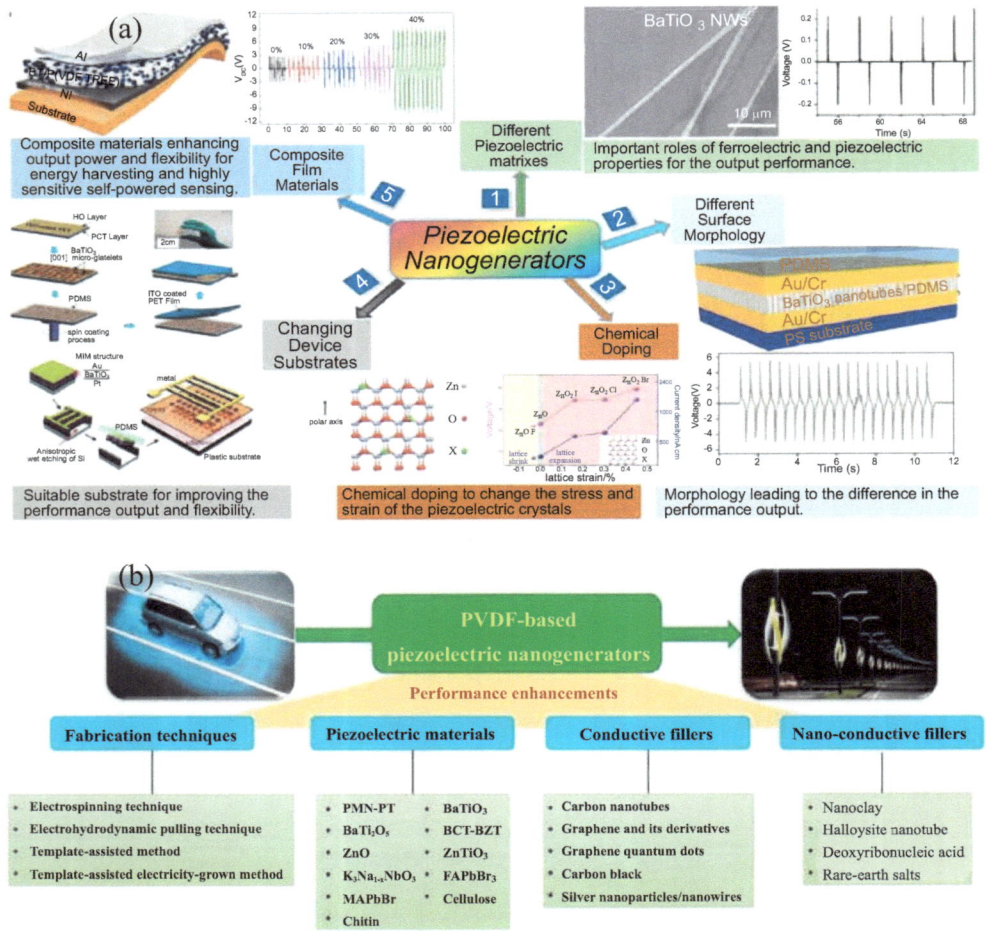

Fig. (11). (a) Various factors affecting PENGs' output performance [46] and (b) performance enhancement of PVDF-based PENGs [77].

Mechanical NGs constructed from PVDF nanofibers prove to be viable possibilities for creating energy by directly exploiting human-produced mechanical energy. The device known as a nanogenerator is used to generate voltage or power. The application is also quite flexible, which is convenient [46, 81, 82]. Applying mechanical force or vibration provides enormous power [2, 58, 80]. PVDF is a polymer and one of the piezoelectric materials. The mechanical energy can be converted into electricity *via* a triboelectric and piezoelectric nanogenerator. However, thermal energy can be obtained from a time-dependent temperature variation using pyroelectric nanogenerators. The PVDF-based PENGs' multi-functionalities, including their high sensitivity, flexibility, stretchability, and multi-shape transformability, were shown to be necessary for the steady and continuous operation of self-powered sensors. The most recent developments in piezoelectric PVDF production techniques for diverse structures, as well as structure-dependent piezoelectricity, are illustrated in Fig. (**12**). PENGs based on PVDF and its composites exhibit varying output performance depending on the film structure, whether it is a thin film, microstructured, nanostructured, or a nanocomposite [2, 10, 58] (Fig. **12**).

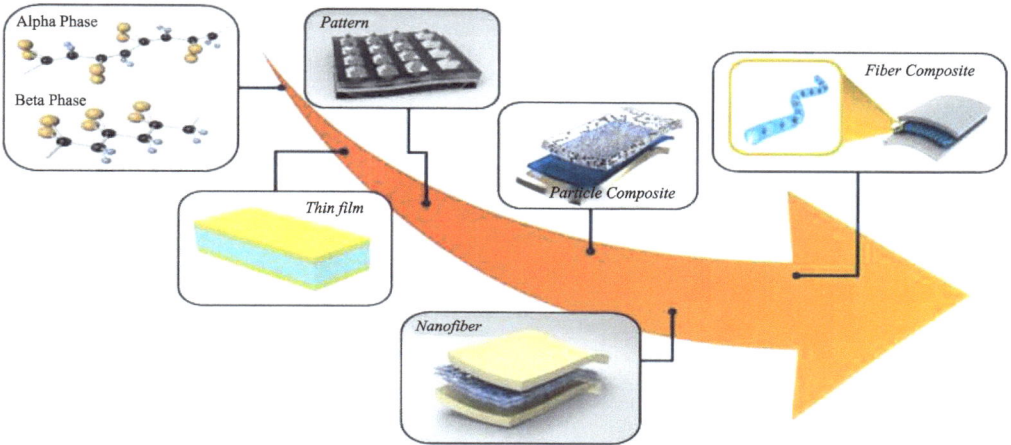

Fig. (12). A graphical illustration of the several piezoelectric nanogenerators based on structured polyvinylidene fluoride (PVDF) [10].

In the preparation of nanogenerators, a variety of materials are utilized, including zinc oxide nanowires, lead zirconate titanate, barium titanate nanowires, and PVDF. The nanogenerator is anticipated to be used in a variety of situations where periodic kinetic energy is present, including large-scale processes like wind and ocean waves, as well as smaller ones like heartbeat-induced muscle contractions and lung inhalation [83, 84]. PENG devices based on PVDF can be engineered to be stretchable, twistable, and foldable, making them perfect for wearable electronics [45]. Applications for PVDF-based nanogenerators include wireless transmission, capacitor charging, and LED light glow, as shown in Fig. (**13**).

Sensors

Sensors are tools that are used to find or sense various kinds of physical quantities in the environment. The input could be vibration, motion, wetness, heat, or light. Piezoelectric polymer materials with flexible properties are employed for developing flexible and wearable pressure sensors for sensing applications [1, 42, 86]. PVDF (Polyvinylidene Fluoride) resolves the drawbacks of PZTs and also has the benefits of a flexible structure and stable performance, especially for EH [58]. The low piezoelectric coefficient of PVDF, which is typically considered a disadvantage, can in fact be advantageous for impact applications as sensors are not susceptible to voltage saturation when detecting an impact load [87]. PVDF films also cost a lot less than PZTs, as shown in Fig. (14). PVDF films offer a wide range of uses, including structural health monitoring. Fig. (14) shows the schematic diagram of a PVDF-based sensor.

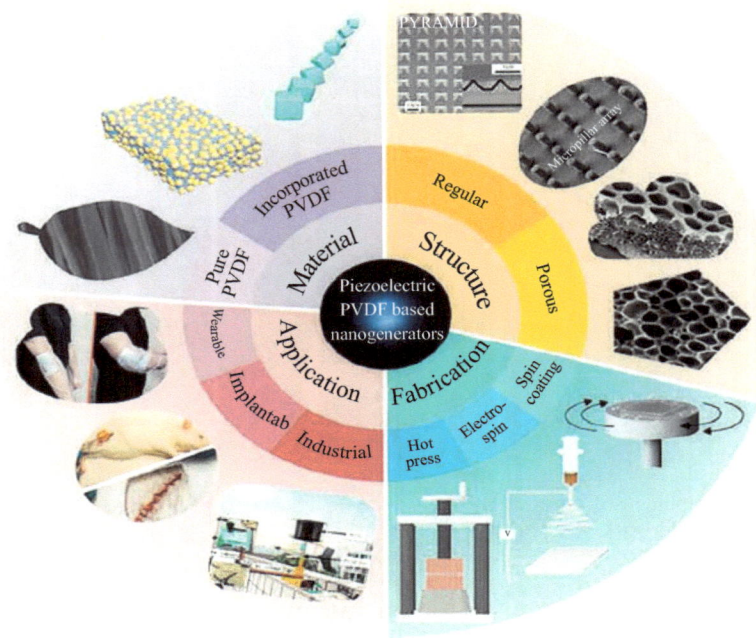

Fig. (13). An overview of PVDF-based piezoelectric nanogenerators [85].

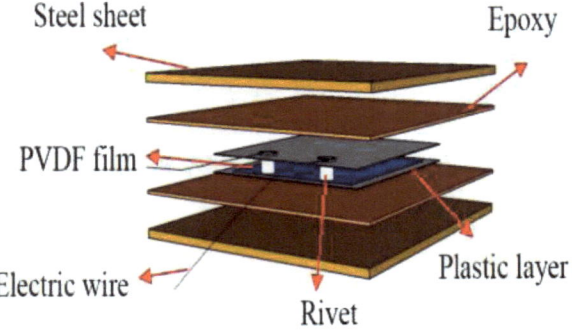

Fig. (14). Schematic diagram of PVDF-based sensor [87].

Various Types of Sensors

Tactile Sensors

A device that measures forces in response to physical contact with the environment is called a tactile sensor. The sense of touch in humans is modelled based on the biological sense of touch provoked by the mechanical stimulation of the skin. Tactile sensing is created by a coordinated set of touch sensors [1, 37]. Touch sensing pertains to the identification and/or quantification of the magnitude of pressure and spatial information at a given location. These sensors' ability to sense minute variations in contact force with extremely high sensitivity is a crucial feature. Numerous applications, such as in the automotive sector, use tactile sensors, such as brakes, clutches, door seals, gaskets, and so on [37]. Electrospun PVDF is frequently utilized in tactile sensing because of its capacity to transform mechanical deformations into electrical signals. To detect any movement or displacement, such as body movements, *etc.*, tactile sensors are used. Tactile sensors come in a variety of forms, including triboelectric, piezoelectric, piezo-resistive, capacitive, optical, and piezoelectric [1]. Due to the limitations of piezoelectric ceramics, such as their high cost of production, difficulty in processing, and lack of flexibility, polymer materials have attracted a lot of attention. Figs. (**15a** and **15b**) show the uses of tactile sensors in human body monitoring. Tactile sensing is generated by a coordinated group of touch sensors. From touch sensing, we can detect the measurement of point contact force and location information. These sensors have very high sensitivities to very small changes in contact force. These sensors have applications in areas like the automobile industry, *e.g.*, brakes, door seals, fuel cells and so on. Tactile sensors do not need any external power. These sensors are utilized in human health monitoring and human-machine interaction. These sensors have the potential to become an integral part of wearable devices due to their biocompatibility, portability, and small volume.

Force Sensors

In the 1970s, Sir Franklin Eventoff discovered that the resistance feedback of some materials can alter when they are pressed. Force-sensing resistors were the name given to these substances. A sensor that can measure force is made from these materials. The force imparted to an object can be measured with the use of a force sensor [1]. The applied force can be determined by measuring how much the resistance readings of force-sensing resistors fluctuate [90]. "The working principle behind the force sensor is based on the property of contact resistance. When the force is applied on the surface of the film, the microsized particle touches the sensor electrode by changing the resistance of the film."Applications for force sensors that use force-sensing resistors include pressure-sensitive buttons, musical instruments, automotive occupancy sensors, *etc* [91].

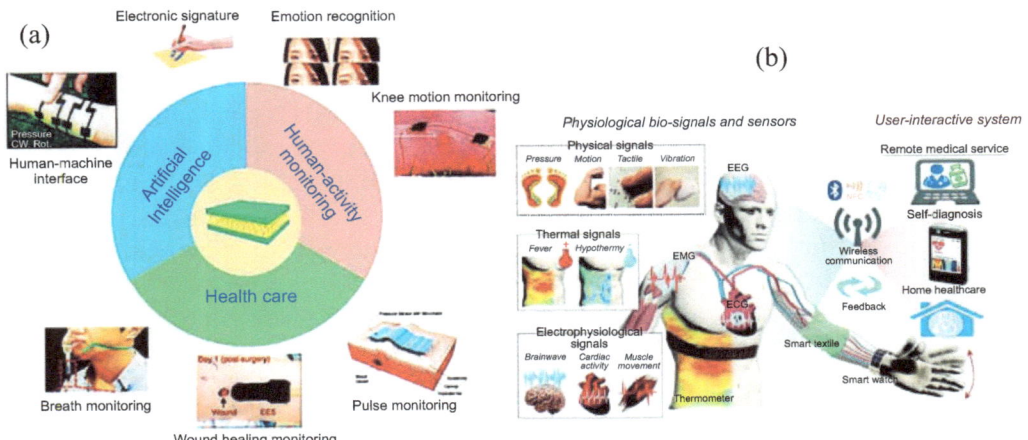

Fig. (15). (a) Various uses of tactile sensors [88] and (b) Application of tactile sensors in human body monitoring [89].

Self-charging Power Cell

A Self-Charging Power Cell (SCPC) is a device that combines the mechanisms for energy storage and conversion into a single system, allowing mechanical energy to be transformed directly into electrochemical energy. The energy may be stored in power cells, and electricity can be produced from a variety of energy sources by generators, utilizing chemical energy storage to produce electrical energy [92]. Energy generation and storage are often thought of as two distinct processes needing two different devices, but recent developments have shown that we can perform both processes with just one device. As power cells grow more efficient and gadgets use less power, research into self-charging power cells and batteries is advancing quickly. In order to operate, these self-charging power cells convert mechanical energy, the motion and position of an object's associated energy, into chemical energy [93]. As an electric current, the transformed energy can be released by the technologies and used to power other devices. Self-charging batteries may improve energy efficiency and reduce the size and weight of our electronic devices. A simple diagram of a self-charging power cell is shown in Fig. (**16**).

Self-charging power cell is a mechanism that works under an electrochemical process. It is inspired by the piezoelectric capacity operated by deformation. A self-charging power cell is composed of basically 3 major components: anode, separator, and cathode. Xinyu *et al*. [93] took $LiCoO_2$ as a cathode, aligned TiO_2

nanotube as an anode, and 1M $LiPF_6$ in 1:1 EC:DMC as an electrolyte. This structure was sealed in a 2016 stainless steel coin-type cell. They found that the stored electric capacity of the power cell was about 0.036 µAh. The self-charging process is governed by the piezoelectric effect by changing parameters such as frequency and applied force.

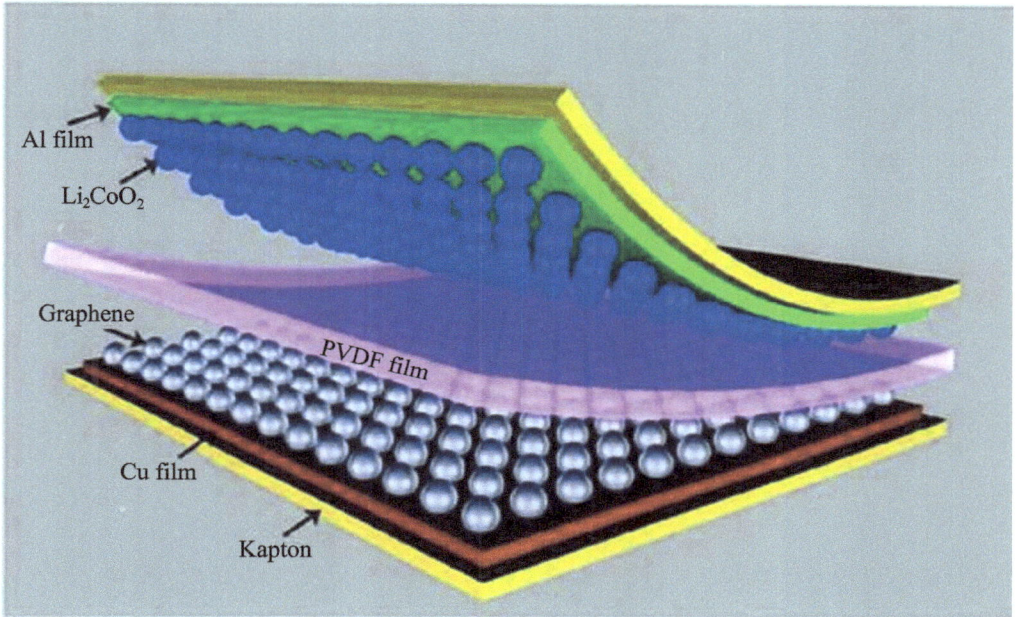

Fig. (16). A schematic illustration of the flexible SCPC design. Anode is made of graphene nanosheet composites on copper foil; separator is a layer of polarized PVDF film; cathode is a mixture of $LiCoO_2$-based materials on aluminium foil; and the shells of flexible SCPCs are made of Kapton boards [92].

Piezoelectric Transducer

A piezoelectric transducer is a device that uses the piezoelectric effect to monitor changes in acceleration, pressure, tension, temperature, or force by converting this energy into an electric charge [94 - 96]. Anything that changes one form of energy into another is a transducer. One type of transducer is made of piezoelectric material. The transducer transforms this energy into voltage when we squeeze or exert any pressure or force on this piezoelectric material. This voltage depends on the pressure or force exerted on it. The voltage measuring equipment can quickly and easily measure the electric voltage generated by a piezoelectric transducer, as shown in Fig. (17). We can determine the force or pressure that was exerted by looking at the voltage readout because this voltage depends on it. Therefore, the utilization of a piezoelectric transducer enables the direct sensing of physical quantities, such as mechanical stress or force.

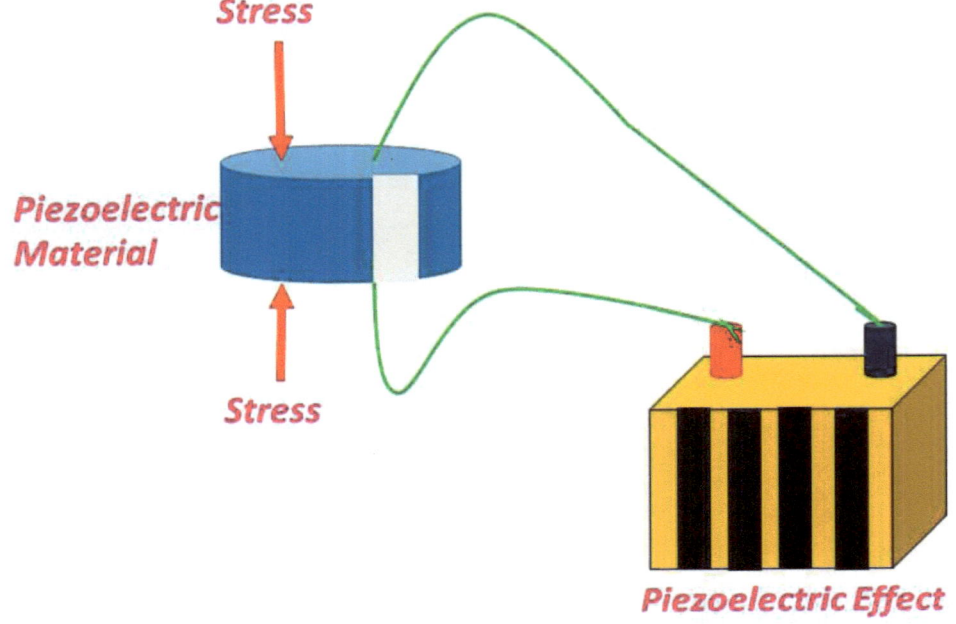

Fig. (17). Piezoelectric transducer.

Application of Piezoelectric Transducers

Piezoelectric transducers can be employed in many different applications by using piezoelectric materials, such as:

1. The process of converting sound pressure into an electric signal in microphones, followed by amplification to produce a louder sound, is a fundamental aspect of audio technology.
2. A piezoelectric substance is also used to lock car seat belts in response to a sudden deceleration.
3. It also functions as a diagnostic tool in medicine.
4. They are used to analyze blast waves and shock waves that travel at high speeds.
5. They are utilized in kitchen electric lighters. An electric signal is produced when pressure is applied to a piezoelectric sensor, and this signal finally triggers the flash.
6. They are used in Inkjet printers.

. It is also utilized at restaurants and airports, where the door opens as a patron approaches it. The idea behind this is that when someone gets close to the door, their weight puts pressure on the sensors, creating an electric reaction that causes the door to open on its own.

Piezoelectric Actuator

The behavior of a piezoelectric actuator is the opposite of that of a piezoelectric sensor. It is the one in which the material deforms, stretches, or bends as a result of the electric influence [2, 4, 92], as depicted in Fig. (**18**). This suggests that a piezoelectric actuator undergoes deformation, stretching, or bending upon the application of an electric potential, in contrast to a piezoelectric sensor that produces an electric potential in response to the application of force to stretch or bend it.

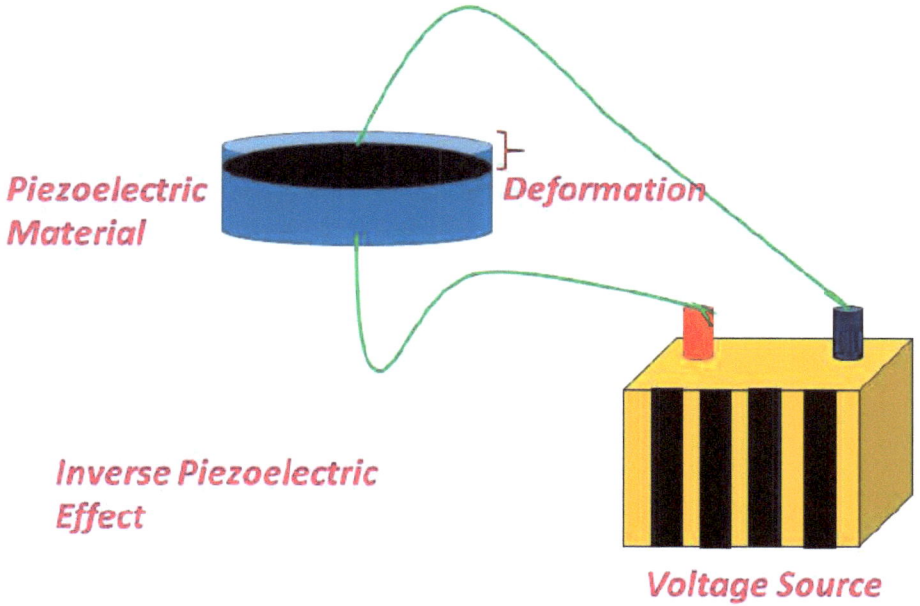

Fig. (18). Piezoelectric Actuator.

CONCLUSION

Despite having a high strength and a high piezoelectric coefficient, inorganic piezoelectric materials are brittle and unsuitable for applications requiring more flexibility. Polyvinylidene Fluoride (PVDF) is a highly advantageous material in the advancement of piezoelectric materials, serving as a prototypical example of flexible piezoelectric polymers. It is extremely adaptable, and doping optimization or other procedures can also enhance its piezoelectric capabilities. The main problem of using traditional techniques to create PVDF films, such as drawing, spin coating, and solution casting, is the requirement for additional processing (poling or mechanical stretching) to enhance the material's piezoelectricity. On the other hand, due to its promising outcomes and straightforward preparation steps, the electrospinning approach is currently frequently utilized to create piezoelectric PVDF. It regulates the polymorphism of the PVDF (α, β, and γ-phase) and promotes the creation of the β-phase, which has exceptional piezoelectric characteristics. As a result of their outstanding physical and mechanical properties, chemically stable nature, and affordable manufacturing costs, PVDF-based piezoelectric materials have much higher potential for application in devices such as sensors, energy generation, and biomedical science, among other fields.

REFERENCES

[1] G. Kalimuldina, N. Turdakyn, I. Abay, A. Medeubayev, A. Nurpeissova, D. Adair, and Z. Bakenov, "A review of piezoelectric PVDF film by electrospinning and its applications", *Sensors (Basel)*, vol. 20, p. 5214, 2020.

[2] S. Rajput, M. Averbukh, and N. Rodriguez, "Energy harvesting and energy storage systems", *Electronics (Basel)*, vol. 11, p. 984, 2022.

[3] Q. Xu, X. Gao, S. Zhao, Y.N. Liu, D. Zhang, K. Zhou, H. Khanbareh, W. Chen, Y. Zhang, and C. Bowen, "Construction of bio-piezoelectric platforms: From structures and synthesis to applications", *Adv. Mater.*, vol. 33, p. 2008452, 2021.

[4] S. Rajput, A. Sharma, V. Jately, M. Ram, Ed., *Recent Advances in Energy Harvesting Technologies*. 1st ed. River Publishers, 2023.

[5] T. Furukawa, "Piezoelectricity and pyroelectricity in polymers", *IEEE Trans. Electr. Insul.*, vol. 24, pp. 375-394, 1989.

[6] L. Persano, A. Catellani, C. Dagdeviren, Y. Ma, X. Guo, Y. Huang, A. Calzolari, and D. Pisignano, "Shear Piezoelectricity in Poly (vinylidenefluoride-co-trifluoroethylene): Full Piezotensor Coefficients by Molecular Modeling, Biaxial Transverse Response, and Use in Suspended Energy-Harvesting Nanostructures", *Adv. Mater.*, vol. 28, pp. 7633-7639, 2016.

[7] M. Smith, Y. Calahorra, Q. Jing, and S. Kar-Narayan, *Direct observation of shear piezoelectricity in poly-l-lactic acid nanowires*. vol. Vol. 5. APL Materials, 2017.

[8] Z. Yang, S. Zhou, J. Zu, and D. Inman, "High-performance piezoelectric energy harvesters and their applications", *Joule*, vol. 2, pp. 642-697, 2018.

[9] M. Kumar, N.D. Kulkarni, and P. Kumari, "Fabrication and characterization of PVDF/BaTiO3 nanocomposite for energy harvesting application", *Mater. Today Proc.*, vol. 56, pp. 1151-1155, 2022.

[10] X. Hu, X. Bao, J. Wang, X. Zhou, H. Hu, L. Wang, S. Rajput, Z. Zhang, N. Yuan, G. Cheng, and J. Ding, "Enhanced energy harvester performance by a tension annealed carbon nanotube yarn at extreme temperatures", *Nanoscale*, vol. 14, pp. 16185-16192, 2022.

[11] N.D. Kulkarni, M. Kumar, and P. Kumari, "PVDF/RGO based piezoelectric nanocomposite films for enhanced mechanical and dielectric properties", *Mater. Today Proc.*, vol. 76, pp. 81-87, 2023.

[12] A.I. Kingon, and S. Srinivasan, "Lead zirconate titanate thin films directly on copper electrodes for

ferroelectric, dielectric and piezoelectric applications", *Nat. Mater.*, vol. 4, pp. 233-237, 2005.

[13] P. Martins, A.C. Lopes, and S. Lanceros-Mendez, "Electroactive phases of poly (vinylidene fluoride): Determination, processing and applications", *Prog. Polym. Sci.*, vol. 39, pp. 683-706, 2014.

[14] L. Wu, Z. Jin, Y. Liu, H. Ning, X. Liu, and N. Alamusi, "Hu, ''Recent advances in the preparation of PVDF-based piezoelectric materials", *Nanotechnol. Rev.*, vol. 11, pp. 1386-1407, 2022.

[15] G.H. Kim, S.M. Hong, and Y. Seo, "Piezoelectric properties of poly (vinylidene fluoride) and carbon nanotube blends: β-phase development", *Phys. Chem. Chem. Phys.*, vol. 11, pp. 10506-10512, 2009.

[16] L. Wu, M. Jing, Y. Liu, H. Ning, X. Liu, S. Liu, L. Lin, N. Hu, and L. Liu, "Power generation by PVDF-TrFE/graphene nanocomposite films", *Compos., Part B Eng.*, vol. 164, pp. 703-709, 2019.

[17] K.S. Ramadan, D. Sameoto, and S. Evoy, "A review of piezoelectric polymers as functional materials for electromechanical transducers", *Smart Mater. Struct.*, vol. 23, p. 033001, 2014.

[18] A.C. Lopes, J. Gutiérrez, and J.M. Barandiarán, "'Direct fabrication of a 3D-shape film of polyvinylidene fluoride (PVDF) in the piezoelectric β-phase for sensor and actuator applications, '", *Eur. Polym. J.*, vol. 99, pp. 111-116, 2018.

[19] S.P. Muduli, L. Lipsa, A. Choudhary, S. Rajput, and S. Parida, "Modulation of electrical characteristics of polymer-ceramic-graphene hybrid composite for piezoelectric energy harvesting", *ACS Appl. Electron. Mater.*, vol. 5, pp. 3023-3037, 2023.

[20] S.P. Muduli, S. Parida, S.K. Behura, S. Rajput, S.K. Rout, and S. Sareen, "Synergistic effect of graphene on dielectric and piezoelectric characteristic of PVDF-(BZT-BCT) composite for energy harvesting applications", *Polym. Adv. Technol.*, vol. 33, pp. 3628-3642, 2022.

[21] R. Guo, H. Zhang, S. Cao, X. Cui, Z. Yan, and S. Sang, "''A self-powered stretchable sensor fabricated by serpentine PVDF film for multiple dynamic monitoring, '", *Mater. Des.*, vol. 182, p. 108025, 2019.

[22] M. Wegener, W. Künstler, and R. Gerhard-Multhaupt, "Piezo-, pyro-and ferroelectricity in poly (vinylidene fluoride-hexafluoropropylene) copolymer films", *Integr. Ferroelectr.*, vol. 60, pp. 111-116, 2004.

[23] G.T. Hwang, V. Annapureddy, J.H. Han, D.J. Joe, C. Baek, D.Y. Park, D.H. Kim, J.H. Park, C.K. Jeong, K.I. Park, and J.J. Choi, "Self-powered wireless sensor node enabled by an aerosol-deposited PZT flexible energy harvester", *Adv. Energy Mater.*, vol. 6, p. 1600237, 2016.

[24] T. Badapanda, R. Harichandan, T.B. Kumar, S. Parida, S. Rajput, P. Mohapatra, S. Anwar, and R. Ranjan, "Improvement in dielectric and ferroelectric property of dysprosium doped barium bismuth titanate ceramic", *J. Mater. Sci. Mater. Electron.*, vol. 27, pp. 7211-7221, 2016.

[25] E.L. Nix, and I.M. Ward, "The measurement of the shear piezoelectric coefficients of polyvinylidene fluoride", *Ferroelectrics*, vol. 67, pp. 137-141, 1986.

[26] R. Fu, S. Chen, Y. Lin, S. Zhang, J. Jiang, Q. Li, and Y. Gu, "Improved piezoelectric properties of electrospun poly (vinylidene fluoride) fibers blended with cellulose nanocrystals", *Mater. Lett.*, vol. 187, pp. 86-88, 2017.

[27] S. Manna, S.K. Batabyal, and A.K. Nandi, "Preparation and characterization of silver− poly (vinylidene fluoride) nanocomposites: formation of piezoelectric polymorph of poly (vinylidene fluoride)", *J. Phys. Chem. B*, vol. 110, pp. 12318-12326, 2006.

[28] S. You, L. Zhang, J. Gui, H. Cui, and S. Guo, "A flexible piezoelectric nanogenerator based on aligned P (VDF-TrFE) nanofibers", *Micromachines (Basel)*, vol. 10, p. 302, 2019.

[29] H Li, C Tian, and Z D Deng, "Energy harvesting from low frequency applications using piezoelectric materials", In: *Appl Phys Rev* vol. 1. , 2014.

[30] M. Smith, and S. Kar-Narayan, "Piezoelectric polymers: Theory, challenges and opportunities", *Int. Mater. Rev.*, vol. 67, pp. 65-88, 2022.

[31] X.H. Du, J. Zheng, U. Belegundu, and K. Uchino, "Crystal orientation dependence of piezoelectric

properties of lead zirconate titanate near the morphotropic phase boundary", *Appl. Phys. Lett.*, vol. 72, pp. 2421-2423, 1998.

[32] H. Kawai, "the piezoelectricity of poly (vinylidene fluoride)", *Jpn. J. Appl. Phys.*, vol. 8, p. 975, 1969.

[33] J.H. McFee, J.G. Bergman Jr, and G.R. Crane, "Pyroelectric and nonlinear optical properties of poled polyvinylidene fluoride films", *Ferroelectrics,* vol. 3, pp. 305-313, 1972.

[34] P. Buchman, "Pyroelectric and switching properties of polyvinylidene fluoride film", *Ferroelectrics,* vol. 5, pp. 39-43, 1973.

[35] R.G. Kepler, and R.A. Anderson, "Ferroelectricity in polyvinylidene fluoride", *J. Appl. Phys.*, vol. 49, pp. 1232-1235, 1978.

[36] J.C. Dubois, "Ferroelectric polymers: Chemistry, physics, and applications", *Adv. Mater.*, vol. 8, p. 542, 1996.

[37] K.K. Sappati, and S. Bhadra, "Piezoelectric polymer and paper substrates: A review", *Sensors (Basel),* vol. 18, p. 3605, 2018.

[38] J. Hillenbrand, and G.M. Sessler, "Quasistatic and dynamic piezoelectric coefficients of polymer foams and polymer film systems", *IEEE Trans. Dielectr. Electr. Insul.*, vol. 11, pp. 72-79, 2004.

[39] S. Bauer, R. Gerhard-Multhaupt, and G.M. Sessler, "Ferroelectrets: Soft electroactive foams for transducers", *Phys. Today,* vol. 57, pp. 37-43, 2004.

[40] R.E. Newnham, D.P. Skinner, and L.E. Cross, "'Connectivity and piezoelectric-pyroelectric composites, '", *Mater. Res. Bull.,* vol. 13, pp. 525-536, 1978.

[41] S.M. Pilgrim, and R.E. Newnham, "3: 0: A new composite connectivity", *Mater. Res. Bull.,* vol. 21, pp. 1447-1454, 1986.

[42] S. Rajput, M. Averbukh, and A. Yahalom, "Electric power generation using a parallel-plate capacitor", *Int. J. Energy Res.*, vol. 43, pp. 3905-3913, 2019.

[43] L. Ruan, X. Yao, Y. Chang, L. Zhou, G. Qin, and X. Zhang, "Properties and applications of the β phase poly (vinylidene fluoride)", *Polymers (Basel),* vol. 10, p. 228, 2018.

[44] Z. He, F. Rault, M. Lewandowski, E. Mohsenzadeh, and F. Salaün, "Electrospun PVDF nanofibers for piezoelectric applications: A review of the influence of electrospinning parameters on the β phase and crystallinity enhancement", *Polymers (Basel),* vol. 13, p. 174, 2021.

[45] W. Zhang, G. Wu, H. Zeng, Z. Li, W. Wu, H. Jiang, W. Zhang, R. Wu, Y. Huang, and Z. Lei, "'The preparation, structural design, and application of electroactive poly (vinylidene fluoride)-based materials for wearable sensors and human energy harvesters", *Polymers (Basel),* vol. 15, p. 2766, 2023.

[46] D. Hu, M. Yao, Y. Fan, C. Ma, M. Fan, and M. Liu, "Strategies to achieve high performance piezoelectric nanogenerators", *Nano Energy,* vol. 55, pp. 288-304, 2019.

[47] C. Lee, and J.A. Tarbutton, "Polyvinylidene Fluoride (PVDF) direct printing for sensors and actuators", *Int. J. Adv. Manuf. Technol.*, vol. 104, pp. 3155-3162, 2019.

[48] S.K. Ghosh, A. Biswas, S. Sen, C. Das, K. Henkel, D. Schmeisser, and D. Mandal, "Yb3+ assisted self-polarized PVDF based ferroelectretic nanogenerator: A facile strategy of highly efficient mechanical energy harvester fabrication", *Nano Energy,* vol. 30, pp. 621-629, 2016.

[49] T. Lei, P. Zhu, X. Cai, L. Yang, and F. Yang, "Electrospinning of PVDF nanofibrous membranes with controllable crystalline phases", *Appl. Phys., A Mater. Sci. Process.*, vol. 120, pp. 5-10, 2015.

[50] A.D. Hussein, "Output piezoelectric layers nanofibers of nanocomposites PVDF/Ceramic," Eurasian Journal of Physics", *Chemistry and Mathematics,* vol. 18, pp. 65-71, 2023.

[51] M. Habib, I. Lantgios, and K. Hornbostel, "'A review of ceramic, polymer and composite piezoelectric materials, '", *J. Phys. D Appl. Phys.*, vol. 55, p. 423002, 2022.

[52] B. Neese, Y. Wang, B. Chu, K. Ren, S. Liu, Q.M. Zhang, and C. Huang, "J. West," Piezoelectric responses in poly (vinylidene fluoride/hexafluoropropylene) copolymers", *Appl. Phys. Lett.*, vol. 90, p. 242917, 2007.

[53] L. Shi, R. Wang, and Y. Cao, "Effect of the rheology of poly (vinylidene fluoride-c--hexafluropropylene)(PVDF–HFP) dope solutions on the formation of microporous hollow fibers used as membrane contactors", *J. Membr. Sci.*, vol. 344, pp. 112-122, 2009.

[54] Y. Huan, Y. Liu, Y. Yang, and Y. Wu, "Influence of extrusion, stretching and poling on the structural and piezoelectric properties of poly (vinylidene fluoride-hexafluoropropylene) copolymer films", *J. Appl. Polym. Sci.*, vol. 104, pp. 858-862, 2007.

[55] D. Mao, B.E. Gnade, and M.A. Quevedo-Lopez, "Ferroelectric properties and polarization switching kinetic of poly (vinylidene fluoride-trifluoroethylene) copolymer", In: *Lallart M., F-P. Effects*, Ed., InTech, 2011, pp. 78-100.

[56] T. Yamada, and T. Kitayama, "Ferroelectric properties of vinylidene fluoride-trifluoroethylene copolymers", *J. Appl. Phys.*, vol. 52, pp. 6859-6863, 1981.

[57] X. Li, J. Pan, F. Macedonio, C. Ursino, M. Carraro, M. Bonchio, E. Drioli, A. Figoli, Z. Wang, and Z. Cui, "Fluoropolymer membranes for membrane distillation and membrane crystallization", *Polymers (Basel)*, vol. 14, p. 5439, 2022.

[58] S. Mohammadpourfazeli, S. Arash, A. Ansari, S. Yang, K. Mallick, and R. Bagherzadeh, "Future prospects and recent developments of polyvinylidene fluoride (PVDF) piezoelectric polymer; fabrication methods, structure, and electro-mechanical properties", *RSC Advances*, vol. 13, pp. 370-387, 2023.

[59] S.J. Choi, L. Persano, A. Camposeo, J.S. Jang, W.T. Koo, S.J. Kim, H.J. Cho, I.D. Kim, and D. Pisignano, "Electrospun nanostructures for high performance chemiresistive and optical sensors", *Macromol. Mater. Eng.*, vol. 302, p. 1600569, 2017.

[60] Y. Fang, Y. Lv, R. Che, H. Wu, X. Zhang, D. Gu, G. Zheng, and D. Zhao, "Two-dimensional mesoporous carbon nanosheets and their derived graphene nanosheets: synthesis and efficient lithium ion storage", *J. Am. Chem. Soc.*, vol. 135, pp. 1524-1530, 2013.

[61] D.H. Kang, and H.W. Kang, "Surface energy characteristics of zeolite embedded PVDF nanofiber films with electrospinning process", *Appl. Surf. Sci.*, vol. 387, pp. 82-88, 2016.

[62] R. Gopal, S. Kaur, Z. Ma, C. Chan, S. Ramakrishna, and T. Matsuura, "Electrospun nanofibrous filtration membrane", *J. Membr. Sci.*, vol. 281, pp. 581-586, 2006.

[63] R.S. Barhate, C.K. Loong, and S. Ramakrishna, "Preparation and characterization of nanofibrous filtering media", *J. Membr. Sci.*, vol. 283, pp. 209-218, 2006.

[64] R.S. Barhate, and S. Ramakrishna, "Nanofibrous filtering media: Filtration problems and solutions from tiny materials", *J. Membr. Sci.*, vol. 296, pp. 1-8, 2007.

[65] J. Ning, M. Yang, H. Yang, and Z. Xu, "Tailoring the morphologies of PVDF nanofibers by interfacial diffusion during coaxial electrospinning", *Mater. Des.*, vol. 109, pp. 264-269, 2016.

[66] S. Aghayari, "PVDF composite nanofibers applications", *Heliyon*, vol. 8, p. e11620, 2022.

[67] D. Lolla, J. Gorse, C. Kisielowski, J. Miao, P.L. Taylor, G.G. Chase, and D.H. Reneker, "Polyvinylidene fluoride molecules in nanofibers, imaged at atomic scale by aberration corrected electron microscopy", *Nanoscale*, vol. 8, pp. 120-128, 2016.

[68] E. Yuliwati, A.F. Ismail, T. Matsuura, M.A. Kassim, and M.S. Abdullah, "Effect of modified PVDF hollow fiber submerged ultrafiltration membrane for refinery wastewater treatment", *Desalination*, vol. 283, pp. 214-220, 2011.

[69] S. Rajput, R. Katoch, K.K. Sahoo, G.N. Sharma, S.K. Singh, R. Gupta, and A. Garg, "Enhanced ferroelectricity in La and Ni co-doped BiFeO3 thin films", *J. Alloys Compd.*, vol. 621, pp. 339-344,

2015.

[70] P. Chaturvedi, A.B. Kanagaraj, A. Alhammadi, H. Al Shibli, and D.S. Choi, "Fabrication of PVDF–HFP-based microporous membranes by the tape casting method as a separator for flexible Li-ion batteries", *Bull. Mater. Sci.,* vol. 44, pp. 1-7, 2021.

[71] Y. Lin, Y. Zhang, F. Zhang, M. Zhang, D. Li, G. Deng, L. Guan, and M. Dong, "Studies on the electrostatic effects of stretched PVDF films and nanofibers", *Nanoscale Res. Lett.,* vol. 16, p. 79, 2021.

[72] Y.H. Liu, F.S. Lin, L.X. Chen, H.Y. Su, Y.C. Hsu, S. Pal, Y.H. Wang, and C.H. Huang, "Wearable transparent PVDF transducer for photoacoustic imager in body sensor network", *2020 IEEE International Ultrasonics Symposium (IUS),* 2020pp. 1-3

[73] R. Dallaev, T. Pisarenko, D. Sobola, F. Orudzhev, S. Ramazanov, and T. Trčka, "Brief review of PVDF properties and applications potential", *Polymers (Basel),* vol. 14, p. 4793, 2022.

[74] S. Rajput, M. Averbukh, A. Yahalom, and T. Minav, "An approval of MPPT based on PV cell's simplified equivalent circuit during fast-shading conditions", *Electronics (Basel),* vol. 8, p. 1060, 2019.

[75] J. Shi, W. Zeng, Z. Dai, L. Wang, Q. Wang, S. Lin, Y. Xiong, S. Yang, S. Shang, W. Chen, and L. Zhao, "Piezocatalytic foam for highly efficient degradation of aqueous organics", *Small Sci.,* vol. 1, p. 2000011, 2021.

[76] X. Xu, D. Cao, H. Yang, and M. He, "Application of piezoelectric transducer in energy harvesting in pavement", *Int. J. Pavement Res. Technol.,* vol. 11, pp. 388-395, 2018.

[77] J. Yan, M. Liu, Y.G. Jeong, W. Kang, L. Li, Y. Zhao, N. Deng, B. Cheng, and G. Yang, " Performance enhancements in poly (vinylidene fluoride)-based piezoelectric nanogenerators for efficient energy harvesting", *Nano Energy,* vol. 56, pp. 662-692, 2019.

[78] C. Dagdeviren, P. Joe, O.L. Tuzman, K.I. Park, K.J. Lee, Y. Shi, Y. Huang, and J.A. Rogers, "Recent progress in flexible and stretchable piezoelectric devices for mechanical energy harvesting, sensing and actuation", *Extreme Mech. Lett.,* vol. 9, pp. 269-281, 2016.

[79] X. Dong, Z. Yang, J. Li, W. Jiang, J. Ren, Y. Xu, T.S. Hu, and M. Li, "Recent advances of triboelectric, piezoelectric and pyroelectric nanogenerators", *Nano-Structures & Nano-Objects,* vol. 35, p. 100990, 2023.

[80] S. Rajput, S. Keshri, V.R. Gupta, N. Gupta, V. Bovtun, and J. Petzelt, "Design of microwave dielectric resonator antenna using MZTO-CSTO composite", *Ceram. Int.,* vol. 38, pp. 2355-2362, 2012.

[81] B. Bera, and M.D. Sarkar, "Piezoelectricity in PVDF and PVDF based piezoelectric nanogenerator: a concept. IOSR", *J. Appl. Phys.,* vol. 9, pp. 95-99, 2017.

[82] L. Koroglu, E. Ayas, and N. Ay, "3D printing of polyvinylidene fluoride based piezoelectric nanocomposites: an overview", *Macromol. Mater. Eng.,* vol. 306, p. 2100277, 2021.

[83] S. Sripadmanabhan Indira, C. Aravind Vaithilingam, K.S. Oruganti, F. Mohd, and S. Rahman, "Nanogenerators as a sustainable power source: state of art, applications, and challenges", *Nanomaterials (Basel),* vol. 9, p. 773, 2019.

[84] L. Joshi, S. Rajput, and S. Keshri, "Structural and magneto-transport properties of LCMO-STO composites", *Phase Transit.,* vol. 83, pp. 482-490, 2010.

[85] L. Lu, W. Ding, J. Liu, and B. Yang, "Flexible PVDF based piezoelectric nanogenerators", *Nano Energy,* vol. 78, p. 105251, 2020.

[86] S.K. Sonika, "Verma, S. Samanta, A. K. Srivastava, S. Biswas, R. M. Alsharabi, S. Rajput, "Conducting polymer nanocomposite for energy storage and energy harvesting systems"", *Adv. Mater. Sci. Eng.,* vol. 1, p. 2266899, 2022.

[87] G. Du, Z. Li, and G. Song, "A PVDF-based sensor for internal stress monitoring of a concrete-filled

steel tubular (CFST) column subject to impact loads", *Sensors (Basel)*, vol. 18, p. 1682, 2018.

[88] B. Zazoum, K.M. Batoo, and M.A. Khan, "Recent advances in flexible sensors and their applications", *Sensors (Basel)*, vol. 22, p. 4653, 2022.

[89] S Rajput, A Lugovskoy, M Averbukh, and A Yahalom, "Porous metal-oxide based electrostatic energy generator", *IEEE CANDO EPE*, 2019pp. 133-136

[90] Available from: https://www.iqsdirectory.com/articles/load-cell/force-sensors.html

[91] S Rajput, S Parida, A Sharma, and Sonika, "Dielectric materials for energy storage and energy harvesting devices", *River Publ*, River Publishers, 2023. Available from: https://www.elprocus.com/force-sensor-working-principle-and-applicationa/

[92] S. Rajput, S. Parida, and A. Sharma, Available from: https://ieeexplore.ieee.org/servlet/opac?bknumber=10177830

[93] X. Xue, S. Wang, W. Guo, Y. Zhang, and Z.L. Wang, "Hybridizing energy conversion and storage in a mechanical-to-electrochemical process for self-charging power cell", *Nano Lett.*, vol. 12, pp. 5048-5054, 2012.

[94] M. Fang, S. Rajput, Z. Dai, Y. Ji, Y. Hao, S. Ren, and X. Ren, "Understanding the mechanism of thermal-stable high-performance piezoelectricity", *Acta Mater.*, vol. 169, pp. 155-161, 2019.

[95] S. Rajput, X. Ke, X. Hu, D. Hu, F. Ye, M. Fang, Y. Hao, and X. Ren, "Critical triple point as the origin of giant piezoelectricity in PbMg1/3Nb2/3O3-PbTiO3 System", *J. Appl. Phys.*, vol. 128, p. 104105, 2020.

[96] X. Hu, S. Rajput, S. Parida, J. Li, W. Wang, L. Zhao, J. Gao, L. Zhong, and X. Ren, "Electrostrain enhancement at rhombohedral region for BaTi1-xHfxO3 Ceramic", *J. Mater. Eng. Perform.*, vol. 29, pp. 5388-5394, 2020.

CHAPTER 5

Innovative, Cutting-Edge Technologies for Energy Harvesting Using Polymer Nanocomposites

Sonika[1,*], Manas Ranjan Nayak[2], Ratikanta Nayak[2], Varatharajan Prasanna Venkadesan[3] and Sushil Kumar Verma[4]

[1] Department of Physics, Rajiv Gandhi University, Rono Hills, Doimukh, Itanagar 791112, Arunachal Pradesh, India

[2] Novel Material Research Laboratory, NIST University, Berhampur 761008, Odisha, India

[3] School of Mechanical and Aerospace Engineering, Queens University of Belfast, Belfast, Northern Ireland

[4] Centre for Sustainable Polymer, Technology Complex, Indian Institute of Technology Guwahati, Guwahati 781039, Assam, India

Abstract: This chapter discusses the cutting-edge polymer nanocomposite energy harvesting technology. Attention has been drawn to polymer nanocomposites, which combine polymers and nanoparticles, for their potential for effective energy conversion. Polymer nanocomposites are used in piezoelectric, thermoelectric, and photovoltaic systems to transform mechanical, thermal, and light energy into electrical energy. Techniques like nanofillers and interface engineering maximize the effectiveness of energy conversion. Improvements in characterization, modeling, and manufacturing techniques are essential for commercialization. Polymer nanocomposites can potentially be used for energy harvesting in self-powered sensors, Internet of Things (IoT) devices, and wearable electronics. In order to advance sustainable energy solutions, this chapter outlines current research initiatives and future uses of polymer nanocomposites.

Keywords: Applications, Challenges, Electronics devices, Energy harvesting, Innovative technologies, Piezoelectric materials, Polymer nanocomposites.

INTRODUCTION

It is possible to "harvest" a small portion of dissipating energy from the environment with minimal effort to utilize it as electrical energy. This concept is referred to as "energy harvesting". Recently, it has attained enormous attention as an alternative/source for achieving Sustainable Development Goals (SDG) [1, 2, 3].

---

* **Corresponding author Sonika:** Department of Physics, Rajiv Gandhi University, Rono Hills, Doimukh, Itanagar 791112, Arunachal Pradesh, India; E-mail: sonika.gupta@rgu.ac.in

Sushil Kumar Verma, Sonika & Arbind Prasad (Eds.)
All rights reserved-© 2026 Bentham Science Publishers

The Internet of Things (IoT) technology provides more sophistication to society, and as a result, it is envisioned to enter the Trillion-Sensor Universe (TSU). Generally, the challenging task is to link these sensors to an individual energy supply, and for this purpose, batteries are used despite their disadvantages, like high power costs. However, it is difficult to change the batteries often, especially for a TSU. So, the communal application of energy harvesting techniques becomes unavoidable.

From the literature, it has been identified that the term harvesting has been explicitly used for photovoltaic applications [4]. Different energy harvesting methodologies were reported by researchers [5, 6]. The overall outline of the technologies is shown in Fig. (**1**) for different environmental targets. In general, energy harvesting may include one or all of the following:

a. Harvesting: Generally in small quantities
b. Conversion: Harvested energy to electric energy
c. Processing: Preferably in power conversion circuits
d. Utilizing: For different applications

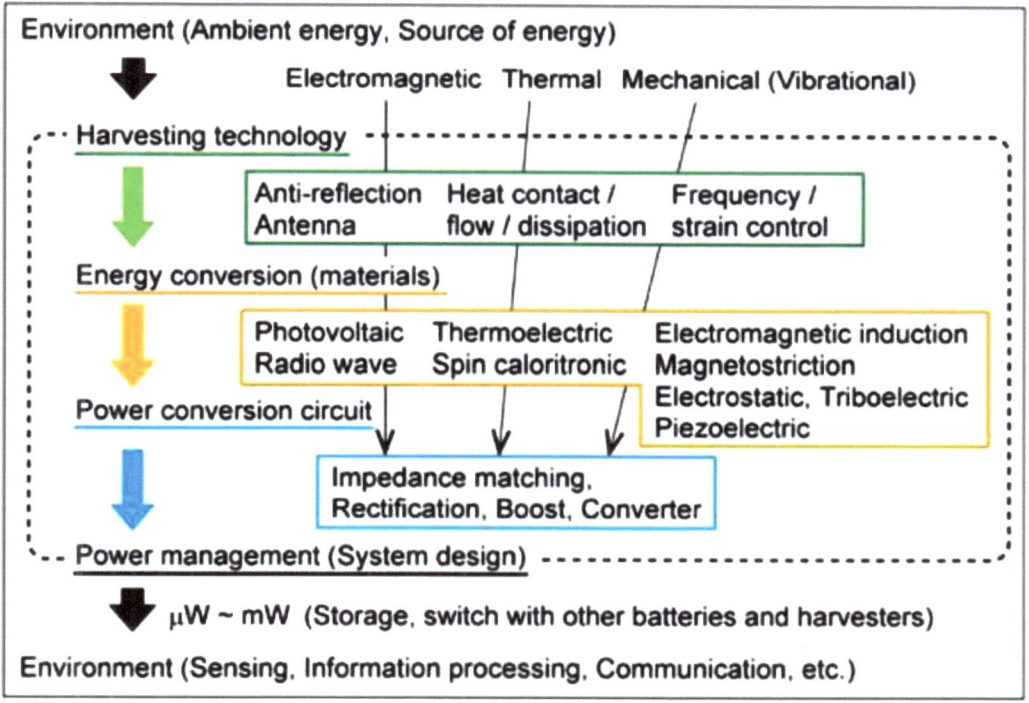

Fig. (1). Energy harvesting techniques [7].

Throughout this chapter, all the above are referred to as energy harvesting. Among the various technologies, solar cells have wide popularity due to their high yield output, and they have already been in practice for a long time. Still, it is necessary to harvest the energy from indoor and outdoor light [8, 9, 10]. Thus, standardization is essential for the different solar cell types for further development.

Besides solar energy, vibrational energy, radio wave energy, and thermoelectric energy are also essential sources/techniques. The factors to be considered while implementing the energy harvesters are reliability, durability, environment, and operational cost.

CLASSIFICATION

According to the matrix, nanocomposites can be categorizedinto ceramic, metal, and polymer nanocomposites.

Ceramic-Matrix Nanocomposites (CMCs)

Industrial ceramics are inorganic, non-metallic compounds composed of borides, oxides, nitrides, silicides, *etc*. Since ceramics are fragile, their toughness is one of the most interesting research areas. For this reason, more pliable metal phases are usually interspersed in a ceramic matrix. To achieve unique nanoscopic qualities, it is ideal for both the metallic and ceramic components to be properly circulated within one another. Moreover, information on improved qualities, including optical, electrical/magnetic, tribological, and corrosion resistance, has been distributed [11].

Since a metallic constituent can freely react with ceramic and lose its metallic properties, protective layers must be formed to prevent chemical reactions between the two components. Finding a metal that is miscible with the ceramic matrix is one of these measures. For instance, Cu is miscible with TiO_2 in the Gibbs' triangle of Cu-O-Ti in vast sections.

Ceramic-matrix nanocomposite films can be produced utilizing gas current sputtering with the resonating cathode method in thin-film (thicknesses from a few nm to tens of m applied to a connecting substrate). High deposition rates and the growth of nanoparticles in the gas phase are associated with this vacuum-based deposition technology. The technique forms nanocomposite layers consisting of TiO_2 and Cu that have admirable corrosion resistance, low friction, and high mechanical hardness.

The commonly used material is lead zirconate titanate (PZT, perovskite structure - $Pb[Zr_xTi_{1-x}]O_3$), which belongs to the category of piezoelectric polycrystalline ceramics. The piezoelectric effect exhibited by PZT arises from both intrinsic and extrinsic factors of the material [12]. The intrinsic effect is ascribed to the relative shift of anions and cations, which maintains the ferroelectric crystal structures. On the other hand, the extrinsic effect is associated with the motion of domain walls. To achieve desired properties, these are often doped with various elements.

PZT exhibits exceptional performance when compared to alternative materials. It exhibits excellent manufacturability, allowing for the fabrication of different forms with various geometries at a minimal cost. With an advanced Curie point than polymers, PZT offers a wide temperature range. Its stiffness makes it susceptible to significant strain from even slight deflections. The relatively high piezoelectric coefficients (d and g) and mechanical coercive field make it appropriate for high-voltage applications.

Metal-Matrix Nanocomposites (MMCs)

In order to produce a superthermite "energetic" nanocomposite, metal-matrix nanocomposites can associate several types of dispersed components, such as carbon nitride and boron nitride nanoparticles, metal oxides, and nano-scale Al powder within a silica-based hybrid sol-gel matrix [13]. The excellent tensile strength and electrical conductivity of carbon nanotubes have been widely explored in metal-matrix nanocomposites as well. The growth of strong interfacial adhesion between the matrix and the carbon nanotubes, as well as consistent dispersion of carbon nanotubes into the metallic matrix, was revealed to be essential.

Polymer-Matrix Nanocomposites (PMCs)

Numerous auxiliary materials, such as ceramics, clays, carbon nanotubes, and nanoparticles, as well as inorganic nanowires and nanorods, can be embedded in polymer matrix nanocomposites [14].

In addition to mechanical properties like stiffness and strength, there are also spatial qualities like crystallisation behavior, flame resistance, biodegradability of the matrix, and implant biocompatibility [15]. In order to effectively make use of the nano-fillers' high specific surface area and aspect ratio, stronger groups of reducing particle size must be split up in order to attain homogenous and stable dispersion within the matrix. Nanoparticle dispersibility is a complex problem that is influenced by a range of elements, such as inter-particle attraction and fusion, surface chemistry and conductivity, as well as processing variables including viscosity and shear mixing intensity, *etc* [16].

The Poly(Vinylidene Fluoride) (PVDF) and its copolymers, such as Poly(Vinylidene Fluoride-trifluoroethylene) (P(VDF-TrFE)), are widely recognized as prominent piezoelectric polymer materials. While their piezoelectric coefficients are lower compared to ceramics, these polymers offer enhanced mechanical flexibility, reduced stiffness, and increased chemical resistance. These characteristics contribute to their ability to prevent fatigue and extend the lifespan of devices [10, 17]. PVDF and its copolymers find applications in sensors, actuators, and energy harvesters. The following examples highlight some key advantages of PVDF materials.

One such example is the development of a backpack energy harvester utilizing a PVDF strap with piezoelectric properties [18]. The conventional fabric strips are substituted with a flexible and durable PVDF polymer strip, capitalizing on the advantages it offers. The authors predicted a maximum instantaneous power of 0.345 W and a normal power of 45.6 mW.

LEAD-FREE MATERIALS IN THE PIEZOELECTRIC ENERGY HARVESTER

In recent decades, environmental concerns have led to increased attention toward lead (Pb)-free materials [19]. Traditional piezoelectric ceramics, such as PZT, commonly contain toxic lead, necessitating the exploration of alternative materials to comply with environmental policies [19]. Organic materials include PVDF and its copolymers, while inorganic Pb-free materials can be further categorized as tungsten bronze, Aurivillius (bismuth-layer-structured ferroelectrics), and perovskite families. Among the inorganic configurations, Barium Titanate (BT), Bismuth Sodium Titanate (BNT), and Sodium Potassium Niobate (KNN) have been widely considered.

A quantitative comparison reveals distinct differences in performance: PZT typically exhibits a high piezoelectric coefficient ($d_{33} \approx$ 300–600 pC/N), Curie Temperature (TC) around 350°C, and a dielectric constant (ε_r) ~1000–1500. BT-based ceramics show lower d_{33} values (~190 pC/N), TC near 120°C, and (ε_r) ~1500. KNN-based materials demonstrate d_{33} values in the range of 80–400 pC/N, depending on composition and processing, with TC around 210–420°C, offering a good balance of performance and thermal stability. BNT-based systems often have d_{33} values of 70–150 pC/N and higher TC (~320°C), though they face challenges related to high coercive fields and poor depolarization behavior. PVDF, as a polymer alternative, shows lower d_{33} (~20–30 pC/N), but offers high flexibility, lightweightness, and biocompatibility.

BT-based ceramics were the first recognized lead-free piezoelectric ceramics, suitable for low-temperature applications. Among them, KNN is considered particularly promising for its relatively high piezoelectric performance and thermal stability, making it a viable candidate for replacing PZT in high-performance applications.

Apart from materials, researchers investigated piezoelectric energy harvesters with respect to orientation as well. Fig. (2) shows the representation of different orientations considered in the development of piezoelectric harvesters.

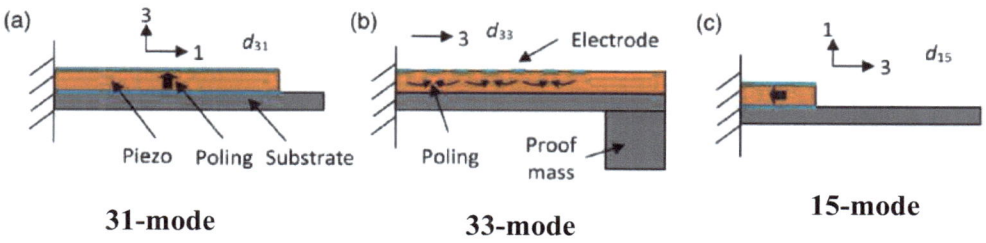

Fig. (2). Orientations considered during the design of piezoelectric harvesters.

PIEZOELECTRIC NANOCOMPOSITE ENERGY HARVESTERS

The limitations of monolithic piezoceramics include their inability to conform to curved surfaces, brittleness that makes them prone to breakage, and their high density resulting from ceramics. To overcome these constraints, researchers have developed composite piezoelectric devices that consist of an active piezoceramic phase embedded in a polymeric matrix phase. These composite materials offer increased strength, flexibility, and robustness due to the protective nature of the polymer matrix. Active devices employing active and macro fiber composites [20, 21], as well as particles known as 0-3 composites, have been developed as a result of research in this area. These materials, sometimes referred to as "piezoelectric paints," are easy to apply to surfaces and can cover large areas. Preliminary research on 0-3 piezoelectric composites, specifically PZT paint sensors, was reported in the literature [22, 23].

In recent years, piezoelectric Nanowires (NWs) have experienced significant growth. Wang and Song [24] reported the fabrication and testing of ZnO NWs by deflecting a single nanowire using atomic force microscopy. The study demonstrated that these NWs lead to the exploration of a novel harvesting technique and surpass many Micro-Electromechanical System (MEMS)-based designs [25].

Due to the semiconductive nature of ZnO, it requires careful consideration. Early reports stressed the importance of a Schottky barrier for establishing electrical connectivity with the nanowire. This barrier played a crucial role [26, 27]. In 2008, Liu *et al.* [26] conducted a study that explored the characteristics of ZnO nanowires. They adjusted the conductivity and carrier density of the nanowires. The study underscored the significant impact on the semiconductive properties of the device. Furthermore, Briscoe *et al.* [27] demonstrated the feasibility of constructing a ZnO nanorod with a p-n junction, instead of a Schottky barrier.

Koka *et al.* [28, 29] compared the $BaTiO_3$ nanowires to ZnO nanowires and found that the $BaTiO_3$ nanogenerator (Fig. **3**) produced 20 times more energy, even without tuning the Schottky barrier or doping. In 2014, Koka and Sodano [30] achieved 155 Hz by growing ultralong $BaTiO_3$ nanowires with reduced stiffness. Various works on nanofibers have been reviewed by Chang *et al.* [31], Espinosa *et al.* [32], and Brisco and Dunn [33]. These articles cover different types of piezoelectric nanofiber materials, their characterization, performance, and device architectures.

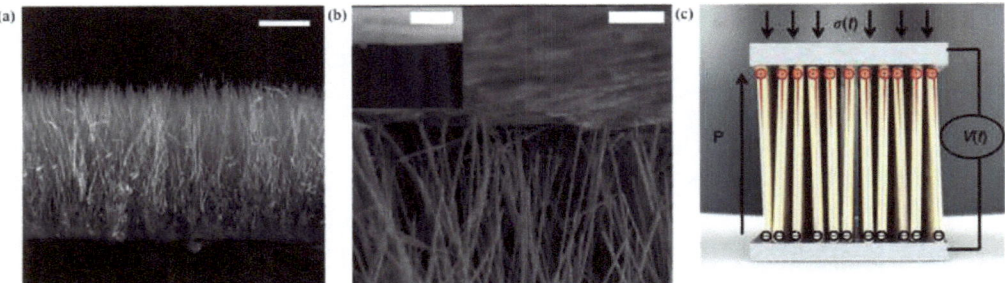

Fig. (3). (a) $BaTiO_3$ NW arrays (cross-section), (b) nanowires contact with the foil, and (c) representation of voltage production from nanowires [28].

INNOVATIVE STRETCHABLE PIEZOELECTRIC ENERGY HARVESTER

Energy Harvesting (EH) refers to the concept of transforming waste energy sources, such as bodily motion or structural vibrations, into electrical energy. This energy can be organized to be used by driving a wireless sensor node or reviving a wearable device's battery. Using piezoelectric materials is one way to convert mechanical energy into electrical energy. When a piezoelectric material is deformed mechanically, it becomes polarized, and the subsequent electric field can be used to transfer charge through an external circuit [34]. Fig. (**4**) illustrates a sample piezoelectric energy harvester. Piezoelectric Energy Harvesting (PEH) is an ingrained field, and several writers have studied the subject in depth [35, 36, 37].

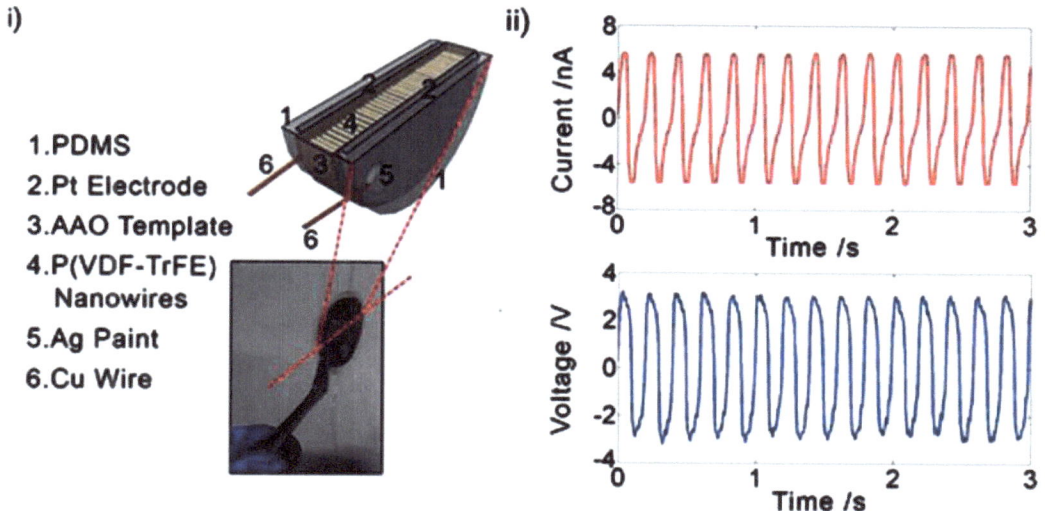

Fig. (4). (i) An EH device contains P(VDF-TrFE) nanowires grown by pattern wetting. (ii) The open-circuit voltage and short-circuit current of this expedient when exposed to cyclic loading [40].

The level of mechanical energy stored in stress-driven situations—deformation resulting from a fixed stress rather than a static strain—is inversely associated with the material's elastic modulus. Compliant piezoelectric polymers can therefore supply significantly more mechanical energy than stiff ceramic materials under these situations. Piezoelectric polymers have a low internal conversion efficiency, but because there is initially a lot of mechanical energy stored, there could still be a lot of electrical energy available for the gearbox [38, 39]. To ensure maximal power transmission, the mechanical resistivity of the EH device should be suited to the source of mechanical energy. This can be done by retaining the benefit of the mechanical properties of polymers.

When referring to EH devices, especially those made using piezoelectric materials, the phrase "nanogenerator" is usually used. The technology's earliest demonstration, which described the piezoelectric voltage generated across zinc oxide nanowires when probed by an AFM tip, gave rise to the name [24]. Even though many of the subsequent devices did not really include any nanomaterials, they were nonetheless referred to as nanogenerators. However, a lot of researchers continue to develop EH devices using nanomaterials. There are two explanations for this. First, a material's effective stiffness can be decreased through nano-structuring. Therefore, the previous explanations for how more flexible piezoelectric materials are a better option for stress-driven nanogenerators still hold true. Second, it is frequently noted that as material dimensions are lowered, a material's piezoelectric capabilities get better.

MAJOR CHALLENGES FOR ENERGY HARVESTING

In piezoelectric materials, the piezoelectric property is induced during the manufacturing/processing stage. During processing, the continuous application of strain to the crystal results in grain boundary defects and sometimes a macroscopic crack [41]. For commercialization, mechanical property evaluation is unavoidable. In this case, the reproducibility of the mechanical test results is crucial. Also, as explained in the earlier section, the combination of different design ideas is considered in the vibrational energy harvester device. Hence, it is advisable to consider a suitable variety of materials to satisfy the design and energy harvesting requirements. Thus, standardization will lead to competition and the practice of making energy harvesters for intended applications. More precisely, in the case of piezoelectric methods, standard protocols for testing the materials are required.

Vibration Method

The energy harvesting from vibrational modes has three categories:

i. Electromagnetic method: The electrostatic induction and inverse magnetostrictive principles are used in the electromagnetic method.
ii. Electrostatic method: MEMS and triboelectric modes are primarily used in the electrostatic method.
iii. Piezoelectric method: This is centered on the piezoelectric effect, *i.e.*, the dielectric polarization. The surface charge will appear when mechanical stress/strain is applied to the piezoelectric materials.

Apart from general design considerations, the perception of vibration must also be taken into account. The vibrational frequency in the environment is around or less than 200 Hz. In the case of humans and bridges, the design is intended for a frequency range of 2–3 Hz. Thus, the features intended to be implemented in the harvesters should aid in efficient harvesting and energy conversion.

Much research has been carried out on vibrational energy harvesters, especially in the electromagnetic and piezoelectric mode harvesting methods. The commonly used magnetostrictive materials considered in the vibrational energy harvesters are alloys like Terfenol-D and Fe-Ga alloy Galfenol [42]. The device built using FeCo alloy was reported in the literature [43]. Also, piezoelectric and perovskite-type materials like $Pb(Zr,Ti)O_3$ (PZT) are used by the researchers. For further improvement in the conversion, techniques like multi-layered and nanostructures are recommended by the researchers.

It is reported in the literature that downsizing materials to nanometers and adjusting charge shielding instead of controlling strain will result in improved performance [44]. However, the utilization of lead is restricted by RoHS due to its harmful consequences [45]. This leads to the advancement of lead (Pb) free materials and the field's expanded momentum [46, 47, 48].

Radio Frequency (RF) Method

Electromagnetic waves that are less than 3000 GHz are classified as radio frequency. The rectification process is unavoidable in the radio frequency energy harvesting technique. Commercial wireless systems for different applications transmit RF energy, similar to RF energy harvesting. The circuit design requirement in this method is more complex than the vibrational system, as the RF method deals with very high frequencies. As the world is heading toward the 5G era, harvesting energy from higher bandwidth is more beneficial. However, the complexity of circuit design for efficient harvesting and handling still exists.

In RF energy harvesting devices, rectennas (antenna + rectifier) are used for converting the harvested energy to direct current (Fig. 5). As rectennas play a crucial role in RF harvesting devices, researchers performed various investigations to evaluate the rectennas using different technologies [49, 50, 51]

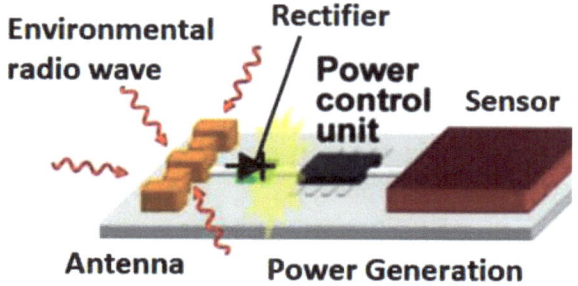

Fig. (5). Energy harvesting from environmental radio waves [7].

Furthermore, significant enhancements have been achieved in the diode properties comprising Silicon-On-Insulator (SOI) FETs, specifically known as p-n junction body-tie SOI-FETs, which exhibit steep current characteristics. These advancements have generated anticipation for their utilization in radio wave energy harvesting [52, 53, 54]

Additionally, researchers have effectively invented flexible diodes using molybdenum (IV) sulfide (MoS2), a two-dimensional (2D) semiconductor. These diodes have demonstrated outstanding potential for implementation in rectennas operating within the 2.4 GHz frequency range [55]. Despite these advancements, there remains ample opportunity to discover novel materials and structures that could revolutionize diode development and the progress of compact, high-efficiency antennas. It is expected that future research and development endeavors will result in substantial breakthroughs in the field of radio wave energy harvesting.

Thermoelectric Method

According to multiple surveys, it has been found that a significant portion, approximately 70%, of the total energy consumed is wasted as heat [56]. This waste heat often remains at temperatures below 100°C. However, this low-grade heat can be effectively harnessed using this technique. This technology has gained attention as it provides an independent source for various IoT (Internet of Things) devices [57, 58].

The phenomenon responsible for this energy conversion is known as the Seebeck effect. When a temperature difference (ΔT) is applied across a conductor, it generates a voltage (ΔV) proportional to the temperature difference. The voltage and temperature difference proportionality coefficient is called the Seebeck coefficient (S), calculated as $S = \Delta V/\Delta T$. Thermoelectric materials are generally categorized using ZT, a dimensionless quantity that depends on the Seebeck coefficient (S), as well as electrical and thermal conductivity.

To be practically viable, the thermoelectric conversion efficiency must surpass 1, which is achieved through an increased ZT value. However, conventional metals typically struggle to attain high thermoelectric conversion efficiency due to the limited variation in the number of free electrons with temperature and the adherence to the Wiedemann-Franz law, which establishes a proportionality between σ and κ. Additionally, the equation for ZT involves S, σ, and κ, all of which are influenced by several factors and are challenging to constrain separately.

Researchers have explored selective enhancement through microstructures to overcome the limitations that hinder the increase in ZT [59, 60, 61, 62]. Furthermore, efforts have been made to control the structure of crystals [63, 64] and employ a μ-fabrication process, including the utilization of nanostructures [65, 66]. Another promising approach is the introduction of magnetic ions [67, 68], which involves techniques such as spin-orbit interaction to form polarons and increase the effective mass of carriers. Additionally, the benefits of two-dimensional materials [69, 70, 71, 72, 73] and band engineering [74, 75] have been extensively investigated to enhance the Power Factor (PF). Notably, recent advancements in band engineering have shown significant progress in improving ZT.

Extensive research has been conducted on carbon nanotube thin films as thermoelectric materials due to their ability to achieve stable n-type doping. Impressive Power Factors (PFs) at room temperature have been documented [76, 77, 78]. Additionally, silicon compounds have garnered interest as they are both human- and environmentally friendly [79, 80, 81].

Conductive oxides have historically received limited attention. However, remarkable thermoelectric properties were reported for Na_2CoO_4 [82], followed by $(Ca_2CoO_3)_xCoO_2$, $SrTiO_3$, and other materials [83, 84]. These findings have sparked significant interest in conductive oxides as potential candidates for energy harvesting applications. One advantage of oxides is that their crystal structure and electron-phonon properties can be easily manipulated through element substitution. Current research trends in conductive oxides focus on (1) enhancing the Seebeck coefficient, (2) mitigating the phonon contribution, and (3) suppressing the thermal conductivity through interface engineering [85, 86].

While numerous thermoelectric materials have been investigated, a significant portion of them exhibit a substantial reduction in their performance around Room Temperature (RT). Consequently, Bi_2Te_3, known for its favorable thermoelectric properties near RT, remains the primary material of choice for practical Internet of Things (IoT) applications. Moreover, it becomes more challenging, necessitating advancements in harvesting technology. However, the advantage of utilizing lower temperatures is the alleviation of degradation in thermoelectric materials and electrode interfaces, thus enhancing reliability. The degradation mechanism of thermoelectric materials, particularly their ionic conductivity, has been investigated with an emphasis on their anticipated use for IoT applications [87, 88]. Unravelling the degradation mechanisms underpinning thermoelectric materials in IoT scenarios will be a crucial area of future research.

To complement the understanding of degradation mechanisms and improve energy harvesting in IoT environments, material-level innovations must be coupled with advancements in thermal energy management. Recent research has increasingly focused on manipulating heat transport at the nanoscale to support more efficient and stable operation of miniaturized thermoelectric systems. This includes exploring methods to enhance thermal conduction control, optimize energy collection efficiency, and mitigate temperature-induced performance loss in integrated devices.

Researchers have made significant advancements in controlling thermal conduction, which possesses periodic nanostructures. Additionally, thermal collection techniques utilizing lens structures with radial hole arrays [89, 90, 91, 92] have been successfully implemented. Another notable development is the generation of heat at the nanometer scale through plasmons [93]. These breakthroughs in heat current control hold immense potential and are expected to drive further progress in the field. However, accurately measuring the properties of miniaturized assemblies poses a significant challenge. Prior studies have utilized a transmission electron microscope to measure the temperature of an illustration with nanometer spatial resolution [94]. Nonetheless, precise measurements require sufficiently sharp plasmon peaks.

CONCLUSION

In electrospun PVDF composite membranes, a wide variety of filler systems that span three different material classes— carbon (CNT and GO), ceramics (BT, ZnO, and two nanoclays, halloysite and bentonite), and cellulose (MCCs and NCCs)—have been explored. Testing and analysis have been performed on the effects of filler type and freight on the crystalline and piezoelectric qualities of the composites. At specific filler loadings, the piezoelectric coefficient (d_{33}) for electrospun PVDF can be exceeded by that of all PVDF composites. It was emphasized in an evaluation of the processing of piezoelectric polymers that the processing essential for each polymer can vary significantly. There was also discussion of certain techniques to enhance the piezoelectric effect in polymers. Piezoelectric polymer applications were presented in the final section. It is hoped that this conversation will assist in advancing knowledge of piezoelectric materials and direct research efforts toward the areas of the science that still require clarification. In addition to exciting intellectual curiosity, a deeper comprehension of piezoelectricity in polymers will also result in innovations in applications for industry, medicine, and engineering.

REFERENCES

[1] "Transforming our world: the 2030 agenda for sustainable development", *United Nations,* United Nations: New York, 2015. Available from: https://sdgs.un.org/2030agenda

[2] "Sustainable development goal 7; ensure access to affordable, reliable, sustainable and modern energy for all", *United Nations,* United Nations: New York, 2015. Available from: https://sustainabledevelopment.un.org/sdg7/

[3] H. Akinaga, H. Fujita, M. Mizuguchi, and T. Mori, "Focus on advanced materials for energy harvesting: prospects and approaches of energy harvesting technologies", *Sci. Technol. Adv. Mater.,* vol. 19, no. 1, pp. 543-544, 2018.
[http://dx.doi.org/10.1080/14686996.2018.1491165]

[4] S. Rajput, M. Averbukh, and N. Rodriguez, "Energy Harvesting and Energy Storage Systems", *Electron.,* vol. 11, no. 7, pp. 2-5, 2022.
[http://dx.doi.org/10.3390/electronics11070984]

[5] S. Rajput, A. Sharma, V. Jately, and M. Ram, *Recent Advances in Energy Harvesting Technologies.* River Publishers: New York, 2023.

[6] J. Krikke, "Sunrise for energy harvesting products", *IEEE Pervasive Comput.,* vol. 4, no. 1, pp. 4-5, 2005.

[http://dx.doi.org/10.1109/MPRV.2005.23]

[7] H. Akinaga, "Recent advances and future prospects in energy harvesting technologies", *Jpn. J. Appl. Phys.,* vol. 59, no. 11, p. 110201, 2020.
[http://dx.doi.org/10.35848/1347-4065/abbfa0]

[8] S. Rajput, M. Averbukh, and A. Yahalom, "Electric power generation using a parallel-plate capacitor", *Int. J. Energy Res.,* vol. 43, no. 8, pp. 3905-3913, 2019.
[http://dx.doi.org/10.1002/er.4492]

[9] Y. Li, N.J. Grabham, S.P. Beeby, and M.J. Tudor, "The effect of the type of illumination on the energy harvesting performance of solar cells", *Sol. Energy,* vol. 111, pp. 21-29, 2015.
[http://dx.doi.org/10.1016/j.solener.2014.10.024]

[10] S.P. Muduli, L. Lipsa, A. Choudhary, S. Rajput, and S. Parida, "Modulation of electrical characteristics of polymer–ceramic–graphene hybrid composite for piezoelectric energy harvesting", *ACS Appl. Electron. Mater.,* vol. 5, no. 6, pp. 3023-3037, 2023.
[http://dx.doi.org/10.1021/acsaelm.3c00078]

[11] C.D. Usurelu, S. Badila, A.N. Frone, and D.M. Panaitescu, "Poly(3-hydroxybutyrate) nanocomposites with cellulose nanocrystals", *Polymers (Basel),* vol. 14, no. 10, p. 1974, 2022.
[http://dx.doi.org/10.3390/polym14101974]

[12] M. Fang, S. Rajput, Z. Dai, Y. Ji, Y. Hao, and X. Ren, "Understanding the mechanism of thermal-stable high-performance piezoelectricity", *Acta Mater.,* vol. 169, pp. 155-161, 2019.
[http://dx.doi.org/10.1016/j.actamat.2019.03.011]

[13] S.R. Bakshi, D. Lahiri, and A. Agarwal, "Carbon nanotube reinforced metal matrix composites - a review", *Int. Mater. Rev.,* vol. 55, no. 1, pp. 41-64, 2010.
[http://dx.doi.org/10.1179/095066009X12572530170543]

[14] S. Zeng, D. Baillargeat, H-P. Ho, and K-T. Yong, "Nanomaterials enhanced surface plasmon resonance for biological and chemical sensing applications", *Chem. Soc. Rev.,* vol. 43, no. 10, p. 3426, 2014.
[http://dx.doi.org/10.1039/c3cs60479a]

[15] T. Subbiah, G.S. Bhat, R.W. Tock, S. Parameswaran, and S.S. Ramkumar, "Electrospinning of nanofibers", *J. Appl. Polym. Sci.,* vol. 96, no. 2, pp. 557-569, 2005.
[http://dx.doi.org/10.1002/app.21481]

[16] J.H. Song, and J.R.G. Evans, "Flocculation after injection molding in ceramic suspensions", *J. Mater. Res.,* vol. 9, no. 9, pp. 2386-2397, 1994.
[http://dx.doi.org/10.1557/JMR.1994.2386]

[17] S.P. Muduli, S. Parida, S.K. Behura, S. Rajput, S.K. Rout, and S. Sareen, "Synergistic effect of graphene on dielectric and piezoelectric characteristic of PVDF BZT-BCT composite for energy harvesting applications", *Polym. Adv. Technol.,* vol. 33, no. 10, pp. 3628-3642, 2022.
[http://dx.doi.org/10.1002/pat.5816]

[18] H.A. Sodano, J. Granstrom, J. Feenstra, and K. Farinholt, *Harvesting of electrical energy from a backpack using piezoelectric shoulder straps.,* Y. Matsuzaki, M. Ahmadian, D.J. Leo, Eds., , 2007, p. 652502.

[19] J. Song, and J. Wang, "Ferroelectric materials for vibrational energy harvesting", *Sci. China Technol. Sci.,* vol. 59, no. 7, pp. 1012-1022, 2016.
[http://dx.doi.org/10.1007/s11431-016-6081-7]

[20] A. Bent, Active fiber composites for structural actuation, PhD thesis, Massachusetts Institute of Technology, 1997.

[21] W.K. Wilkie, *Low-cost piezocomposite actuator for structural control applications.,* J.H. Jacobs, Ed., , 2000, pp. 323-334.

[22] K.A. Klein, A. Safari, R.E. Newnham, and J. Runt, "Composite piezoelectric paints", *Sixth IEEE International Symposium on Applications of Ferroelectrics,* 1986pp. 285-287

[23] K.A. Hanner, A. Safari, R.E. Newnham, and J. Runt, "Thin film 0–3 polymer/piezoelectric ceramic composites: Piezoelectric paints", *Ferroelectrics,* vol. 100, no. 1, pp. 255-260, 1989.
[http://dx.doi.org/10.1080/00150198908007920]

[24] Z L Wang, and J Song, "Piezoelectric nanogenerators based on zinc oxide nanowire arrays", *Science,* vol. 312, no. 5771, pp. 242-246, 2006.
[http://dx.doi.org/10.1126/science.1124005]

[25] S. Xu, Y. Qin, C. Xu, Y. Wei, R. Yang, and Z.L. Wang, "Self-powered nanowire devices", *Nat. Nanotechnol.,* vol. 5, no. 5, pp. 366-373, 2010.
[http://dx.doi.org/10.1038/nnano.2010.46]

[26] J. Liu, "Carrier Density and Schottky Barrier on the Performance of DC Nanogenerator", *Nano Lett.,* vol. 8, no. 1, pp. 328-332, 2008.
[http://dx.doi.org/10.1021/nl0728470]

[27] J. Briscoe, M. Stewart, M. Vopson, M. Cain, P.M. Weaver, and S. Dunn, "Nanostructured p-n Junctions for Kinetic-to-Electrical Energy Conversion", *Adv. Energy Mater.,* vol. 2, no. 10, pp. 1261-1268, 2012.
[http://dx.doi.org/10.1002/aenm.201200205]

[28] A. Koka, and H.A. Sodano, "High-sensitivity accelerometer composed of ultra-long vertically aligned barium titanate nanowire arrays", *Nat. Commun.,* vol. 4, no. 1, p. 2682, 2013.
[http://dx.doi.org/10.1038/ncomms3682]

[29] A. Koka, Z. Zhou, and H.A. Sodano, "Vertically aligned BaTiO 3 nanowire arrays for energy harvesting", *Energy Environ. Sci.,* vol. 7, no. 1, pp. 288-296, 2014.
[http://dx.doi.org/10.1039/C3EE42540A]

[30] A. Koka, and H.A. Sodano, "A low-frequency energy harvester from ultralong, vertically aligned BaTiO 3 nanowire arrays", *Adv. Energy Mater.,* vol. 4, no. 11, 2014.
[http://dx.doi.org/10.1002/aenm.201301660]

[31] J. Chang, M. Dommer, C. Chang, and L. Lin, "Piezoelectric nanofibers for energy scavenging applications", *Nano Energy,* vol. 1, no. 3, pp. 356-371, 2012.
[http://dx.doi.org/10.1016/j.nanoen.2012.02.003]

[32] H.D. Espinosa, R.A. Bernal, and M. Minary-Jolandan, "A review of mechanical and electromechanical properties of piezoelectric nanowires", *Adv. Mater.,* vol. 24, no. 34, pp. 4656-4675, 2012.
[http://dx.doi.org/10.1002/adma.201104810]

[33] J. Briscoe, and S. Dunn, "Piezoelectric nanogenerators – a review of nanostructured piezoelectric energy harvesters", *Nano Energy,* vol. 14, pp. 15-29, 2015.
[http://dx.doi.org/10.1016/j.nanoen.2014.11.059]

[34] C.R. Bowen, H.A. Kim, P.M. Weaver, and S. Dunn, "Piezoelectric and ferroelectric materials and structures for energy harvesting applications", *Energy Environ. Sci.,* vol. 7, no. 1, pp. 25-44, 2014.
[http://dx.doi.org/10.1039/C3EE42454E]

[35] H. Liu, J. Zhong, C. Lee, S-W. Lee, and L. Lin, "A comprehensive review on piezoelectric energy harvesting technology: Materials, mechanisms, and applications", *Appl. Phys. Rev.,* vol. 5, no. 4, 2018.
[http://dx.doi.org/10.1063/1.5074184]

[36] K. Uchino, "Piezoelectric Energy Harvesting Systems—Essentials to Successful Developments", *Energy Technol. (Weinheim),* vol. 6, no. 5, pp. 829-848, 2018.
[http://dx.doi.org/10.1002/ente.201700785]

[37] L. Joshi, S. Singh Rajput, and S. Keshri, "Structural and magneto-transport properties of LCMO–STO composites", *Phase Transit.,* vol. 83, no. 7, pp. 482-490, 2010.

[http://dx.doi.org/10.1080/01411594.2010.492466]

[38] J.I. Roscow, H. Pearce, H. Khanbareh, S. Kar-Narayan, and C.R. Bowen, "Modified energy harvesting figures of merit for stress- and strain-driven piezoelectric systems", *Eur. Phys. J. Spec. Top.*, vol. 228, no. 7, pp. 1537-1554, 2019.
[http://dx.doi.org/10.1140/epjst/e2019-800143-7]

[39] D.B. Deutz, J-A Pascoe, B. Schelen, S. van der Zwaag, D.M. de Leeuw, and P. Groen, "Analysis and experimental validation of the figure of merit for piezoelectric energy harvesters", *Mater. Horiz.*, vol. 5, no. 3, pp. 444-453, 2018.
[http://dx.doi.org/10.1039/C8MH00097B]

[40] S.S. Rajput, S. Keshri, V.R. Gupta, N. Gupta, V. Bovtun, and J. Petzelt, "Design of microwave dielectric resonator antenna using MZTO–CSTO composite", *Ceram. Int.*, vol. 38, no. 3, pp. 2355-2362, 2012.
[http://dx.doi.org/10.1016/j.ceramint.2011.10.088]

[41] J. Glaum, and M. Hoffman, "Electric fatigue of lead-free piezoelectric materials", *J. Am. Ceram. Soc.*, vol. 97, no. 3, pp. 665-680, 2014.
[http://dx.doi.org/10.1111/jace.12811]

[42] S. Palumbo, P. Rasilo, and M. Zucca, "Experimental investigation on a Fe-Ga close yoke vibrational harvester by matching magnetic and mechanical biases", *J. Magn. Magn. Mater.*, vol. 469, pp. 354-363, 2019.
[http://dx.doi.org/10.1016/j.jmmm.2018.08.085]

[43] Z. Yang, K. Nakajima, R. Onodera, T. Tayama, D. Chiba, and F. Narita, "Magnetostrictive clad steel plates for high-performance vibration energy harvesting", *Appl. Phys. Lett.*, vol. 112, no. 7, 2018.
[http://dx.doi.org/10.1063/1.5016197]

[44] T. Yamada, "Charge screening strategy for domain pattern control in nano-scale ferroelectric systems", *Sci. Rep.*, vol. 7, no. 1, p. 5236, 2017.
[http://dx.doi.org/10.1038/s41598-017-05475-x]

[45] "Restriction of hazardous substances in electrical and electronic equipment (rohs)", *RoHS*, 2020. Available from: https://environment.ec.europa.eu/topics/waste-and-recycling/rohs-directive_en

[46] H. Wei, "An overview of lead-free piezoelectric materials and devices", *J. Mater. Chem. C Mater. Opt. Electron. Devices*, vol. 6, no. 46, pp. 12446-12467, 2018.
[http://dx.doi.org/10.1039/C8TC04515A]

[47] S. Rajput, "Critical triple point as the origin of giant piezoelectricity in PbMg1/3Nb2/3O3-PbTiO3 system", *J. Appl. Phys.*, vol. 128, no. 10, 2020.
[http://dx.doi.org/10.1063/5.0021765]

[48] X. Hu, "Electrostrain enhancement at tricritical point for BaTi1−xHfxO3 ceramics", *J. Mater. Eng. Perform.*, vol. 29, no. 8, pp. 5388-5394, 2020.
[http://dx.doi.org/10.1007/s11665-020-05003-5]

[49] C.H.P. Lorenz, "Breaking the efficiency barrier for ambient microwave power harvesting with heterojunction backward tunnel diodes", *IEEE Trans. Microw. Theory Tech.*, vol. 63, no. 12, pp. 4544-4555, 2015.
[http://dx.doi.org/10.1109/TMTT.2015.2495356]

[50] S. Mizojiri, and K. Shimamura, Wireless power transfer *via* subterahertz-wave., *Appl. Sci. (Basel)*, vol. 8, no. 12, p. 2653, 2018.
[http://dx.doi.org/10.3390/app8122653]

[51] X. Gu, S. Hemour, and K. Wu, "Low thermally activated Schottky barrier rectifier: A new class of energy harvester", *Proc. IEEE Int. Conf. RFID Technol. Appl. (RFID-TA)*, vol. vol. 1, 2019p. 76 Pisa, Italy

[52] T. Mori, J. Ida, S. Momose, K. Itoh, K. Ishibashi, and Y. Arai, "Diode characteristics of a super-steep

[52] ...subthreshold slope PN-body tied SOI-FET for energy harvesting applications", *IEEE J. Electron Devices Soc.,* vol. 6, pp. 565-570, 2018.
[http://dx.doi.org/10.1109/JEDS.2018.2824344]

[53] S. Momose, "First experimental confirmation of ultralow voltage rectification by super steep subthreshold slope 'PN-body tied SOI-FET' for high efficiency RF energy harvesting and ultralow voltage sensing", In: *In 2018 IEEE SOI-3D-Subthreshold Microelectronics Technology Unified Conference (S3S).* IEEE, 2018, pp. 1-3.

[54] I.J. Yasumaru, N. Nakanishi, K. Itoh, S. Tsuchimoto, T. Yamada, and T. Mori, "1 MHz band rectenna with several rectifier devices in nW operation", *Proc. IEEE MTT-S Wireless Power Transfer Conf. (WPTC),* 2019p. 1 London

[55] X. Zhang, "Two-dimensional MoS2-enabled flexible rectenna for Wi-Fi-band wireless energy harvesting", *Nature,* vol. 566, no. 7744, pp. 368-372, 2019.
[http://dx.doi.org/10.1038/s41586-019-0892-1]

[56] C. Forman, I.K. Muritala, R. Pardemann, and B. Meyer, "Estimating the global waste heat potential", *Renew. Sustain. Energy Rev.,* vol. 57, pp. 1568-1579, 2016.
[http://dx.doi.org/10.1016/j.rser.2015.12.192]

[57] T. Mori, "Novel principles and nanostructuring methods for enhanced thermoelectrics", *Small,* vol. 13, no. 45, 2017.
[http://dx.doi.org/10.1002/smll.201702013]

[58] Y. Shi, C. Sturm, and H. Kleinke, "Chalcogenides as thermoelectric materials", *J. Solid State Chem.,* vol. 270, pp. 273-279, 2019.
[http://dx.doi.org/10.1016/j.jssc.2018.10.049]

[59] K. Biswas, "High-performance bulk thermoelectrics with all-scale hierarchical architectures", *Nature,* vol. 489, no. 7416, pp. 414-418, 2012.
[http://dx.doi.org/10.1038/nature11439]

[60] "Effect of A-site modification on structural and microwave dielectric properties of calcium titanate", *J. Met. Mater. Miner.,* vol. 32, no. 3, pp. 118-125, 2022.
[http://dx.doi.org/10.55713/jmmm.v32i3.1525]

[61] T. Kanno, "Enhancement of average thermoelectric figure of merit by increasing the grain-size of Mg3.2Sb1.5Bi0.49Te0.01", *Appl. Phys. Lett.,* vol. 112, no. 3, 2018.
[http://dx.doi.org/10.1063/1.5016488]

[62] S. Saini, "Porosity-tuned thermal conductivity in thermoelectric Al-doped ZnO thin films grown by mist-chemical vapor deposition", *Thin Solid Films,* vol. 685, pp. 180-185, 2019.
[http://dx.doi.org/10.1016/j.tsf.2019.06.010]

[63] L-D. Zhao, "Ultralow thermal conductivity and high thermoelectric figure of merit in SnSe crystals", *Nature,* vol. 508, no. 7496, pp. 373-377, 2014.
[http://dx.doi.org/10.1038/nature13184]

[64] S.S. Rajput, and S. Keshri, Structural and microwave properties of (Mg,Zn/Co)TiO$_3$ dielectric ceramics., *J. Mater. Eng. Perform.,* vol. 23, no. 6, pp. 2103-2109, 2014.
[http://dx.doi.org/10.1007/s11665-014-0950-7]

[65] M. Tomita, "Modeling, simulation, fabrication, and characterization of a mu W/cm class si-nanowire thermoelectric generator for IoT applications", *IEEE Trans. Electron Dev.,* vol. 65, no. 11, pp. 5180-5188, 2018.
[http://dx.doi.org/10.1109/TED.2018.2867845]

[66] S. Rajput, A. Lugovskoy, M. Averbukh, and A. Yahalom, "Porous Metal-Oxide Based Electrostatic Energy Generator", *2019 International IEEE Conference and Workshop in Óbuda on Electrical and Power Engineering (CANDO-EPE),* 2019pp. 133-136

[67] F. Ahmed, N. Tsujii, and T. Mori, "Thermoelectric properties of CuGa 1−x Mn x Te 2: power factor

enhancement by incorporation of magnetic ions", *J. Mater. Chem. A Mater. Energy Sustain.*, vol. 5, no. 16, pp. 7545-7554, 2017.
[http://dx.doi.org/10.1039/C6TA11120C]

[68] H. Takahashi, S. Ishiwata, R. Okazaki, Y. Yasui, and I. Terasaki, "Enhanced thermopower via spin-state modification", *Phys. Rev. B,* vol. 98, no. 2, p. 024405, 2018.
[http://dx.doi.org/10.1103/PhysRevB.98.024405]

[69] N.T. Hung, E.H. Hasdeo, A.R.T. Nugraha, M.S. Dresselhaus, and R. Saito, "Quantum effects in the thermoelectric power factor of low-dimensional semiconductors", *Phys. Rev. Lett.,* vol. 117, no. 3, p. 036602, 2016.
[http://dx.doi.org/10.1103/PhysRevLett.117.036602]

[70] Y. Zhang, "Double thermoelectric power factor of a 2D electron system", *Nat. Commun.,* vol. 9, no. 1, p. 2224, 2018.
[http://dx.doi.org/10.1038/s41467-018-04660-4]

[71] H. Ohta, S.W. Kim, S. Kaneki, A. Yamamoto, and T. Hashizume, "High thermoelectric power factor of high-mobility 2D electron gas", *Adv. Sci.,* vol. 5, no. 1, 2018.
[http://dx.doi.org/10.1002/advs.201700696]

[72] C Chang, "3D charge and 2D phonon transports leading to high out-of-plane zt in n-type snse crystals", *Science,* vol. 360, no. 6390, pp. 778-783, 2018.
[http://dx.doi.org/10.1126/science.aaq1479]

[73] S. Shimizu, "Giant thermoelectric power factor in ultrathin FeSe superconductor", *Nat. Commun.,* vol. 10, no. 1, p. 825, 2019.
[http://dx.doi.org/10.1038/s41467-019-08784-z]

[74] J P Heremans, "Enhancement of thermoelectric efficiency in pbte by distortion of the electronic density of states", *Science,* vol. 321, no. 5888, pp. 554-557, 2008.
[http://dx.doi.org/10.1126/science.1159725]

[75] A. Nomura, "Chalcopyrite ZnSnSb 2: A promising thermoelectric material", *ACS Appl. Mater. Interfaces,* vol. 10, no. 50, pp. 43682-43690, 2018.
[http://dx.doi.org/10.1021/acsami.8b16717]

[76] Y. Nonoguchi, "Simple salt-coordinated n-type nanocarbon materials stable in air", *Adv. Funct. Mater.,* vol. 26, no. 18, pp. 3021-3028, 2016.
[http://dx.doi.org/10.1002/adfm.201600179]

[77] I. Amiel, S. Rajput, and M. Averbukh, "Capacitive reactive power compensation to prevent voltage instabilities in distribution lines", *Int. J. Electr. Power Energy Syst.,* vol. 131, p. 107043, 2021.
[http://dx.doi.org/10.1016/j.ijepes.2021.107043]

[78] S. Rajput, E. Lockshin, A. Schochet, and M. Averbukh, "Reactance regulation using coils with perpendicular magnetic field in the tubular core design", *Appl. Sci. (Basel),* vol. 10, no. 21, p. 7645, 2020.
[http://dx.doi.org/10.3390/app10217645]

[79] T. Taniguchi, T. Ishibe, H. Miyamoto, Y. Yamashita, and Y. Nakamura, "Thermoelectric properties of epitaxial Ge thin films on Si(001) with strong crystallinity dependence", *Appl. Phys. Express,* vol. 11, no. 11, p. 111301, 2018.
[http://dx.doi.org/10.7567/APEX.11.111301]

[80] S. Rajput, A. Kuperman, A. Yahalom, and M. Averbukh, "Studies on dynamic properties of ultracapacitors using infinite r–C chain equivalent circuit and reverse fourier transform", *Energies,* vol. 13, no. 18, p. 4583, 2020.
[http://dx.doi.org/10.3390/en13184583]

[81] S. Rajput, I. Amiel, M. Sitbon, I. Aharon, and M. Averbukh, "Control the voltage instabilities of distribution lines using capacitive reactive power", *Energies,* vol. 13, no. 4, p. 875, 2020.

[http://dx.doi.org/10.3390/en13040875]

[82] I. Terasaki, Y. Sasago, and K. Uchinokura, "Large thermoelectric power in NaCo2O4 single crystals", *Phys. Rev. B Condens. Matter,* vol. 56, no. 20, pp. R12685-R12687, 1997.
[http://dx.doi.org/10.1103/PhysRevB.56.R12685]

[83] K. Koumoto, "Thermoelectric ceramics for energy harvesting", *J. Am. Ceram. Soc.,* vol. 96, no. 1, pp. 1-23, 2013.
[http://dx.doi.org/10.1111/jace.12076]

[84] I. Terasaki, "Research Update: Oxide thermoelectrics: Beyond the conventional design rules", *APL Mater.,* vol. 4, no. 10, 2016.
[http://dx.doi.org/10.1063/1.4954227]

[85] A. Sharma, "Opposition-based tunicate swarm algorithm for parameter optimization of solar cells", *IEEE Access,* vol. 9, pp. 125590-125602, 2021.
[http://dx.doi.org/10.1109/ACCESS.2021.3110849]

[86] S. Rajput, E. Farber, and M. Averbukh, "Optimal selection of asynchronous motor-gearhead couple fed by VFD for electrified vehicle propulsion", *Energies,* vol. 14, no. 14, p. 4346, 2021.
[http://dx.doi.org/10.3390/en14144346]

[87] P. Qiu, "Suppression of atom motion and metal deposition in mixed ionic electronic conductors", *Nat. Commun.,* vol. 9, no. 1, p. 2910, 2018.
[http://dx.doi.org/10.1038/s41467-018-05248-8]

[88] P. Jafarzadeh, "Thermoelectric properties and stability of Ba3Cu16 − x Se11 − y Te y", *J. Appl. Phys.,* vol. 126, no. 2, 2019.
[http://dx.doi.org/10.1063/1.5110043]

[89] C. O'Dwyer, R. Chen, J-H. He, J. Lee, and K.M. Razeeb, "Scientific and technical challenges in thermal transport and thermoelectric materials and devices", *ECS J. Solid State Sci. Technol.,* vol. 6, no. 3, pp. N3058-N3064, 2017.
[http://dx.doi.org/10.1149/2.0091703jss]

[90] J. Maire, R. Anufriev, R. Yanagisawa, A. Ramiere, S. Volz, and M. Nomura, "Heat conduction tuning by wave nature of phonons", *Sci. Adv.,* vol. 3, no. 8, 2017.
[http://dx.doi.org/10.1126/sciadv.1700027]

[91] R. Hu, "Machine-learning-optimized aperiodic superlattice minimizes coherent phonon heat conduction", *Phys. Rev. X,* vol. 10, no. 2, p. 021050, 2020.
[http://dx.doi.org/10.1103/PhysRevX.10.021050]

[92] R. Anufriev, A. Ramiere, J. Maire, and M. Nomura, "Heat guiding and focusing using ballistic phonon transport in phononic nanostructures", *Nat. Commun.,* vol. 8, no. 1, p. 15505, 2017.
[http://dx.doi.org/10.1038/ncomms15505]

[93] S Rajput, S Parida, A Sharma, and Sonika, "Dielectric materials for energy storage and energy harvesting devices", In: *River Publ* River Publishers: New York, 2023.
[http://dx.doi.org/10.1201/9781032630816]

[94] M Mecklenburg, "Nanoscale temperature mapping in operating microelectronic devices", *Science,* vol. 347, no. 6222, pp. 629-632, 2015.
[http://dx.doi.org/10.1126/science.aaa2433]

CHAPTER 6

Methodologies Exploited in the Synthesis of Conducting Polymer Nanocomposites: Concept, Strategies, and Development

Anju Dhillon[1,*], **Manoj Kumar Srivastava**[2], **Vasudha Agarwal**[3] and **Raman Sankar**[4]

[1] *Department of Applied Sciences, Maharaja Surajmal Institute of Technology Affiliated to GGSIP University, New Delhi 110058, India*

[2] *Department of Physics, DAV PG College, DDU Gorakhpur University, Gorakhpur 273001, Uttar Pradesh, India*

[3] *Department of Physics, Maitreyi College (University of Delhi), Chanakyapuri, Delhi, India*

[4] *Institute of Physics, Academia Sinica, 128 Academia Road, Section 2, Nankang, Taipei 11529, Taiwan*

Abstract: This book chapter provides a comprehensive overview of conducting polymer nanocomposites, including their types and potential applications. The article highlights six types of conducting polymer nanocomposites, which include metal nanoparticles, carbon nanotubes, graphene, metal oxide nanoparticles, nanocellulose, and clay nanoparticles. Additionally, the chapter discusses the various methods available for the synthesis of conducting polymer nanocomposites, which include *in situ* polymerization, solution blending, *in situ* reduction, electrochemical deposition, and layer-by-layer assembly. Overall, this chapter serves as a valuable resource for researchers interested in the synthesis and application of conducting polymer nanocomposites.

Keywords: Conducting polymer nanocomposites, Functional polymer, Fabrication techniques, Hybrid nanomaterials.

INTRODUCTION

Conducting polymers, also known as conjugated polymers, are a class of organic materials that have the ability to conduct electricity. They are formed by linking together small molecules called monomers, which form long chains. The unique structure of conducting polymers allows for the flow of electrons, making them attractive for a variety of applications, such as electronics, optoelectronics, and

---

[*] **Corresponding author Anju Dhillon:** Department of Applied Sciences, Maharaja Surajmal Institute of Technology affiliated to GGSIP University, New Delhi 110058, India; E-mail: anju.dhillon@msit.in

energy storage. The properties of conducting polymers can be tailored to suit specific applications by modifying their chemical structure, leading to the development of different types of conducting polymers.

Types of Conducting Polymer Nanocomposites

Conducting polymer nanocomposites are a type of composite material that combines conducting polymers with nanoparticles or nanofibers to enhance their electrical and mechanical properties. The addition of nanoparticles to conducting polymers can result in increased electrical conductivity, improved thermal stability, and enhanced mechanical strength. There are different types of polymer nanocomposites, such as

1. Polymer nanocomposites with metal nanoparticles

Conducting polymers can be combined with metal nanoparticles such as silver, gold and copper to create materials with high electrical conductivity and antimicrobial properties. These materials have potential applications in electronic devices, sensors, and medical devices.

Polymer nanocomposites incorporating metal nanoparticles exhibit antibacterial activity through a synergistic combination of physical and chemical mechanisms:

- a. Physical membrane disruption: The high surface area-to-volume ratio of nanoparticles facilitates strong interactions with bacterial cell membranes. These interactions may involve direct contact, surface adhesion, or even partial envelopment of bacterial cells, ultimately leading to membrane destabilization and increased permeability.
- b. Metal ion release: Certain metal oxide nanoparticles, such as Zinc Oxide (ZnO) and Copper Oxide (CuO), release metal ions into the surrounding environment. These ions can compromise membrane integrity, disrupt vital enzymatic processes, interfere with nutrient uptake, and inhibit DNA replication, collectively impairing bacterial viability.
- c. Generation of Reactive Oxygen Species (ROS): Noble metal nanoparticles, including Silver (Ag) and Gold (Au), can catalyze the formation of reactive oxygen species. ROS such as superoxide anions, hydroxyl radicals, and hydrogen peroxide induce oxidative stress, damaging bacterial DNA, proteins, and lipid membranes, ultimately resulting in cell death.

Polymer Nanocomposites With Carbon Nanotubes

Conducting polymers can also be combined with Carbon Nanotubes (CNTs) to create nanocomposites with high electrical conductivity, thermal stability, and mechanical strength. CNTs are one of the strongest materials known to man, and their addition to conducting polymers can significantly improve their mechanical properties. These materials have potential applications in flexible electronics, energy storage devices, and aerospace applications.

Polymer Nanocomposites with Graphene

Graphene is a two-dimensional material consisting of a single layer of carbon atoms arranged in a hexagonal lattice. Conducting polymers can be combined with graphene to create nanocomposites with high electrical conductivity, thermal conductivity, and mechanical strength. These materials have potential applications in energy storage devices, electronic devices, and sensors.

Polymer Nanocomposites with Metal Oxide Nanoparticles

Conducting polymers can also be combined with metal oxide nanoparticles such as titanium oxide, zinc oxide, and iron oxide to create nanocomposites with improved electrical conductivity, thermal stability, and mechanical strength. These materials have potential applications in sensors, electronic devices, and energy storage devices.

Polymer Nanocomposites with Nanocellulose

Nanocelluloses are biodegradable materials that can be extracted from plant-based sources such as wood, cotton, and hemp. Conducting polymers can be combined with nanocellulose to create nanocomposites with high electrical conductivity, thermal conductivity, and mechanical strength. These materials have potential applications in flexible electronics, biomedical devices, and packaging materials.

Polymer Nanocomposites with Clay Nanoparticles

Clay nanoparticles such as montmorillonite and kaolinite can be combined with conducting polymers to create nanocomposites with improved mechanical strength, thermal stability, and flame retardancy. These materials have potential applications in electronic devices, coatings, and flame-retardant materials.

Advantages of Conducting Polymer Nanocomposites

Conducting polymer nanocomposites offer several advantages over pure conducting polymers, including:

- Enhanced electrical conductivity: The addition of nanoparticles to conducting polymers can significantly enhance the electrical conductivity of the material by providing additional charge carriers.

- Improved mechanical properties: Nanoparticles can improve the mechanical properties of conducting polymers, such as strength, stiffness, and toughness, making them more suitable for a wider range of applications.
- Tailored properties: Conducting polymer nanocomposites can be tailored to exhibit specific properties by adjusting the type and concentration of nanoparticles used. This allows for the development of materials with desired electrical, mechanical, and optical properties.
- Increased stability: Conducting polymer nanocomposites can exhibit increased stability compared to pure conducting polymers. The presence of nanoparticles can help prevent degradation and improve the material's resistance to environmental factors such as moisture, heat, and UV light.
- Lower cost: Nanoparticles are often less expensive than the conducting polymers themselves, making nanocomposites a more cost-effective solution for certain applications.

These benefits make nanocomposites a promising solution for a wide range of electronic, optoelectronic, and energy-related applications.

Methods for the Synthesis of Conducting Polymer Nanocomposites

Conducting polymer nanocomposites are a versatile class of materials that have the potential to transform a variety of industries. There are several methods available for the synthesis of conducting polymer nanocomposites. In this chapter, we will discuss some of the commonly used methods.

IN SITU POLYMERIZATION

In situ polymerization involves the simultaneous synthesis of conducting polymers and the incorporation of nanoparticles or nanofibers into the polymer matrix. This method involves adding the monomer, initiator, and nanoparticles or nanofibers to a reaction vessel and then heating or irradiating the mixture to initiate the polymerization process. The resulting nanocomposite consists of a conducting polymer matrix with uniformly dispersed nanoparticles or nanofibers. In situ polymerization is a commonly employed method for the synthesis of conducting polymer nanocomposites, where the polymerization of the conducting monomer and the dispersion of nanofillers occur simultaneously within a single reaction system.

The process of conducting polymer nanocomposite in situ polymerization typically involves the following steps (Fig. **1**) [1 - 4]:

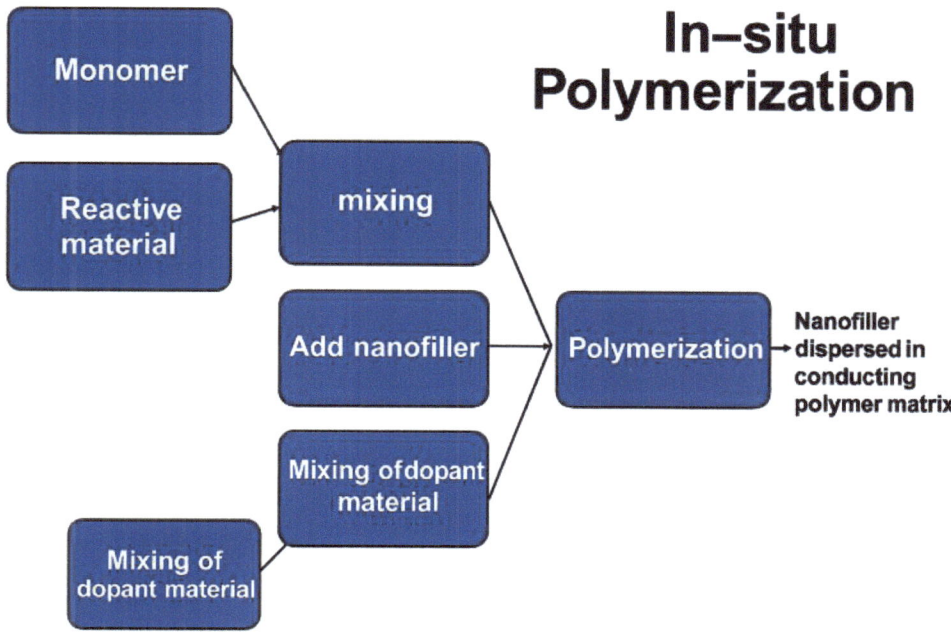

Fig. (1). Systematic flow chart of in situ polymerization.

- **Selection of Monomer:** The choice of monomer is crucial as it determines the conductivity and other properties of the conducting polymer. Commonly used monomers include aniline, pyrrole, and thiophene, among others.
- **Selection of Nanofillers:** Various nanofillers can be incorporated into the conducting polymer matrix to enhance its properties. These nanofillers can include carbon-based materials like Carbon Nanotubes (CNTs) and graphene, metal nanoparticles, metal oxide nanoparticles, and polymer nanoparticles.
- **Dispersion of Nanofillers:** Prior to polymerization, the nanofillers need to be uniformly dispersed within the monomer solution. Techniques such as sonication, mechanical stirring, and surfactant-assisted methods are often employed to achieve good dispersion and prevent agglomeration of the nanofillers.
- **Initiator and Catalyst:** Initiators or oxidants are used to initiate the polymerization process. Commonly used initiators include Ammonium Persulfate (APS), Iron Chloride ($FeCl_3$), and Hydrogen Peroxide (H_2O_2), while catalysts such as acids or bases may be added to enhance the polymerization rate and control the properties of the resulting nanocomposite.
- **Polymerization:** The monomer solution, containing the dispersed nanofillers, is then subjected to polymerization conditions. Typically, the reaction is carried

out under controlled temperature and reaction time to facilitate the formation of a conducting polymer network around the nanofillers.
- **Washing and Drying:** After the polymerization, the resulting nanocomposite is typically washed with appropriate solvents to remove any unreacted monomer or impurities. The nanocomposite is then dried to obtain the final product.

Conducting polymer nanocomposites synthesized *via in situ* polymerization exhibit several advantages, including good dispersion of nanofillers, strong interfacial adhesion between the polymer matrix and nanofillers, and improved electrical, mechanical, and thermal properties compared to the pure conducting polymer.

The interfacial interactions between polymer matrices and nanofillers play a pivotal role in determining the functional performance of polymer nanocomposites. These interactions, governed by factors such as surface chemistry, filler morphology, and polymer chain dynamics, directly influence both charge transport and mechanical coupling within the composite.

In terms of charge transport, strong interfacial adhesion facilitates effective electron or ion transfer pathways across the polymer–nanofiller interface. Functionalization of nanofillers (*e.g.*, carbon nanotubes, graphene, metal oxides) enhances interfacial compatibility, promoting better dispersion and reducing interfacial resistance. This leads to the formation of conductive networks, improving overall electrical conductivity.

For mechanical coupling, robust interfacial bonding ensures efficient stress transfer from the polymer matrix to the nanofiller. This enhances the mechanical strength, modulus, and toughness of the composite. Conversely, weak interactions can lead to filler agglomeration, stress concentration, and interfacial debonding, which degrade both mechanical integrity and functional performance.

Sahoo *et al.*, in 2019, employed an in situ polymerization method to synthesize the Polypyrrole (PPy)/graphene nanocomposites. In this approach, Graphene Oxide (GO) was first prepared and then mixed with pyrrole monomers. The mixture was subjected to a chemical reaction, resulting in the simultaneous reduction of GO and polymerization of pyrrole, leading to the formation of PPy/graphene nanocomposites. The synthesized nanocomposites were characterized using various techniques, and results confirmed the successful incorporation of PPy onto the graphene sheets [5]. The studies also evaluated the electrochemical performance of the conducting polymer/graphene nanocomposites for flexible supercapacitor applications. The researchers fabricated flexible electrodes where the nanocomposites were deposited onto a flexible substrate. The electrodes were then assembled into symmetric supercapacitor cells [6 - 14]. Wanh *et al.* [15] used the facile *in situ* polymerization method to prepare the PANI/rGO nanocomposites. They combined aniline monomers and reduced graphene oxide, and then polymerized the mixture to form a composite material. The resulting nanocomposites were characterized using various techniques to evaluate their properties. The study found that the incorporation of reduced graphene oxide into the polyaniline matrix significantly improved the electrical conductivity of the nanocomposites. The presence of reduced graphene oxide facilitated charge transfer and enhanced the overall conductivity of the material. Additionally, the mechanical properties of the nanocomposites were also improved. The presence of reduced graphene oxide enhanced the tensile strength and modulus of the composite material, making it more mechanically robust.

The researchers attributed [15, 16] these improvements in electrical conductivity and mechanical properties to the unique structure and properties of reduced graphene oxide. The two-dimensional structure of reduced graphene oxide provided a conductive network within the conducting polymer matrix, resulting in improved electrical conductivity. Furthermore, the high aspect ratio and strong interactions between reduced graphene oxide and conducting polymer contributed to the enhanced mechanical properties. Zhang *et al.* [17] used an *in situ* polymerization method to fabricate the (Polyaniline) PANI/CNT (Carbon Nanotubes) nanocomposites. In this approach, the polymerization of aniline monomers occurred directly on the surface of the carbon nanotubes, resulting in a uniform and intimate integration of PANI with CNTs. The use of carbon nanotubes as a support material helps to enhance the electrical conductivity and structural stability of the resulting nanocomposites, exhibiting excellent electrochemical performance with high specific capacitance and good cycling stability. The incorporation of carbon nanotubes [18] effectively improved the electrical conductivity and mechanical strength of the nanocomposites, leading to enhanced energy storage capabilities. The researchers attributed these improvements to the synergistic effects between PANI and CNTs. An *in situ* polymerization technique has been used to fabricate the PPy/MoS2 nanocomposites [19]. This method involved the polymerization of pyrrole monomers in the presence of MoS2, resulting in the formation of a hybrid material with PPy and MoS2 well-integrated at the nanoscale. MoS2, a two-dimensional layered material, serves as a supporting matrix that enhances the electrical conductivity and structural stability of the nanocomposites [20]. The results suggest that these nanocomposites have promising potential for energy storage applications due to their favorable electrochemical performance and the unique properties of MoS2 as a supporting material.

The *in situ* polymerization method holds significant importance in the fabrication of conducting polymer nanocomposites due to the following reasons [1 - 20]:

Control over Nanocomposite Structure: *In situ* polymerization allows for the simultaneous formation of the conducting polymer and the incorporation of nanomaterials within the polymer matrix. This method provides precise control over the composition, morphology, and dispersion of the nanomaterials in the polymer matrix, leading to tailored nanocomposite structures. The ability to control the nanocomposite structure is crucial for optimizing the material properties and achieving desired functionalities.

Improved Electrical Conductivity: The *in situ* polymerization method enables the uniform dispersion of nanomaterials, such as carbon nanotubes, graphene, or metal nanoparticles, within the conducting polymer matrix. This intimate contact between the conducting polymer chains and nanomaterials facilitates efficient charge transport pathways, resulting in enhanced electrical conductivity. Improved electrical conductivity is vital for applications such as sensors, energy storage devices, and electronic devices.

Synergistic Property Enhancement: The incorporation of nanomaterials into the conducting polymer matrix through *in situ* polymerization often leads to synergistic property enhancements. The combination of the unique properties of nanomaterials and conducting polymers can result in superior mechanical strength, thermal stability, electrochemical performance, and optical properties compared to the individual components. Synergistic effects enable the development of high-performance nanocomposites with enhanced functionality for various applications.

Tailored Properties: *In situ* polymerization offers the flexibility to tune the properties of conducting polymer nanocomposites by adjusting the polymerization conditions, monomer ratios, and nanomaterial loading. This control allows for the tailoring of properties such as conductivity, mechanical strength, thermal stability, and electrochemical performance according to specific application requirements. The ability to customize the properties makes conducting polymer nanocomposites versatile and adaptable for a wide range of applications.

Scalability and Cost-effectiveness: *In situ* polymerization is a scalable and cost-effective method for the production of conducting polymer nanocomposites. The process can be easily adapted for large-scale manufacturing, making it suitable for industrial applications. The use of simple and readily available precursors, along with a straightforward synthesis process, contributes to the cost-effectiveness of this method.

These advantages make the *in situ* polymerization method highly valuable for the fabrication of conducting polymer nanocomposites with enhanced performance and tailored functionalities for various applications.

SOLUTION BLENDING

Solution blending involves dissolving the conducting polymer and nanoparticles or nanofibers in a common solvent and then mixing the two solutions together. The resulting mixture is then dried to remove the solvent, leaving behind a solid nanocomposite material. The advantage of this method is that it allows for precise control over the nanoparticle or nanofiber content in the resulting nanocomposite. The synthesis of conducting polymers using the solution blending method involves the combination of a conductive polymer and a non-conductive polymer in a solution to create a composite material with enhanced electrical conductivity. This method allows for the fabrication of conductive polymer films, coatings, fibers, and other forms.

Here is a step-by-step explanation of the synthesis process (Fig. **2**):

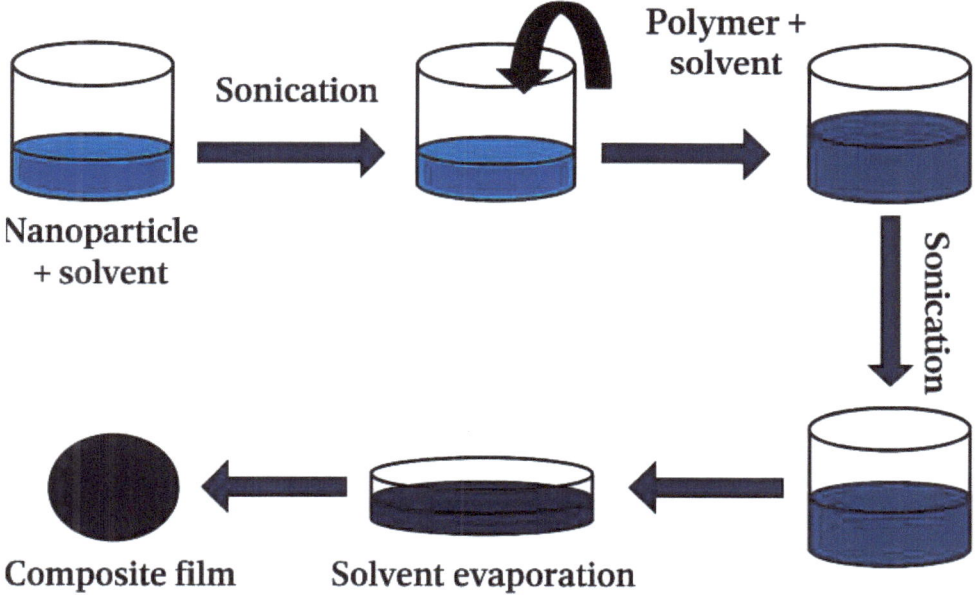

Fig. (2). Synthesis of conducting polymer nanocomposites by the solution blending method.

- **Polymer Selection:** The first step is to select the appropriate conducting polymer and non-conductive polymer. Conducting polymers, such as

Polyaniline (PANI), Polypyrrole (PPy), or Polythiophene (PTh), are chosen for their ability to conduct electricity. The non-conductive polymer can be any compatible polymer, such as Polystyrene (PS), Polyethylene (PE), or Polyvinyl Chloride (PVC).
- **Solvent Selection:** A suitable solvent is chosen that can dissolve both the conducting polymer and the non-conductive polymer. The choice of solvent depends on the polymers used and their solubility properties. Common solvents include Tetrahydrofuran (THF), chloroform, toluene, or a mixture of solvents.
- **Polymer Dissolution:** The conducting polymer and the non-conductive polymer are separately dissolved in the chosen solvent. Each polymer is typically weighed and added to the solvent gradually while stirring to ensure complete dissolution.
- **Blending:** The two polymer solutions are then combined in a predetermined ratio. The conductive polymer solution is slowly added to the non-conductive polymer solution while stirring continuously. The blending process should be carried out carefully to ensure uniform distribution.

Several studies have demonstrated the efficacy of solution blending as a versatile and scalable technique for the fabrication of conducting polymer nanocomposites with enhanced electrical properties. Liu *et al.* [21] reported the synthesis of conducting polymer–Carbon Nanotube (CNT) composites *via* solution blending, emphasizing the critical role of processing parameters—such as solvent selection and mixing conditions—in achieving uniform dispersion of CNTs within the polymer matrix. The study systematically evaluated the morphology and electrical conductivity of the composites, revealing significant improvements attributable to the presence of well-dispersed CNTs.

Kumar *et al.* [22] presented a simplified and effective strategy to enhance the electrical conductivity of Polyaniline (PANI) by incorporating Multi-Walled Carbon Nanotubes (MWCNTs) through solution blending. Their findings highlight the impact of MWCNT content and blending conditions on the conductive performance and microstructure of the composites, reinforcing the reliability of this method.

Further investigations [23, 24] extended the approach to the incorporation of graphene nanosheets into conducting polymer matrices, demonstrating notable enhancements in both electrical conductivity and mechanical flexibility. These studies underline the importance of nanofiller loading, interfacial interactions, and dispersion uniformity in tailoring composite properties.

Complementary work by other researchers [25] confirmed the broader applicability of the solution blending technique for the development of high-performance conducting polymer-based nanocomposites. This method enables precise control over nanofiller distribution and offers insights into the relationship between nanostructure and macroscopic properties. Collectively, these contributions underscore the utility of solution blending in engineering advanced materials for applications in flexible electronics, sensors, and energy storage systems.

Simplicity and Versatility: Solution blending polymerization is a straightforward and versatile method for preparing conducting polymer nanocomposites. It involves dissolving the monomers and nanomaterials in a suitable solvent and then polymerizing the monomers in the presence of the nanomaterials. This simplicity allows for easy control of the nanocomposite composition, morphology, and properties.

Homogeneous Nanomaterial Dispersion: Solution blending facilitates the homogeneous dispersion of nanomaterials within the conducting polymer matrix. The intimate mixing at the molecular level ensures uniform distribution and strong interactions between the nanomaterials and the polymer chains. This uniform dispersion enhances the electrical conductivity and other properties of the resulting nanocomposite.

Property Enhancement: The incorporation of nanomaterials into the conducting polymer matrix through solution blending can lead to significant property enhancements. For instance, the addition of carbon-based nanomaterials (*e.g.*, carbon nanotubes, graphene) can improve electrical conductivity, mechanical strength, and thermal stability. Similarly, the inclusion of metal nanoparticles can enhance catalytic activity and electrochemical performance.

Diverse Applications: Solution blending polymerization enables the production of conducting polymer nanocomposites with tailored properties suitable for a wide range of applications. These applications include energy storage devices (such as supercapacitors and batteries), sensors, actuators, electronic devices, electromagnetic shielding, and biomedical devices. The tunable properties of the nanocomposites make them suitable for specific requirements in these fields.

Scalability and Cost-effectiveness: Solution blending polymerization can be easily scaled up for industrial production. It offers a cost-effective approach as it does not require specialized equipment or complex processing steps. The availability of precursors and the simplicity of the method contribute to its cost-effectiveness.

Material Compatibility: Solution blending polymerization allows for the combination of different nanomaterials with various conducting polymers. This compatibility enables the creation of hybrid nanocomposites with tailored properties by selecting appropriate combinations. It opens up opportunities to explore synergistic effects and optimize the properties for specific applications.

IN SITU REDUCTION

In situ reduction involves the simultaneous synthesis of conducting polymers and the reduction of metal ions to form metal nanoparticles. This method involves adding the monomer, metal salt, and reducing agent to a reaction vessel and then heating or irradiating the mixture to initiate the polymerization process and the reduction of metal ions. The resulting nanocomposite consists of a conducting polymer matrix with uniformly dispersed metal nanocomposites. The synthesis of conducting polymer nanocomposites using the *in situ* reduction method involves (Fig. **3**) [26] the incorporation of conductive nanoparticles within a polymer matrix through a chemical reduction process.

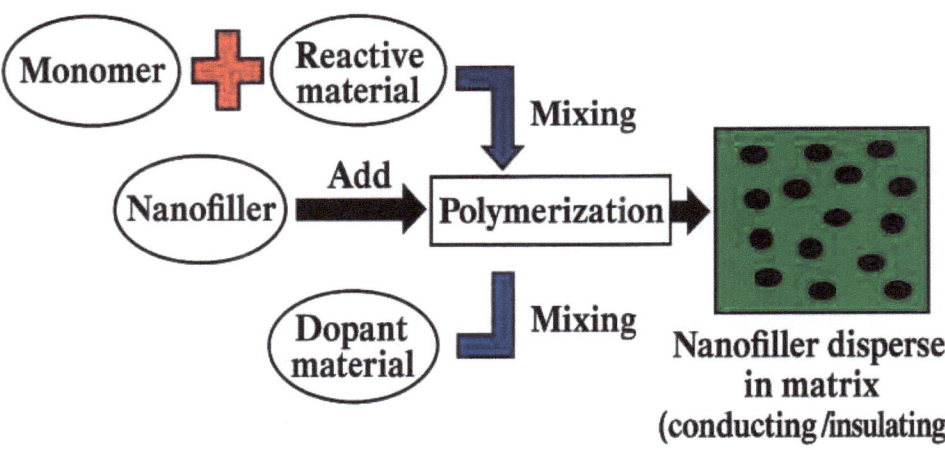

Fig. (**3**). *In situ* reduction polymerization method of preparation of conductive polymers.

This method allows for the simultaneous formation of conducting polymer and the dispersion of nanoparticles, resulting in a composite material with enhanced electrical and mechanical properties.

Here is a step-by-step explanation of the synthesis process:

- **Selection of conducting polymer:** The choice of conducting polymer depends on the desired properties of the nanocomposite. Commonly used conducting

polymers include Polyaniline (PANI), Polypyrrole (PPy), and Polythiophene (PTh). These polymers possess inherent conductivity and can be easily synthesized through chemical oxidation methods.
- Selection of nanoparticles: Various types of nanoparticles can be incorporated into conducting polymers to improve their properties. Commonly used nanoparticles include carbon-based materials (carbon nanotubes, graphene), metal nanoparticles (gold, silver, copper), and metal oxide nanoparticles (titania, zinc oxide). The selection of nanoparticles depends on the intended application and the desired properties of the nanocomposite.
- **Preparation of precursor solution:** A precursor solution is prepared by dissolving the monomer(s) of the conducting polymer in a suitable solvent. The solvent should be compatible with both the monomer and the nanoparticles.
- **Addition of nanoparticles:** The nanoparticles are added to the precursor solution and mixed thoroughly. The concentration of nanoparticles can vary depending on the desired nanoparticle-to-polymer ratio and the target properties of the nanocomposite.
- *In situ* **reduction:** In this step, the reduction of the monomer and the simultaneous formation of the conducting polymer occur in the presence of nanoparticles. Typically, a reducing agent is added to the precursor solution to initiate the *in situ* reduction process. The reducing agent helps in the polymerization of the monomer and the reduction of the nanoparticles, resulting in the formation of conducting polymer chains and the dispersion of nanoparticles throughout the polymer matrix.
- **Polymerization and composite formation:** The precursor solution containing the monomer, nanoparticles, and reducing agent is subjected to suitable conditions for polymerization, such as elevated temperature or a chemical initiator. Polymerization occurs simultaneously with the reduction process, leading to the formation of conducting polymer chains that encapsulate the nanoparticles within the polymer matrix.
- **Washing and drying:** After the completion of the reaction, the conducting polymer nanocomposite is usually washed with a suitable solvent to remove any unreacted monomers, byproducts, or impurities. The nanocomposite is then dried to obtain a solid material.

The *in situ* reduction method allows for the synthesis of conducting polymer nanocomposites by simultaneously forming conducting polymer chains and dispersing nanoparticles within a polymer matrix. This method offers advantages such as easy scalability, control over nanoparticle dispersion, and the ability to tailor the properties of the nanocomposite for specific applications. Zeggai *et al.* in 2017 [27] highlighted the importance of incorporating conducting polymers with nanomaterials to enhance their properties by the *in situ* preparation of conducting polymer/copper(II)-Maghnite clay nanocomposites. The paper describes the experimental procedure for the *in situ* preparation of the nanocomposites. The paper highlights the electrical conductivity, thermal stability, and morphological characteristics of the nanocomposites, providing evidence of the successful incorporation of conducting polymers within the clay matrix. Yang *et al*. [28] focused on the *in situ* reduction of Graphene Oxide (GO) within a conducting polymer matrix to create high-performance thermoelectric nanocomposites. The authors demonstrate a facile method for synthesizing these nanocomposites by reducing GO using a conducting polymer, specifically poly(3,4-ethylenedioxythiophene) (PEDOT). The *in situ* reduction process results in the formation of a highly conductive PEDOT/graphene composite. The nanocomposites exhibit enhanced electrical conductivity and improved thermoelectric properties compared to pure PEDOT. Chen *et al*. [29] explored the *in situ* reduction of Graphene Oxide (GO) and the fabrication of graphene/Polyaniline (PANI) nanocomposites for supercapacitor applications. They presented a method for the simultaneous reduction of GO and polymerization of aniline in the presence of GO. The *in situ* reduction process leads to the formation of graphene/PANI nanocomposites with a well-defined structure. The resulting nanocomposites exhibit enhanced electrical conductivity and improved electrochemical performance compared to pure PANI. The study suggests that the *in situ* reduction of GO enables the effective integration of graphene into the PANI matrix, leading to improved supercapacitor performance. The authors presented [30 - 36] the *in situ* reduction of graphene oxide and its utilization as a scaffold for the synthesis of conducting ploymer in supercapacitor applications. The authors demonstrate a method where GO acts as a support structure for the growth of conducting polymer through *in situ* reduction. The resulting graphene/CP nanocomposites exhibit improved electrochemical performance, such as enhanced electrical conductivity and capacitance, making them promising for supercapacitor applications.

The *in situ* reduction method plays a crucial role in producing conducting polymer nanocomposites [26 - 36], which are promising for supercapacitor applications due to several important reasons.

Enhanced Electrical Conductivity: *In situ* reduction allows for the incorporation of conductive nanomaterials, such as graphene or graphene oxide, into the conducting polymer matrix. This integration significantly enhances the electrical conductivity of the resulting nanocomposite. Improved conductivity is essential for supercapacitors as it enables efficient charge transfer and higher power delivery capabilities.

Improved Electrochemical Performance: The *in situ* reduction process facilitates the formation of well-defined interfaces and interactions between the conducting polymer and nanomaterials. This enhances the electrochemical performance of the nanocomposite, resulting in higher capacitance, faster charge/discharge rates, and improved cycling stability. These factors are critical for the effective functioning of supercapacitors.

Increased Surface Area: The presence of nanomaterials in conducting polymer nanocomposites obtained through *in situ* reduction provides a significantly increased surface area. This expanded surface area allows for more active sites for charge storage, leading to enhanced energy storage capacity and improved overall performance of the supercapacitor.

Structural Integrity and Stability: *In situ* reduction enables the integration of nanomaterials into the conducting polymer matrix at a molecular level. This homogeneous distribution ensures good structural integrity and stability of the nanocomposite. It prevents aggregation and delamination of nanomaterials, maintaining their beneficial properties over time and enhancing the durability and reliability of the supercapacitor.

Scalability and Process Simplification: The *in situ* reduction method offers a scalable and simplified approach for the fabrication of conducting polymer nanocomposites. It allows for the direct synthesis of nanocomposites without the need for additional post-treatment steps or complicated processing techniques. This advantage makes the method more attractive for the large-scale production of supercapacitors.

ELECTROCHEMICAL DEPOSITION

Electrochemical deposition involves the electrodeposition of conducting polymers and nanoparticles or nanofibers onto a substrate. This method involves immersing a conductive substrate in a solution containing the monomer, nanoparticles or nanofibers, and an electrolyte. A voltage is then applied to the substrate, causing the monomer to polymerize and the nanoparticles or nanofibers to be deposited onto the substrate. The resulting nanocomposite consists of a conducting polymer matrix with nanoparticles or nanofibers deposited onto the substrate. Electrochemical deposition polymerization is a method used for the synthesis of conducting polymer nanocomposites. It involves the electrochemical growth of conducting polymers on a conductive substrate, typically an electrode, in the presence of nanomaterials. This process allows for the controlled deposition of the conducting polymer and the incorporation of nanomaterials into the polymer matrix. Here is a more detailed explanation (Fig. **4**) of the electrochemical deposition polymerization method [37]:

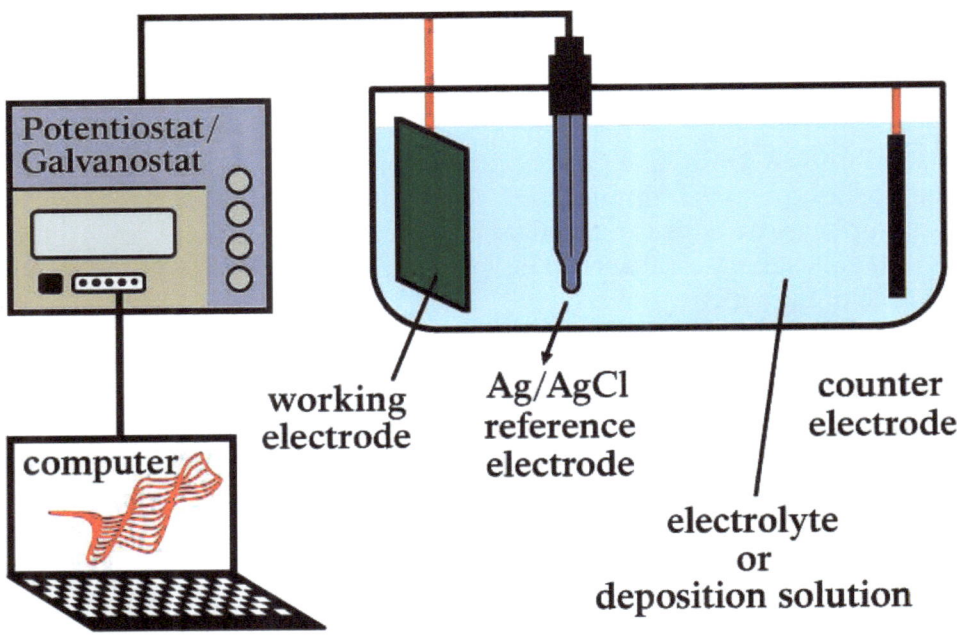

Fig. (4). Schematic illustration of the electrochemical deposition system used in this work.

- **Experimental Setup:** The electrochemical deposition polymerization setup typically consists of a working electrode, a counter electrode, and a reference electrode immersed in an electrolyte solution. The working electrode is the substrate on which the conducting polymer will be deposited. The counter electrode completes the electrical circuit, while the reference electrode helps monitor the electrochemical potential during the polymerization process.
- **Polymerization Process:** The process starts with the application of a voltage or current to the working electrode. The electrochemical potential promotes the oxidation or reduction of monomers in the electrolyte solution. As a result, conducting polymer chains are formed and deposited onto the surface of the working electrode.
- **Nanomaterial Incorporation:** During the electrochemical deposition polymerization, nanomaterials can be introduced into the electrolyte solution. The nanomaterials may include nanoparticles, nanowires, or nanosheets of metals, metal oxides, or carbon-based materials. As the conducting polymer grows, the nanomaterials become embedded within the polymer matrix, leading to the formation of conducting polymer nanocomposites.
- **Control of Nanocomposite Properties:** The properties of the conducting polymer nanocomposites synthesized by electrochemical deposition

polymerization can be controlled by various parameters. These include the polymerization voltage or current, the deposition time, the concentration and type of monomers, and the concentration and characteristics of the nanomaterials. Optimizing these parameters allows for the tuning of the nanocomposite's composition, morphology, and properties.

In electrochemical devices, the performance and durability of electrode coatings produced *via* electrochemical deposition are strongly influenced by the adhesion and thickness uniformity of the deposited layers. Adhesion between the active layer and the substrate is critical to maintaining mechanical and electrical integrity during repeated charge–discharge cycles. Poor adhesion can lead to delamination, cracking, or detachment of the coating under electrochemical or mechanical stress, thereby reducing cycle life. Thickness uniformity is equally important, as non-uniform deposition can result in localized current density variations, uneven charge distribution, and thermal hotspots. These inconsistencies can accelerate degradation mechanisms and compromise the power density and overall efficiency of the device. Controlling deposition parameters such as current density, electrolyte composition, temperature, and deposition time is essential for achieving uniform and adherent coatings. Thus, optimizing the electrochemical deposition process to ensure strong interfacial bonding and homogeneous layer formation is vital for enhancing the long-term performance and reliability of electrochemical energy storage and conversion systems.

Advantages and Applications: Electrochemical deposition polymerization offers several advantages for the synthesis of conducting polymer nanocomposites. It provides a well-controlled and uniform deposition of conducting polymers onto the substrate, enabling precise control over the nanocomposite structure and properties. The incorporation of nanomaterials enhances the electrical, mechanical, and electrochemical properties of the resulting nanocomposites. These materials find applications in various fields, including energy storage devices (batteries, supercapacitors), electrochromic devices, sensors, actuators, and electronic devices.

Challenges: Electrochemical deposition polymerization also has some challenges. It requires careful selection of electrolyte solutions and deposition conditions to ensure the stability and adherence of the deposited polymer film. The uniformity of nanomaterial dispersion within the polymer matrix and their interactions with the polymer chains need to be carefully controlled to achieve the desired nanocomposite properties.

Basically, electrochemical deposition polymerization is a controlled method for synthesizing conducting polymer nanocomposites. It allows for the electrochemical growth of conducting polymers on a substrate while incorporating nanomaterials within the polymer matrix. This technique offers tunable properties and controlled nanocomposite structure and has diverse applications in various fields. Mao et al. [38] conducted a study on the in situ electrochemical polymerization of Poly(3,4-Ethylenedioxythiophene) (PEDOT) and reduced Graphene Oxide (rGO) nanocomposite for high-performance supercapacitors. They employed a facile one-step electrochemical method to synthesize the nanocomposite film. The incorporation of rGO into the PEDOT matrix enhanced the electrical conductivity and surface area of the nanocomposite, resulting in improved electrochemical performance for supercapacitor applications. Anjli Gupta et al. [39] focused on the electrochemical synthesis of Polypyrrole (PPy)/Graphene Oxide (GO) nanocomposite films for supercapacitor applications. They utilized an electrochemical deposition method to fabricate the nanocomposite films. The integration of GO nanosheets into the PPy matrix improved the specific capacitance and rate capability of the nanocomposite. The study highlighted the potential of PPy/GO nanocomposites as electrode materials for high-performance supercapacitors. Cahuhan et al. [40] developed a facile electrochemical synthesis method for Polyaniline (PANI)/Graphene Oxide (GO) nanocomposites with enhanced electrochemical performance. The researchers employed cyclic voltammetry to fabricate the nanocomposites. The incorporation of GO nanosheets in the PANI matrix improved the electrochemical properties, including specific capacitance and cycling stability. The study demonstrated the feasibility of using PANI/GO nanocomposites for applications requiring high-performance electrochemical devices. Dhibar et al. [41] conducted a study on the electrochemical deposition of Polypyrrole (PPy)/Graphene Oxide (GO) nanocomposite films for supercapacitor applications. They employed a two-step electrochemical deposition process to fabricate the nanocomposite films. The incorporation of GO nanosheets into the PPy matrix improved the specific capacitance and cycling stability of the nanocomposite, making it a promising material for supercapacitor applications. Wang et al. [42] focused on the electrochemical synthesis of Polypyrrole (PPy)/reduced Graphene Oxide (rGO) nanocomposites and evaluated their electrochemical performance. They utilized a facile electrochemical method to prepare the nanocomposites. The incorporation of rGO into the PPy matrix enhanced the specific capacitance, rate capability, and cycling stability of the nanocomposite. The study highlighted the potential of PPy/rGO nanocomposites as electrode materials for high-performance supercapacitors. Rahman et al. [43] investigated the electrochemical deposition of Polyaniline (PANI)/graphene nanocomposite films for high-performance supercapacitors. They employed an electrochemical deposition technique to fabricate the nanocomposite films. The incorporation of graphene nanosheets into the PANI matrix significantly improved the specific capacitance and cycling stability of the nanocomposite. The study demonstrated the suitability of PANI/graphene nanocomposites for high-performance supercapacitor applications.

LAYER-BY-LAYER ASSEMBLY

Layer-by-layer assembly involves the sequential deposition of conducting polymers and nanoparticles or nanofibers onto a substrate. This method involves alternately immersing the substrate in solutions containing the conducting polymer and nanoparticles or nanofibers. The resulting nanocomposite consists of multiple layers of conducting polymers and nanoparticles or nanofibers, with each layer providing a different function.

Here is an explanation of the layer-by-layer assembly method for conducting polymer nanocomposites (Fig. **5**) [44]:

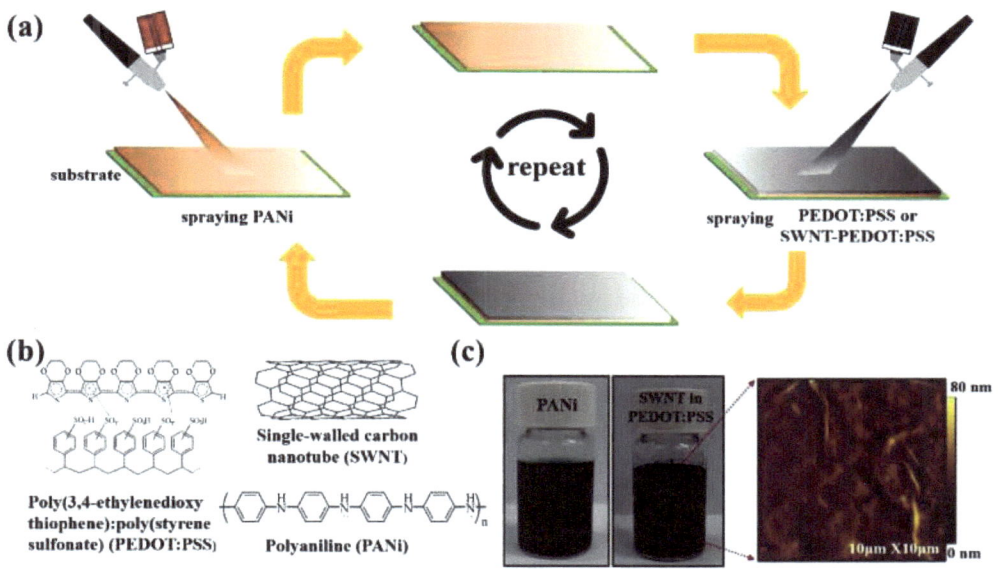

Fig. (5). (a) Illustration for fabrication process of LbL-thermoelectric nanocomposites *via* spray approach, (**b**) chemical structures of PEDOT:PSS, SWNT, and PANi used in this study, and (**c**) photograph of PANi and SWNT, stabilized in PEDOT:PSS, along with a corresponding AFM image that shows uniformly dispersed SWNT in PEDOT:PSS solution [44]..

- **Substrate Preparation:** The LbL assembly process starts with the preparation of a suitable substrate. This substrate can be a solid surface, such as a glass slide or silicon wafer, or a template with a desired shape or pattern. The substrate is typically cleaned and chemically modified to promote adhesion and facilitate layer formation.

- **Polymer Deposition:** The first step in the LbL assembly is the deposition of a conducting polymer layer onto the substrate. This is achieved by immersing the substrate in a solution containing the monomer or pre-formed polymer. The substrate is then rinsed to remove any excess material. This process is repeated to obtain the desired thickness or number of polymer layers.
- **Nanomaterial Deposition:** After the polymer layer deposition, the next step is the deposition of nanomaterials onto the polymer layer. Nanomaterials can include nanoparticles, nanowires, or nanosheets of metals, metal oxides, or carbon-based materials. The nanomaterials are typically dispersed in a solution and then layered onto the polymer-coated substrate using techniques such as dipping, spin coating, or spraying. The substrate is rinsed to remove any unbound nanomaterials.
- **Repeat Layering:** The process of sequential layering is continued by repeating steps 2 and 3. Additional layers of conducting polymer and nanomaterials are deposited one by one, creating a multilayered structure. The number of layers can be controlled to tailor the thickness and properties of the nanocomposite.

Properties: The LbL assembly method allows for precise control over the composition and properties of the conducting polymer nanocomposite. The thickness and order of the layers can be adjusted to achieve desired electrical, mechanical, or optical properties. The choice of conducting polymer and nanomaterials, as well as their concentration and arrangement, can further influence the properties of the nanocomposite.

Applications: Conducting polymer nanocomposites synthesized using the LbL assembly method find applications in various fields. They are used in electronic devices, sensors, actuators, energy storage systems (such as batteries and supercapacitors), and biomedical devices. The precise control over layer thickness and composition enables the customization of nanocomposites for specific applications.

Overall, the layer-by-layer assembly method offers a versatile and controlled approach for synthesizing conducting polymer nanocomposites. It allows for the precise layering of conducting polymers and nanomaterials, enabling the design of nanocomposites with tailored properties for various applications.

Decher's [45] paper titled "Fuzzy nanoassemblies: Toward layered polymeric multicomposites" discusses the concept of layer-by-layer assembly for creating polymeric multicomposites. The paper highlights the potential of this technique for fabricating organized and functional thin films with tailored properties. It emphasizes the importance of controlling the thickness, composition, and functionality of each layer to achieve desired characteristics. Decher's [46] article, "Layer-by-layer assembled multicomposite films," focuses on the fabrication of multimaterial films using the layer-by-layer assembly method. Lvov *et al.* [47] present a study on the assembly, structural characterization, and thermal behavior of layer-by-layer deposited ultrathin films made of poly(allylamine hydrochloride) and DNA. The paper discusses the layer-by-layer deposition technique for creating ultrathin films and explores the structural properties and thermal stability of these films. The study provides valuable insights into the potential applications of layer-by-layer assembled films in areas such as microelectronics and biosensors. Hammond's [48] paper, "Form and function in multilayer assembly: New applications at the nanoscale," discusses the advances and applications of layer-by-layer assembly at the nanoscale. It highlights the diverse range of materials and functionalities that can be incorporated into layer-by-layer assembled films. The paper also showcases various applications of these films in fields such as optics, electronics, and biomedical engineering, emphasizing the importance of tailoring the film properties for specific functionalities.

Richardson *et al.* [49] explored the layer-by-layer assembly of peptides and proteins in their paper. They discussed the techniques, challenges, and opportunities associated with assembling peptides and proteins into multilayer films. The review emphasizes the potential of these assembled structures for various applications, including tissue engineering, biosensors, and drug delivery.

CONCLUSION

In conclusion, the synthesis of conducting polymer nanocomposites has emerged as a promising field in materials science and engineering. Various methods have been developed to fabricate these advanced materials, each offering unique advantages and challenges.

The choice of synthesis method for conducting polymer nanocomposites depends on the desired properties, scalability, and targeted applications. Advances in nanotechnology and polymer science continue to drive the development of new synthesis strategies, enabling the creation of conducting polymer nanocomposites with enhanced electrical conductivity, mechanical strength, and multifunctional capabilities. These materials hold great promise for a wide range of applications, including electronics, energy storage, sensors, and biomedical devices, and are expected to contribute significantly to the advancement of next-generation technologies.

REFERENCES

[1] S. Ghosh, *Conjugated Polymer Nanostructures for Energy Conversion and Storage Applications.* Wiley, 2021.

[2] L. Yang, C.L. Toh, and X. Lu, *In situ preparation of conducting polymer nanocomposites, Synthesis Techniques for Polymer Nanocomposites.* Wiley, 2014, pp. 211-240.

[3] M. Abbas, A. Hachemaoui, A. Yahiaoui, A. Mourad, A.H.I. Belfedal, and N. Cherupurakal, "Chemical synthesis of nanocomposites via in-situ polymerization of aniline and iodoaniline using exchanged montmorillonite", *Polym. Polymer Compos.*, vol. 29, no. 7, 2021.

[4] R. Allen, L. Pan, G.G. Fuller, and Z. Bao, Using *in-situ* polymerization of conductive polymers to enhance the electrical properties of solution-processed carbon nanotube films and fibers, *ACS Appl. Mater. Interfaces,* vol. 6, no. 13, 2014.

[5] S. Sahoo, G. Karthikeyan, G.C. Nayak, and C.K. Das, Electrochemical characterization of *in situ* polypyrrole coated graphene nanocomposites, *Synth. Met.*, vol. 161, no. 15–16, pp. 1713-1719, 2011.

[6] A. Ehsani, A.A. Heidari, and H.M. Shiri, "Electrochemical pseudocapacitors based on ternary nanocomposite of conductive polymer/graphene/metal oxide: An introduction and review to it in recent studies", *Chem. Rec.*, vol. 19, no. 5, pp. 908-926, 2019.

[7] M. Moussa, M.F. El-Kady, S. Abdel-Azeim, R.B. Kaner, P. Majewski, and J. Ma, *Compact, flexible conducting polymer/graphene nanocomposites for supercapacitors of high volumetric energy density.* vol. 160. Compos. Sci. Technol, 2018.

[8] A.E.S. Etman, A.M. Ibrahim, F.A.Z.M. Darwish, and K.F. Qasim, "A 10 years-developmental study on conducting polymer composites for supercapacitors electrodes: A review for extensive data interpretation", *J. Ind. Eng. Chem.*, vol. 122, pp. 27-45, 2023.

[9] M.O. Yanik, E.A. Yigit, Y.E. Akansu, and E. Sahmetlioglu, *Magnetic conductive polymer-graphene nanocomposites based supercapacitors for energy storage.* vol. 138. Energy, 2017.

[10] V. Siva, A. Murugan, A. Shameem, S. Thangarasu, S. Kannan, and A. Raja, "Gel combustion synthesized NiMoO4 anchored polymer nanocomposites as a flexible electrode material for solid-state asymmetric supercapacitors", *Int. J. Hydrogen Energy,* vol. 48, no. 49, 2023.

[11] S. Palsaniya, H.B. Nemade, and A.K. Dasmahapatra, "Synthesis of polyaniline/graphene/MoS 2 nanocomposite for high performance supercapacitor electrode", *Polymer (Guildf.),* vol. 150, pp. 150-158, 2018.

[12] S. Ahmed, A. Ahmed, D.B. Basha, S. Hussain, I. Uddin, and M.A. Gondal, *Critical review on recent developments in conducting polymer nanocomposites for supercapacitors.* vol. 295. Synth. Met, 2023.

[13] S. Sardana, K. Aggarwal, S. Malik, A. Saini, S. Dahiya, R. Punia, A.S. Maan, K. Singh, and A. Ohlan, *Unveiling the surface dominated capacitive properties in flexible ternary polyaniline/NiFe2O4/reduced graphene oxide nanocomposites hydrogel electrode for supercapacitor applications.* vol. 434. Electrochim. Acta, 2022.

[14] R.B. Ambade, and B. Patil, "Conducting polymer-based nanocomposites as electrode materials for supercapacitors", In: *Adv Electron Mater Clean Energy Convers Storage Appl*, 2023.

[15] Y. Wang, X. Wu, W. Zhang, and S. Huang, *Facile synthesis of Ni/PANI/RGO composites and their excellent electromagnetic wave absorption properties.* vol. 210. Synth. Met, 2015.

[16] S.P. Muduli, S. Parida, S.K. Behura, S. Rajput, S.K. Rout, and S. Sareen, "Synergistic effect of graphene on dielectric and piezoelectric characteristic of PVDF-(BZT-BCT) composite for energy harvesting applications", *Polym. Adv. Technol.*, vol. 33, pp. 3628-3642, 2022.

[17] W. Zhang, I.Y. Phang, and T. Liu, "Growth of carbon nanotubes on clay: Unique nanostructured filler for high-performance polymer nanocomposites", *Adv. Mater.*, vol. 18, no. 1, 2006.

[18] S. Kumar, M. Nehra, D. Kedia, N. Dilbaghi, K. Tankeshwar, and K.H. Kim, "Carbon nanotubes: A

potential material for energy conversion and storage", *Pror. Energy Combust. Sci.*, vol. 64, pp. 219-253, 2018.

[19] G Ma, H Peng, J Mu, H Huang, X Zhou, and Z Lei, "In situ intercalative polymerization of pyrrole in graphene analogue of mos2 as advanced electrode material in supercapacitor", *J Power Sources*, vol. 229, .

[20] J. Wang, L. Sun, Y. Gong, L. Wu, C. Sun, X. Zhao, X. Shi, Y. Lin, K. Wang, and Y. Zhang, "A CNT/MoS2@PPy composite with double electron channels and boosting charge transport for high-rate lithium storage", In: *Appl. Surf. Sci* vol. 566. , 2021.

[21] Y. Liu, G. Dong Wang, Y. Shen, E. Blackie, and L. He, *Mode-II fracture toughness of carbon fiber reinforced polymer composites interleaved with polyethersulfone (PES)/carbon nanotubes (CNTs)*. vol. 320. Compos. Struct, 2023.

[22] A. Kumar, V. Kumar, M. Kumar, and K. Awasthi, "Synthesis and characterization of hybrid PANI/MWCNT nanocomposites for EMI applications", *Polym. Compos.*, vol. 39, no. 11, 2018.

[23] S.S. Al-Abbas, R.A. Ghazi, A.K. Al-Shammari, N.R. Aldulaimi, A.R. Abdulridha, S.H. Al-Nesrawy, and E. Al-Bermany, "Influence of the polymer molecular weights on the electrical properties of Poly(vinyl alcohol) - Poly(ethylene glycols)/Graphene oxide nanocomposites", In: *Mater. Today Proc* vol. 42. , 2021.

[24] X. Su, R. Wang, X. Li, S. Araby, H. C. Kuan, M. Naeem, and J. Ma, "A comparative study of polymer nanocomposites containing multi-walled carbon nanotubes and graphene nanoplatelets", *Nano Mater Sci*, vol. 4, no. 3, .

[25] S. Sharma, P. Sudhakara, A.A.B. Omran, J. Singh, and R.A. Ilyas, "Recent trends and developments in conducting polymer nanocomposites for multifunctional applications", *Polymers (Basel)*, vol. 13, no. 17, 2021.

[26] V. Shukla, "Review of electromagnetic interference shielding materials fabricated by iron ingredients", *Nanoscale Adv.*, vol. 1, no. 5, 2019.

[27] F.Z. Zeggai, M. Belbachir, and A. Hachmaoui, *In-situ* preparation of conducting polymers/copper(II)-maghnite clay nanocomposites, *Material Science Research India*, vol. 14, no. 2, pp. 204-211, 2017.

[28] L. Yang, C.L. Toh, and X. Lu, *In situ* preparation of conducting polymer nanocomposites, *Synthesis Techniques for Polymer Nanocomposites*. Wiley, 2014, pp. 211-240.

[29] N. Chen, Y. Ren, P. Kong, L. Tan, H. Feng, and Y. Luo, *In situ* one-pot preparation of reduced graphene oxide/polyaniline composite for high-performance electrochemical capacitors, *Appl. Surf. Sci* vol. 392. , 2017.

[30] Z. Wang, L. Jiang, Y. Wei, and C. Zong, *In-situ* polymerization to prepare reduced graphene oxide/polyaniline composites for high-performance supercapacitors, *J Energy Storage*, vol. 32, .

[31] Z.F. Li, H. Zhang, Q. Liu, L. Sun, L. Stanciu, and J. Xie, "Fabrication of high-surface-area graphene/polyaniline nanocomposites and their application in supercapacitors", *ACS Appl. Mater. Interfaces*, vol. 5, no. 7, pp. 2685-2691, 2013.

[32] Z.F. Li, and J. Xie, "Covalently-functionalized graphene for supercapacitor application", *ECS Meet Abstr*, 2015no. 9,

[33] H. Gul, A.U.H.A. Shah, U. Krewer, and S. Bilal, "Study on direct synthesis of energy efficient multifunctional polyaniline–graphene oxide nanocomposite and its application in aqueous symmetric supercapacitor devices", *Nanomaterials (Basel)*, vol. 10, no. 1, 2020.

[34] L. Wang, Y. Ye, X. Lu, Z. Wen, Z. Li, H. Hou, and Y. Song, "Hierarchical nanocomposites of polyaniline nanowire arrays on reduced graphene oxide sheets for supercapacitors", In: *Sci. Rep* vol. 3. , 2013.

[35] Z. Zheming, Z. Ling, and L. Chunzhong, "Emulsion polymerization: A new approach to prepare graphite oxide coated with polyaniline", *J. Macromol. Sci. Part B Phys.*, vol. 48, no. 2, 2009.

[36] X. Wang, D. Wu, X. Song, W. Du, X. Zhao, and D. Zhang, "Review on carbon/polyaniline hybrids: Design and synthesis for supercapacitor", *Molecules,* vol. 24, no. 12, 2019.

[37] Y.E. Firat, and A. Peksoz, "Electrochemical synthesis of polyaniline/inorganic salt binary nanofiber thin films for electrochromic applications", *J. Mater. Sci. Mater. Electron.,* vol. 28, no. 4, pp. 3515-3522, 2017.

[38] X. Mao, X. He, J. Xu, W. Yang, H. Liu, Y. Yang, and Y. Zhou, "Three-Dimensional Reduced Graphene Oxide/Poly(3,4-Ethylenedioxythiophene) Composite Open Network Architectures for Microsupercapacitors", *Nanoscale Res. Lett.,* vol. 14, no. 1, 2019.

[39] A. Gupta, S. Sardana, S. Dahiya, R. Punia, A.S. Maan, K. Singh, R. Tripathi, and A. Ohlan, "Binder-free polypyrrole/fluorinated graphene nanocomposite hydrogel as a novel electrode material for highly efficient supercapacitors", In: *Appl. Surf. Sci. Adv.* vol. 11. , 2022.

[40] N.P.S. Chauhan, M. Mozafari, N.S. Chundawat, K. Meghwal, R. Ameta, and S.C. Ameta, "High-performance supercapacitors based on polyaniline–graphene nanocomposites: Some approaches, challenges and opportunities", *J. Ind. Eng. Chem.,* vol. 36, pp. 13-29, 2016.

[41] S. Dhibar, A. Roy, and S. Malik, "Nanocomposites of polypyrrole/graphene nanoplatelets/single walled carbon nanotubes for flexible solid-state symmetric supercapacitor", In: *Eur. Polym. J.* vol. 120. , 2019.

[42] W. Wang, O. Sadak, J. Guan, and S. Gunasekaran, "Facile synthesis of graphene paper/polypyrrole nanocomposite as electrode for flexible solid-state supercapacitor", *J Energy Storage,* vol. 30, .

[43] M.M. Rahman, M.R. Shawon, M.H. Rahman, I. Alam, M.O. Faruk, M.M.R. Khan, and O. Okoli, "Synthesis of polyaniline-graphene oxide based ternary nanocomposite for supercapacitor application", *J. Energy Storage,* vol. 67, p. 107615, 2023.

[44] S. Kim, Y.Y. Byun, I.Y. Lee, W. Cho, G. Kim, M. Culebras, J. Jang, and C. Cho, "Organic thermoelectric nanocomposites assembled via spraying layer-by-layer method", *Nanomaterials (Basel),* vol. 13, no. 5, 2023.

[45] G. Decher, "Fuzzy nanoassemblies: Toward layered polymeric multicomposites", *Science,* vol. 277, no. 5330, 1997.

[46] G. Decher, M. Eckle, J. Schmitt, and B. Struth, "Layer-by-layer assembled multicomposite films", *Curr. Opin. Colloid Interface Sci.,* vol. 3, no. 1, pp. 32-39, 1998.

[47] Y. Lvov, G. Decher, and M. Mohwald, "Assembly, structural characterization, and thermal behavior of layer-by-layer deposited ultrathin films of poly(vinyl sulfate) and poly(allylamine)", *Langmuir,* vol. 9, no. 2, 1993.

[48] P.T. Hammond, "Form and function in multilayer assembly: New applications at the nanoscale", *Adv. Mater.,* vol. 16, no. 15, 2004.

[49] J.J. Richardson, M. Björnmalm, and F. Caruso, "Technology-driven layer-by-layer assembly of nanofilms", *Science,* vol. 348, no. 6233, 2015.

CHAPTER 7

Biodegradable Nanocomposites for Energy Harvesting Devices, Piezoelectric Sensors, and Fuel Cells

Pankaj Verma[1], Nitin Kumar[2,*], Nripesh Kumar[2,3], Ratnamala Ganjir[4], Sushil Kumar Verma[5] and Sonika[6]

[1] *Department of Applied Sciences, Galgotias College of Engineering and Technology, Knowledge Park-II, Greater Noida, Uttar Pradesh, India*

[2] *Department of Physics, National Institute of Technology Mizoram, Aizawl 796012, Mizoram, India*

[3] *Department of Physics, Bharat Institute of Engineering and Technology, Hyderabad 501510, Telangana, India*

[4] *Department of Physics, Government J. Yojanandam Chhatisgarh College, Raipur 492001, Chhatisgarh, India*

[5] *Department of Chemical Engineering, CoE-Suspol, Indian Institute of Technology Guwahati, Guwahati 781039, Assam, India*

[6] *Department of Physics, Rajiv Gandhi University, Rono Hills, Doimukh, Itanagar 791112, Arunachal Pradesh, India*

Abstract: The growing demand for sustainable and eco-friendly technologies has spurred significant interest in the development of biodegradable materials with multifunctional capabilities. Biodegradable nanocomposites have emerged as a promising solution to address various environmental challenges while enabling advanced applications in piezoelectric sensors, fuel cells, and energy harvesting devices. This comprehensive review delves into the recent advancements and key considerations in the design, fabrication, and application of biodegradable nanocomposites for these vital technologies. In conclusion, the development of biodegradable nanocomposites presents a transformative opportunity to merge sustainability with cutting-edge technologies. This review underscores the pivotal role of these materials in piezoelectric sensors, fuel cells, and energy harvesting devices.

Keywords: Biodegradable nanocomposites, Energy harvesting, Environmental technology, Fuel cells, Piezoelectric sensors.

---

* **Corresponding author Nitin Kumar:** Department of Physics, National Institute of Technology Mizoram, Aizawl 796012, Mizoram, India; E-mail: nitinphysicskushawaha@gmail.com

INTRODUCTION

The pressing environmental challenges have spurred substantial research endeavors in both industrial and academic sectors, with the aim of developing "green materials" derived from natural resources. These efforts encompass the exploration of biodegradable and biopolymer-based materials as sustainable alternatives. Polylactic Acid (PLA) stands out as the most promising biopolymer due to its biodegradability and its derivation from agricultural resources. Nanocomposite materials offer several advantages over conventional composites, including superior barrier and thermal and mechanical properties, even with minimal reinforcement levels. Additionally, nanocomposites exhibit enhanced recyclability, transparency, and reduced weight compared to their counterparts [1]. Nanocomposites can be categorized into various types, including polymer nanocomposites, metal nanocomposites, bio-nanocomposites, and more. Polymer nanocomposites consist of a polymer matrix containing uniformly dispersed organic or inorganic fillers that serve as additives [2]. In recent years, biopolymers have garnered significant attention as polymer matrices of interest. Commonly used conventional polymers such as polypropylene, polyethylene, polystyrene, and polymethylmethacrylate are non-biodegradable, posing significant challenges for their recycling and reusability [3]. Consequently, an enormous amount of non-biodegradable waste is generated worldwide, including the composites derived from these polymers. The utilization of biodegradable polymers sourced from alternative renewable origins in polymer composite technology is captivating. On the other hand, substituting them for commercial synthetic polymers poses significant challenges. A few of these challenges include low thermal stability, brittleness, and inadequate barrier characteristics [4]. The incorporation of nano-fillers into polymer matrices not only results in selective developments in performance and characteristics but also impacts the biodegradability of the polymer within the nanocomposite. Hence, it is crucial to assess the biodegradability of the nanocomposite. However, when it comes to specific applications such as biomedical applications, no individual biodegradable polymer can accomplish every single one of the necessary requirements. Consequently, a trend has emerged in recent years, focusing on the fabrication of multi-component polymer systems. This approach aims to develop innovative biomaterials that are multifunctional in nature [5].

Biodegradable nanocomposites have captured considerable interest across diverse domains such as piezoelectric devices, fuel cells, sensors, and energy harvesting systems. These materials present the advantage of both functionality and environmentally friendly properties. Piezoelectric substances generate an electric charge under mechanical stress. It is feasible to design biodegradable nanocomposites to display piezoelectric characteristics, enabling them to transform mechanical power into electrical power. The application of biodegradable nanocomposites extends to the construction of environmentally friendly sensors. These sensors possess the capability to identify and quantify different variables, encompassing temperature, humidity, pressure, and chemical analytes [6]. Through the integration of functional nanoparticles or nanowires into biodegradable matrices, these sensors acquire the ability for real-time monitoring, all the while maintaining their eco-friendly nature. Fuel cells transform chemical energy into electrical energy by means of electrochemical reactions. Biodegradable nanocomposites can serve as electrolyte membranes or electrode materials in fuel cells. These materials must exhibit commendable proton or ion conductivity, substantial surface area, and exceptional chemical stability to enhance the fuel cell's efficiency and productivity [7]. Biodegradable nanocomposites can be utilized in energy harvesting systems to capture and convert surrounding vibrations into electrical energy. This technology exhibits a range of applications, including self-sustaining wearable devices, wireless sensors, and remote monitoring systems. The development and commercialization of biodegradable nanocomposites for these applications are still in early stages, with scientists focused on enhancing their characteristics, performance, and scalability for specific needs. This review offers a synopsis of notable progressions in biodegradable nanocomposites that have taken place in recent years. This content encapsulates the authors' contributions to the development and characterization techniques, accompanied by the presentation of typical examples that demonstrate the improved material properties [8]. These developments possess significant promise for a wide range of applications in areas of human interest.

Fig. (**1**) provides an outline categorizing bio-polymers into four distinct groups. The initial three classifications contain bio-based polymers, while the polymers in the final group are ecological but not of bio-based origin. Additionally, the polymers belonging to the last three groups share the common feature of being aliphatic polyesters in the structure.

Polylactic Acid (PLA)-Based Nanocomposites

Polylactic Acid (PLA) has garnered significant attention from researchers due to its exceptional biodegradability, compatibility, and high strength. Furthermore, as an aliphatic polyester derived entirely from renewable resources, PLA has emerged as a highly capable substance for synthesizing biodegradable nanocomposites. By incorporating various nanoparticles into the PLA matrix, the material's mechanical and thermal properties can be significantly enhanced, expanding its range of applications. This key aspect explains the considerable focus of several studies on this procedure [10 - 12]. Zhuang *et al.* [13] employed an *in situ* polymerization technique to fabricate PLA/TiO$_2$ nanocomposites with changing TiO$_2$ Conformation. Notably, titanium dioxide (TiO$_2$) nanoparticles characterized by a prominent aspect ratio exhibit numerous favorable properties, including excellent photocatalytic activity, magnetic characteristics, intriguing hydrophilicity, and antibacterial characteristics. The investigation noted that when the nano-TiO$_2$ content was below 3%, the nanoparticles dispersed uniformly throughout the PLA matrix, resulting in significant enhancements in both mechanical and thermal characteristics. (Fig. 2) Furthermore, as the amount of TiO$_2$ increased, the composite displayed enhanced degradation, as confirmed through exposure to UV light and solution degradation experiments (Fig.3). Remarkably, the amalgamation of TiO$_2$ nanoparticles in PLA/TiO$_2$ nanocomposites exhibited notable bacteriostatic activity. Nekhamanurak *et al.* [14] conducted a study to explore the impact of incorporating CaCO$_3$ nanoparticles on the fracture appearances and mechanical characteristics of PLA. Due to the inherent properties of synthetic silica, such as its high surface area and strong absorbability, the development of a robust interface between the two materials is hindered. To overcome this challenge, the surface of CaCO$_3$ nanoparticles was made more hydrophilic by applying a silica coating through the sol-gel method. The study exposed that a growth in the content of SiO$_2$ on the nanofiller resulted in a consolidation effect of nano-CaCO$_3$, leading to instant improvements in elongation at break, notched impact strength, and elastic modulus. This development confirmed the higher interaction between nanoparticles and the PLA matrix after coating in the nanocomposites.

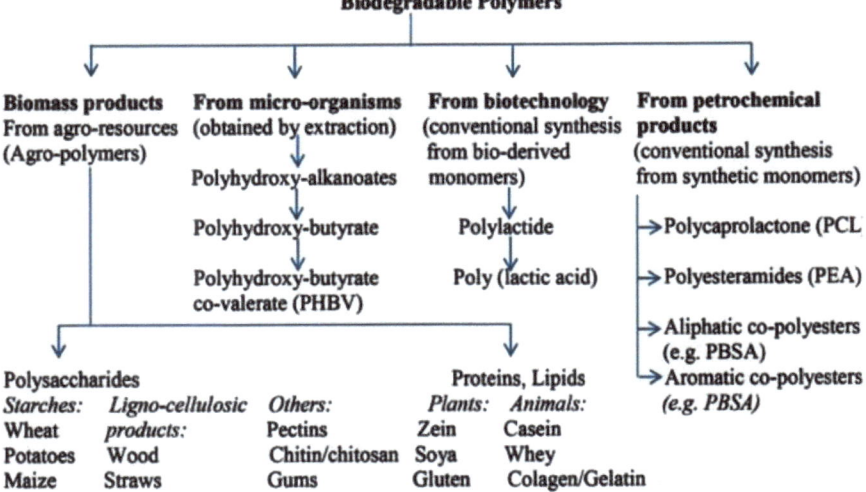

Fig. (1). Categorization of biodegradable polymers [9].

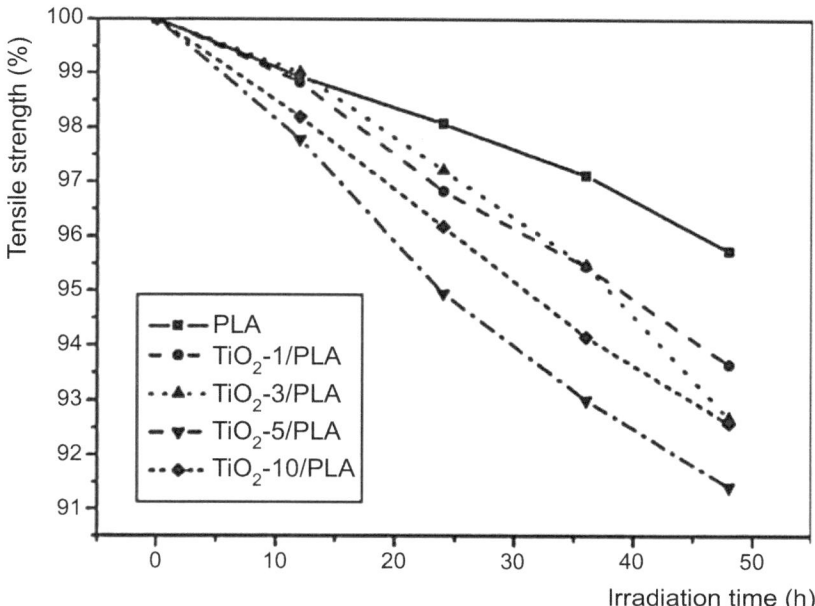

Fig. (2). Consequence of UV exposure duration on the tensile strength of PLA and TiO$_2$/PLA nanocomposites [17].

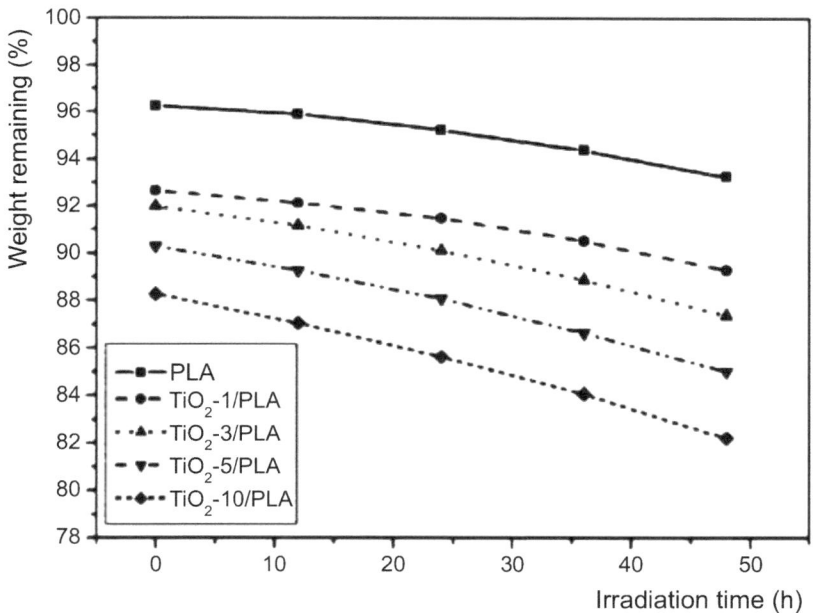

Fig. (3). Outcome of UV exposure duration on the tensile strength of PLA and TiO$_2$/PLA nanocomposites in soil-extraction solution [17].

After processing the nanosilica with two different organosilane agents, Basilissi and his colleagues [15] produced PLA-silica nanocomposites using the bulk Ring-Opening Polymerization (ROP) technique. This behavior is dangerous because silica particles and the PLA matrix have poor compatibility and adhesion, which would otherwise have led to particle collection and diminished material properties. Samples with an extensive loading of silane-coated silica showed better crystallization properties in the nanocomposites. Thermal evaluations further revealed that PLA/silica nanocomposites had more thermal strength than plain PLA, which improved the processing compatibility of the nanocomposite. A study on the antibacterial properties of PLA-cellulose-Ag nanocomposite films for uses in food packaging and sanitation was carried out by Fortunati and their research team [16]. Using sulfuric acid hydrolysis to produce cellulose nanocrystals, the researchers subsequently transformed their surfaces using an acid phosphate ester of ethoxylated nonylphenol, which performed as a surfactant to improve spreading throughout the matrix. After that, a melt technique was used for blending the customized cellulose nanocrystals with PLA and silver nanopowder. The consequent nanocomposite films demonstrated extended antibacterial activity against Staphylococcus aureus and Escherichia coli cells, enhanced tensile strength, and excellent transparency. These characteristics build the nanocomposite films appropriate for multifunctional applications in various industries [15].

Polycaprolactone Nanocomposites (PCL)

Polycaprolactone is acknowledged as a notable aliphatic polyester with potential applications in both the biomedical and environmental fields. Its synthesis usually involves the Ring-Opening Polymerization (ROP) of ε-caprolactone under optimal conditions [18]. PCL shows several characteristic properties, including a low glass conversion temperature, modest melting point, substantial permeability, notable crystallinity, high elongation at break, diminished modulus, and good flexibility. However, further enhancement of these properties is expected to enable their suitability for meeting the diverse requirements of different applications [19]. In pursuit of this objective, researchers have developed PCL nanocomposites by incorporating various nanoparticles, which has proven to be an efficient approach for significantly modifying the properties of PCL. Liu et al. [20] examined the consequence of nano-$CaCO_3$ on the mechanical and thermal characteristics of the PCL matrix. The addition of nano-$CaCO_3$ results in higher mechanical characteristics of the Polycaprolactone (PCL) matrix, with nano-$CaCO_3$ acting as a nucleating agent for crystallization. The researchers utilized the chemical foaming technique to prepare $CaCO_3$/PCL nanocomposites and analyzed parameters related to cell-like characteristics, such as cell wall thickness, mean cell density, and cell size. They noted differences in the cellular configuration of the nanocomposite foams with different concentrations of $CaCO_3$. With higher $CaCO_3$ concentrations, there was an observed increase in cell wall thickness, and the mean cell size reached its lowest value at a $CaCO_3$ concentration of 5% (Fig. 4). Moreover, through the alteration of the cell structure, the mechanical characteristics of $PCL/CaCO_3$ nanocomposite foams were enhanced. As the $CaCO_3$ content increased, the compressive strength of $PCL/CaCO_3$ nanocomposite foams with comparable density was improved as well.

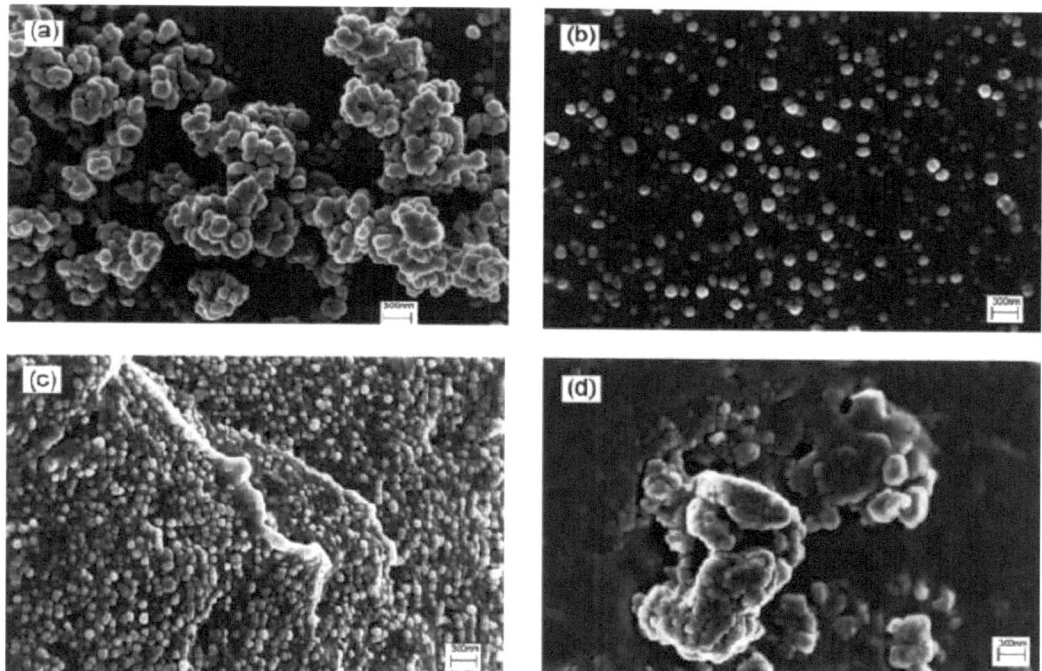

Fig. (4). Scanning electron microscopy (SEM) images of TiO_2 and TiO_2-x/PLA composites (a) TiO_2, (b) TiO_2-1/PLA (1.0 wt%), (c) TiO_2-3/PLA (3.0 wt%), and (d) TiO_2-10/PLA (10.0 wt%).

Wu et al. [21] investigated the mechanical properties and biodegradability of PCL omposites incorporating the phosphate-solubilizing bacterium Bacillus species PG01. The researchers observed that approximately all cell-loaded tablets composed of PCL and PCL composites were degraded by the PG01 strain, resulting in a continuous release of cells. After an incubation period of 15-20 days, simple interruptions to the capsule structure were observed. Similarly, the tensile strength and elongation at break exhibited a similar decrease pattern. These findings demonstrate the strong relationship between the mechanical appearances and biodegradability of the PCL nanocomposites. The addition of starch enhanced the biodegradability of the capsules, while the incorporation of clay resulted in relatively lower biodegradability. The increased biodegradability and decreased mechanical strength of the PCL-based capsules led to higher amounts and faster rates of cell release from the encapsulated cells.

Muñoz-Bonilla et al. [22] conducted a study where the melt processing method was used to produce PCL/TiO_2 nanocomposites. They inserted TiO_2 nanoparticles with a mean diameter of around 10 nm, produced by the microemulsion technique, to a biodegradable PCL matrix at concentrations ranging from 0.5 to 5%. Although some aggregated submicrometer-sized moieties were also present, the researchers were able to disperse the oxide component of the nanoparticles inside the polymeric matrix in a way that was very homogeneous. The TiO_2 nanoparticles were in the anatase form, which, upon UV activation, produced very energetic electron-hole pairs. Even at the nanoscale, these charge carriers can interact with microbes and have bactericidal effects. By evaluating the antibacterial and optical properties, it was determined that the higher loading of nano-TiO_2 content contributed to the enhanced bactericidal properties of the nanocomposites against Gram-positive Staphylococcus aureus and Gram-negative Escherichia coli.

To examine the influence of POSS loading on the PCL crystallization characteristics, Pan and others synthesized PCL/POSS nanocomposites using the solution processing method [23]. The dual impact of POSS on PCL crystallization was identified through the researchers' isothermal and non-isothermal melt crystallization studies. First of all, POSS acted as a nucleating agent, enabling a variety of heterogeneous nucleation sites for PCL crystallization to promote further crystallization. On the other hand, it emerged that POSS aggregates had a regulating effect that inhibited the growth of large-sized crystals, hence restricting the crystallization of PCL.

Poly (p-dioxanone) (PPDO) Nanocomposites

Poly (p-dioxanone) (PPDO) stands out among other biodegradable aliphatic polyesters like PLA and PCL due to its remarkable mechanical properties, in addition to its durability, biodegradability, and biocompatibility. This unique polymer belongs to a rare category of biodegradable aliphatic polyesters that exhibit a combination of high tensile strength and remarkable flexibility [23]. Over the past few decades, an increasing fascination has emerged regarding the creation, properties, and uses of PPDO. Several research endeavors have been dedicated to delving into its miscellaneous facets [24]– [32]. PPDO shows great potential in medical applications, such as devices for fixing bones or tissues, as well as systems for delivering drugs [33, 34]. Moreover, scientists have investigated the utilization of PPDO in producing a range of items, such as adhesives, molded sheets, laminates, films, nonwoven materials, coatings, and foams. However, PPDO encounters notable limitations, including weak melt strength, slow crystallization rate, and hydrophobic nature, all of which considerably diminish its market appeal [35]. Hence, it is important to address these constraints by adapting and enhancing the processing techniques. Blending and copolymerization are commonly employed approaches to overcome these challenges.

Polybutylene Succinate (PBS) Nanocomposites

PBS is a biopolymer with various distinctive characteristics, including high biodegradability, melt processability, and thermo-chemical resistance. Usually, 1,4-butanediol and succinic acid are polycondensed to produce an aliphatic polyester, which is then used to generate PBS. The insertion of two types of clay nanoparticles, TFC and C25A, into PBS systems has produced nanocomposites with improved desired properties, similar to other biodegradable polymers in this category [36 - 39].

In order to develop an array of Poly (Butyl Succinate) (PBS)/Organically Modified Layered Silicates (OMLS) nanocomposites, Ray and his colleagues [36, [37] used a melt intercalation method. The OMLS is comprised of two distinct kinds: quaternary hexadecyl tri-n-butylphosphonium bromide-modified saponite and octadecyl ammonium chloride-modified Montmorillonite (MMT). They revealed that the flocculated OMLS structure, where the intercalated stacked silicate layers flocculated as a result of hydroxylated edge-edge connections, significantly affected the mechanical and other properties of the nanocomposites, whether in the solid or molten stage.

Chen and Evans [39] developed a novel technique to improve the interactions between Cloisite 25A (C25A), a commercially available organoclay, and Polybutylene Succinate (PBS). They developed a method in which (glycidoxypropyl) trimethoxysilane was used to graft epoxy groups onto C25A, resulting in dual-functionalized clay (TFC). After that, PBS and TFC nanocomposites were produced using the melt mixing method. The presence of epoxy groups on the clay surface substantially increased the degree of exfoliation of the silicate layers in PBS/TFC compared to PBS/C25A. The mechanical properties of the nanocomposites were significantly improved as a result of this simultaneous modification. Furthermore, they showed that pure Poly(Butylene Succinate) (PBS) and PBS/clay nanocomposites exhibited non-isothermal crystallization kinetics. At a specific cooling rate, the crystallization rate fell in the following order: PBS/TFC > PBS/C25A > pure PBS. The research revealed that TFC has greater nucleation activity than C25A during PBS crystallization.

This discovery emphasizes the distinct effects of TFC and C25A on the PBS crystallization behaviors [39].

Someya [38] established PBS/layered silicate nanocomposites utilizing the melt intercalation technique. Even though non-modified montmorillonite and five other organo-modified MMTs were used in the studies, no apparent change in mechanical properties was observed. Choi and his team's subsequent studies [40 [41, lined up a series of novel nanocomposites (BAP/OMMT) based on biodegradable aliphatic polyesters derived from diols (1,4-butanediol and ethylene glycol) and dicarboxylic acids (succinic acid and adipic acid) along with O-MMT using a combination of melt intercalation and solvent-casting techniques. BAP/OMMT nanocomposites comprising 6% OMMT [40] were successfully developed *via* the melt intercalation technique. It was determined that the presence of OMMT improved the mechanical strength of both types of composites. The outcomes additionally showed that the OMMT content significantly influenced the rheological properties of the nanocomposites.

Poly (Hydroxyalkanoates) (PHAs) Nanocomposites

A biodegradable polymer known as poly(hydroxyalkanoates) (PHAs) is produced by different microorganisms, generally through extraction. However, PHAs have limitations such as brittleness and melt instability, which have significant effects on their applicability in various sectors. Poly(3-hydroxybutyrate-co- 3-hydroxyvalerate) (PHBV), a partial substitute for Polyhydroxybutyrate (PHB), has been developed to solve these constraints. The potential of PHBV to replace traditional plastics in an eco-friendly way is well known. The limitations of PHBV, however, also include their high cost, complex processing, low crystallization rate, and high degree of crystallinity. Researchers have tried multiple times to change the characteristics of PHBV by adding nanoparticles to the polymer matrix in an effort to get around these challenges [42 - 45].

Choi [43] produced PHBV/MMT nanocomposites using Cloisite 30B as an organoclay using the melt intercalation technique. An intercalated structural morphology produced by this technique resulted in improved tensile strength and thermal stability. The researchers also found that the PHBV crystallization temperature and rate were substantially raised by the non-dispersed organoclay's effective nucleation, which eventually improved the thermal stability and tensile properties of the nanocomposites.

Wang *et al.*'s group [42] examined the biodegradability of PHBV/OMMT nanocomposites in a different analysis. They observed that the biodegradability of PHBV/OMMT nanocomposites in soil suspension diminished as the OMMT level rose.

In a study, Sridhar and his coworkers [42] used the solution casting technique to produce PHBV/graphene nanocomposites. They observed substantial improvements in the mechanical properties of the nanocomposites using dynamic mechanical and static tensile tests. Their research incorporated soil degradation studies and Differential Thermal Analysis (DTA), which showed that the evenly dispersed graphene nanoparticles enhanced the thermal degradation temperature of PHBV without affecting its biodegradability.

METHODS FOR THE FABRICATION OF BIODEGRADABLE NANOCOMPOSITES

Biodegradable polymers have been used in the development of nanocomposites utilizing an array of techniques, primarily solvent-based and melt-mixing techniques. The level of dispersion of the nanofillers inside the polymers is greatly affected by the process technology employed, which also affects the functional characteristics and micro- or nanostructure of films or 3D structures. For optimal use of nanofillers' outstanding capabilities, it is essential to attain their uniform dispersion throughout the continuous polymeric matrix. When selecting the appropriate process technology for nanocomposite production, key considerations include the desired product shape, required performance, manufacturing cost, and ease of implementation [46]. It is significant to recognize that diverse processing techniques usually produce non-equivalent results, necessitating thorough evaluation. Depending on the specific application and desired properties, a range of approaches exists for the fabrication of biodegradable nanocomposites.

Melt Mixing

The melt mixing method is commonly used for the fabrication of biodegradable nanocomposites. The process involves the incorporation of nanoparticles or nanofillers into a polymer matrix by physically mixing them in the molten state. This technique is comparatively straightforward and scalable, making it suitable for industrial production of nanocomposite materials.

These methods are eco-friendly and provide both industrial and economic feasibility. The optimization of molding time, temperature, and pressure is crucial for achieving the desired nanocomposite properties. The extrusion process has been extensively used in the polymer industry for many years as a high-volume manufacturing technique to produce various items like plastic sheets, frames, tubes, and films [47]. It is crucial to emphasize that these processes involve high temperatures, necessitating measures to prevent thermal degradation of the polymers. Even with the involvement of elevated operating temperatures, pressures, and shear rates, extrusion remains a popular and extensively utilized process in the industry due to its straightforward, efficient, continuous, and versatile nature, enabling the transformation of raw materials into finished products.

Moreover, this method is environmentally responsive since it avoids the use of organic solvents. It is also well-suited to prevailing industrial developments [48, 49]. For the fabrication of each polymer system, specific processing conditions are required, considering the desired processing efficiency and product properties.

Solvent-based Process

Solvent casting is an easy technique for producing nanocomposites. A solvent is chosen for manufacturing nanocomposites. The chosen solvent should enable total polymer dissolution and facilitate uniform nanostructure dispersion inside the matrix of the polymer. The viscosity of the solution and the rate of solvent evaporation are both affected by the ratio of polymer to solvent. The properties of the resultant nanocomposites and the surface morphology can be affected by the external aspects, such as temperature, pressure, and humidity, during the evaporation process. Before using the finished nanocomposite, it is essential to ensure that all the solvent has dissipated, as any remaining solvent could prove detrimental and affect the properties of the material. Organic solvents like chloroform are typically not used in industrial applications or scientific research because of the potential risks associated. As a result, using a water-soluble polymer becomes an important strategy. It is important to note that the solvent-based technique has limitations on the types of polymer matrices that can be used, which render it non-industrial and economically unviable. Therefore, this method is often saved for specialized objectives. When dispersing a tiny amount of synthetic nanoparticles in an organic-soluble polymer and for solvent-stable soluble nanoparticles, we mostly used the solvent-based approach in our research [50 - 60].

Electrospinning Technique

The method of electrospinning is extensively employed in the production and fabrication of biodegradable nanocomposites. It encompasses the creation of nanofibers by using an electric field to draw a polymer solution or melt through a spinneret. The process starts with a polymer solution or melt being introduced through the spinneret, and as the solution emerges from the spinneret, an electric field is applied, causing the polymer to be drawn into thin fibers that solidify as they travel to a collector. For the fabrication of biodegradable nanocomposites through electrospinning, the process can be modified by incorporating nanoparticles or nanofillers into the polymer solution or melt before electrospinning. These nanoparticles can be inorganic substances like metal oxides or carbon nanotubes, or they can be biodegradable materials such as cellulose or chitosan nanoparticles. Throughout the electrospinning procedure, the nanoparticles get distributed within the polymer nanofibers, leading to the formation of biodegradable nanocomposites. The incorporation of nanoparticles can enhance a range of properties of the nanofibers, such as mechanical strength, surface area, and bioactivity, making them suitable for specific applications.

The electrospinning technique provides control over the composition, fiber morphology, and alignment, making it a versatile and efficient method for producing biodegradable nanocomposites with tailored properties to meet specific requirements in various fields.

3D Printing

3D printing, recognized as additive manufacturing, is an innovative method used for the manufacture of biodegradable nanocomposites. This technique involves the layer-by-layer deposition of materials to construct three-dimensional objects based on digital designs. For the production of biodegradable nanocomposites through 3D printing, the process can be adapted to incorporate nanoparticles or nano-fillers into the printing material. The initial stage in the 3D printing of biodegradable nanocomposites involves preparing a printable material containing the biodegradable polymer and the desired nanoparticles. This can be accomplished by uniformly dispersing the nanoparticles within the polymer matrix, either through melt mixing or by using solvent-based techniques. Upon the completion of material preparation, it is loaded into the 3D printer, and the printing process begins. The 3D printer follows the digital design, systematically placing the material layer upon layer to construct the desired object. Throughout the printing process, nanoparticles are uniformly distributed within the biodegradable polymer, resulting in the formation of the biodegradable nanocomposite structure.

The advantage of employing 3D printing for fabricating biodegradable nanocomposites lies in its capability to generate complex and customized shapes with precise control over the distribution of nanoparticles. This enables the manufacturing of nanocomposites with properties customized for specific applications, such as drug delivery devices, implants for tissue engineering, and biodegradable components for medical devices.

Furthermore, 3D printing facilitates rapid prototyping and scalability, making it a valuable asset for research and development in the field of biodegradable nanocomposites. With the ongoing advancement of technology, 3D printing holds great potential for revolutionizing the fabrication of biodegradable materials, thus playing a pivotal role in fostering more sustainable and environmentally friendly manufacturing processes.

In Situ Polymerization

In situ polymerization is a technique employed to synthesize polymer materials directly within a matrix or substrate, where the monomers react and form polymer chains *in situ* (in place). This technique involves introducing the monomers, along with any necessary catalysts or initiators, into the matrix or substrate and initiating the polymerization process under specific conditions, such as pressure, temperature, or irradiation. The advantage of *in situ* polymerization is that it allows for the generation of polymer materials that exhibit enhanced compatibility and adhesion to the surrounding matrix or substrate. This method is frequently used to create polymer composites, where the polymer is formed within the structure of another material, such as films, fibers, or coatings. The resultant composites frequently exhibit enhanced mechanical properties, increased strength, and improved interfacial bonding between the polymer and the matrix.

In situ polymerization finds various applications, including the development of biodegradable nanocomposites, where the polymerization of monomers occurs within the biodegradable matrix, leading to the incorporation of nanoparticles or nanofillers directly into the polymer structure. Consequently, this yields biodegradable nanocomposites with customized properties, making them suitable for a wide range of applications in fields such as medicine, packaging, and environmental conservation.

Electrochemical Deposition

This technique involves using an electrochemical process to deposit nanoparticles onto the surface of biodegradable polymers, resulting in the development of nanocomposite materials with improved properties. Electrochemical deposition is a method employed for the fabrication of biodegradable nanocomposites, involving the placement of nanoparticles or nanofillers onto the surface of a substrate or electrode. This approach employs electrochemical processes to control the nucleation and growth of nanoparticles, allowing for precise control over the nanocomposite's composition and properties.

The procedure initiates with the preparation of a solution containing the biodegradable polymer and the desired nanoparticles or precursor ions. The substrate, which can be a conductive material or an electrode, is submerged within the solution and acts as the cathode or anode, depending on the specific deposition process. Upon the application of an electric current, electrochemical reactions occur at the substrate's surface, leading to the reduction or oxidation of the precursor ions. Consequently, this gives rise to the initiation and expansion of nanoparticles directly on the substrate surface, leading to the formation of the biodegradable nanocomposite layer.

The electrochemical deposition technique offers numerous benefits for fabricating biodegradable nanocomposites. It allows for excellent control over the structure and composition of the nanocomposite, as the deposition parameters can be tuned to achieve specific properties. The technique also enables the incorporation of a wide range of nanoparticles with different sizes and functionalities, thereby enhancing the nanocomposite's performance for various applications [61, 62].

Moreover, electrochemical deposition is a comparatively simple and cost-effective approach, making it suitable for large-scale production of biodegradable nanocomposites. The process can be conducted under ambient temperature and pressure, reducing the need for harsh chemical conditions and energy-intensive processes. The biodegradable nanocomposites fabricated through electrochemical deposition find versatile applications, including biomedical implants, drug delivery systems, and sensors. The ability to precisely regulate the nanocomposite's properties and tailor them to the desired application makes electrochemical deposition an attractive technique for the development of advanced and sustainable materials.

Each of these methods has its advantages and limitations, and the selection of the fabrication technique is contingent on the specific necessities of the application and the desired characteristics of the biodegradable nanocomposite.

Biodegradation

In the pursuit of novel materials, the biodegradation characteristics of nanocomposites derived from biodegradable polymers are crucial. Depending on the proposed requirements and post-use circumstances of the newly advanced material, an array of biodegradation processes can be considered, such as composting, hydrolytic, and enzymatic processes.

In the case of biodegradation of PLA, the polymer degradation process is a challenging interaction that involves several different events. These include oligomer destruction resulting from ester cleavage, water absorption, the subsequent dispersion of soluble oligomers aided by bacteria, and ester cleavage leading to the production of oligomer fragments [63]. The PLA disintegration process starts with surface hydrolysis, which causes the polymer to break down into components with lower molecular weights. The differences in molecular weight provide important information on the rate of division of polymers and the time of fragmentation inside a polymer. Microorganisms primarily target the polymer's amorphous phase during the initial phase of the disintegration process, which causes a loss of transparency. It was determined that adding nanostructures to a biodegradable polymer matrix had an effect on several properties of the nanocomposite, notably the rate of disintegration. Furthermore, the type of filler used has an impact on this effect. Additionally, research has shown that clays can have an impact on the bacterial breakdown of polymers based on their chemical composition and the attraction of bacteria the clay. The use of nanoclays has been seen to accelerate PLA decomposition in the context of composting. The hydroxyl groups found in the silicate layers of these clays are liable for this action [63, 64]. Furthermore, due to the hydrophilic properties of nanocellulose, it was found that Cellulose Nanocrystals (CNCs) enhanced the disintegrability rate of PLA [65].

APPLICATION OF BIODEGRADABLE NANOCOMPOSITES

Piezoelectric Devices

Biodegradable nanocomposites for piezoelectric applications have gained significant consideration in recent years due to their potential for sustainable and environmentally friendly energy harvesting. These nanocomposites combine biodegradable polymers with piezoelectric nanoparticles or fibers to enable the generation of electrical energy from mechanical vibrations or stress. One of the key advantages of biodegradable nanocomposites for piezoelectric applications is their compatibility with eco-friendly and biocompatible materials. Biodegradable polymers such as Polyhydroxyalkanoates (PHA), Polylactic Acid (PLA), and cellulose derivatives serve as the matrix material, providing biocompatibility and natural degradation over time [66]– [68]. This characteristic makes them appropriate for applications where sustainability and minimal environmental impact are essential.

Incorporating piezoelectric fillers, like Barium Titanate ($BaTiO_3$), Lead Zirconate Titanate (PZT), or Zinc Oxide (ZnO), into the polymer matrix imparts piezoelectric properties to the composite. These fillers play a vital role in converting mechanical energy into electrical energy and are crucial for efficient energy harvesting. The selection of piezoelectric filler depends on factors such as desired properties, performance requirements, and processing method [69]. Several processing methods can be utilized to fabricate biodegradable piezoelectric nanocomposites, including solution casting, electrospinning, melt blending, or 3D printing. The choice of the method depends on the desired shape, scalability, and specific requirements of the application. These fabrication methods enable the production of complex geometries and structures, expanding the potential applications of biodegradable piezoelectric nanocomposites [70 - 72].

Enhancing the performance of biodegradable piezoelectric nanocomposites frequently requires significant interface modifications. Surface modifications and the application of coupling agents enhance the compatibility between the polymer matrix and the piezoelectric fillers. This promotes good dispersion and bonding between the components, resulting in enhanced overall performance of the nanocomposites [73, [74].

Considerations related to biodegradability play an imperative role in the development of these nanocomposites. Customizing the degradation properties of the nanocomposites to the specific application is essential. Through the careful selection of suitable biodegradable polymers and controlling the nanocomposite composition, it is possible to achieve desired degradation rates while upholding piezoelectric performance. This aspect guarantees the environmentally friendly nature of the nanocomposites throughout their lifecycle [46].

In conclusion, biodegradable nanocomposites designed for piezoelectric applications hold promise for sustainable energy harvesting and other related fields. Ongoing research endeavors are focused on refining their characteristics, exploring novel materials, enhancing their performance, and exploring novel materials and fabrication techniques. These advancements strive to develop viable alternatives to traditional piezoelectric materials while minimizing environmental impact [75].

The field of biodegradable nanocomposites for piezoelectric applications is a burgeoning area of study that focuses on creating sustainable materials that can generate electrical energy from mechanical stress, all while maintaining an environmentally friendly nature. Piezoelectric materials can transform mechanical energy, such as vibration or pressure, into electrical energy, and they have applications in various fields, including energy harvesting, actuators, and sensors [67, 68].

Scientists utilize a range of strategies to develop biodegradable piezoelectric nanocomposites by combining biodegradable polymers with piezoelectric nanoparticles or fibers. Biodegradable matrix materials such as Polylactic Acid (PLA), Polyhydroxyalkanoates (PHA), or cellulose derivatives, providing biocompatibility, degrade naturally over time. Inorganic piezoelectric materials like Barium Titanate ($BaTiO_3$), Lead Zirconate Titanate (PZT), or Zinc Oxide (ZnO) can be incorporated into the polymer matrix to impart piezoelectric properties to the composite [68]. Fabrication methods such as melt blending, electrospinning, solution casting, or 3D printing are employed according to the desired shape, properties, and scalability. Interface modification techniques, such as surface modifications and coupling agents, are used to advance compatibility between the polymer matrix and piezoelectric fillers, improving dispersion and bonding. Biodegradability considerations involve tailoring the degradation properties of nanocomposites to the specific application, which is achieved by selecting appropriate biodegradable polymers and controlling the composition [73 - 76].

Despite being in the early stages of development, ongoing efforts focus on optimizing the properties of biodegradable piezoelectric nanocomposites, enhancing their performance, and exploring novel materials and fabrication techniques. These advancements strive to establish sustainable alternatives to conventional piezoelectric materials for various applications while minimizing their environmental impact [77].

Sensors

Biodegradable nanocomposites offer a promising foundation for the advancement of sensors characterized by enhanced sustainability and biocompatibility. These nanocomposites combine biodegradable polymers, such as polyhydroxyalkanoates (PHA), cellulose derivatives, or Polylactic Acid (PLA) with functional nanoparticles or nanofillers to make sensor materials that can detect and respond to a range of stimuli. The incorporation of different sensing mechanisms within biodegradable nanocomposites is dependent upon the type of nanoparticles or nanofillers used. For example, conductive nanoparticles allow for electrical sensing through changes in conductivity, functionalized nanoparticles enable the detection of target molecules, and variations in optical properties can be employed for optical sensing applications [5].

Biodegradable nanocomposites can be designed to respond to specific stimuli, such as moisture, temperature, pH, or mechanical stress. By incorporating responsive nanoparticles or fillers, these nanocomposites exhibit changes in properties like optical properties, electrical conductivity, or mechanical behavior when subjected to the intended stimulus. This responsiveness facilitates real-time monitoring and detection of environmental or biological changes. Fabrication methods like solution casting, electrospinning, electrochemical deposition, and 3D printing offer flexibility in designing sensor structures and formats, permitting the creation of sensors with tailored properties and geometries [78]. The biodegradability of these nanocomposites offers a significant advantage, particularly in applications where the sensor may need to be implanted in the body or used in environmentally sensitive areas. Biodegradable polymers undergo natural degradation over time, ensuring that the sensors do not contribute to long-term environmental pollution and can be safely disposed of following their use.

Current research in the field of biodegradable nanocomposites for sensors focuses on improving sensitivity and selectivity, optimizing sensor performance, and exploring new materials and fabrication techniques. The ultimate goal is to create sustainable sensing solutions for various applications, including bioengineering, medical diagnostics, and environmental monitoring.

Fuel Cells

The application of biodegradable nanocomposites in fuel cells can support the progress of sustainable and environmentally friendly energy conversion devices. Fuel cells are electrochemical devices that transform the chemical energy stored in fuels, like hydrogen or methanol, into electrical power [79]. An application of biodegradable nanocomposites in fuel cells is as a catalyst support. Catalysts, commonly constructed from precious metals such as platinum (Pt), are essential for facilitating fuel cell reactions. By incorporating these catalysts into biodegradable nanocomposites, it becomes possible to enhance their stability, their surface, and their catalytic performance [80]. Another application lies in the advancement of membranes and electrolytes. Fuel cells necessitate ion-conducting membranes or electrolytes to enable the transport of ions between the fuel and the electrode. Biodegradable nanocomposites offer the potential for fabricating membranes and electrolytes with improved ion conductivity and mechanical properties. Through the amalgamation of polymers like Poly(Lactic Acid) (PLA), Poly(Hydroxyalkanoates) (PHA), or cellulose derivatives with nanofillers, nanocomposite membranes can be produced, exhibiting enhanced fuel cell performance [81]. Biodegradable nanocomposites can also be used as gas diffusion layers in fuel cells. These layers facilitate the distribution of reactant gases, such as oxygen and hydrogen, across the surfaces of the electrodes. Creating nanocomposite materials with optimal porosity, effective gas permeability, and excellent electrical conductivity, efficient gas flow and electron transfer within the fuel cell can be achieved [82].

Moreover, researchers are investigating the feasibility of employing fully biodegradable materials for fuel cell components, including current collectors, electrodes, and other structural elements. Incorporating nanocomposites based on biodegradable polymers enables the creation of fuel cells with reduced environmental repercussions and can be disposed of in a more sustainable manner [83].

Recognizing the ongoing nature of research, it is crucial to acknowledge that the advancement of biodegradable nanocomposites for fuel cells is an active area of research, and several challenges need to be addressed. These challenges include optimizating nanocomposite properties, ensuring their long-term stability, and maintaining acceptable performance levels compared to conventional non-biodegradable materials. Nonetheless, the exploration of biodegradable nanocomposites in fuel cells offers the potential to advance the field of sustainable energy conversion.

Energy Harvesting

Biodegradable nanocomposites have the potential to be utilized in energy harvesting systems, which transform ambient energy from the surrounding environment into usable electrical power. These nanocomposites provide sustainable and eco-friendly materials for energy harvesting applications [84] [66]. They can be applied in various ways. One methodology involves piezoelectric energy harvesting. Biodegradable nanocomposites can be engineered to show piezoelectric properties, converting mechanical vibrations or stress into electrical energy. By incorporating piezoelectric nanoparticles or fibers into a biodegradable polymer matrix, these nanocomposites can generate electrical power from sources such as footstep vibrations and motion-induced oscillations, among others. This makes them suitable for wearable devices, self-powered sensors, and low-power electronics [85]. Triboelectric energy harvesting presents an alternate choice. Triboelectric nanocomposites utilize the electrostatic charge separation induced by contact between two materials to generate electricity. Through the amalgamation of biodegradable polymers with triboelectric materials like nanoparticles or fillers with high electron affinity, nanocomposites can harvest energy from friction or contact between surfaces. This application finds relevance in scenarios such as human-body motion, wind-induced vibrations, or water droplet impacts [86, [87]. Moreover, biodegradable nanocomposites find utility in solar energy harvesting. Through the inclusion of photovoltaic nanoparticles such as organic semiconductors or perovskite materials into biodegradable polymers, flexible and lightweight solar cells can be formed. These cells transform sunlight into electrical energy and are suitable for portable electronics, outdoor applications, and remote areas where sustainable and biodegradable energy sources are needed [88]. Thermoelectric energy harvesting offers an alternative pathway. Enhanced thermoelectric properties can be achieved in biodegradable nanocomposites by incorporating conductive fillers or nanoparticles into a biodegradable polymer matrix. These nanocomposites can harvest waste heat from sources such as industrial processes or body heat, transferring it into usable electrical power [66]. Furthermore, biodegradable nanocomposites find utility in bioenergy harvesting systems, which extract energy from biological sources. For example, they can be used to create biodegradable sensors or devices that harness energy from body fluids like glucose or sweat to power medical implants or wearable electronics [89]. The utilization of biodegradable nanocomposites in energy harvesting is an active research area, with ongoing efforts focused on improving their performance, stability, and biodegradability. These materials hold the potential to play a role in the advancement of sustainable energy solutions by utilizing renewable sources and reducing environmental impact.

Summary

i. Biodegradable nanocomposites are emerging as multifunctional materials with broad applications in advanced technologies. Their eco-friendliness, tunable properties, and nanoscale reinforcement make them suitable for piezoelectric devices, fuel cells, sensors, and energy harvesting systems. Piezoelectric devices offer sustainable alternatives to traditional ceramics, supporting the development of flexible, green electronics. In fuel cells, they function as biodegradable membranes and structural components, improving ionic conductivity while reducing environmental impact. Their biocompatibility and responsiveness to various stimuli make them ideal for sensors in healthcare, environmental monitoring, and food safety. In energy harvesting, they enable the creation of sustainable, self-powered systems.

ii. These applications share common goals such as sustainability, miniaturization, and multifunctionality, positioning biodegradable nanocomposites as a unifying platform for eco-innovative solutions. However, further integrative research is essential to better understand cross-disciplinary synergies and support practical implementation.

CONCLUSION

Biodegradable Polymer Nanocomposites (BPNs) possess a wide array of properties that make them highly valuable as high-performance materials across various industrial sectors. In particular, biopolymers are poised to play a vital role in the coming decades, as concerns grow over the depletion of fossil fuel resources and the environmental impact of conventional synthetic polymers, which are non-renewable and non-biodegradable. In response to these challenges, the past decade has seen a surge in research focused on biodegradable polymers and strategies to enhance their functional properties. One promising and widely adopted approach involves the incorporation of organic or inorganic nanomaterials into biodegradable polymer matrices to form nanocomposites. These eco-friendly materials not only align with global sustainability goals but also offer significantly improved characteristics, including mechanical strength, barrier efficiency, thermal stability, crystallization behavior, degradation rate, and melt processability, when compared to unmodified polymers. As such, biodegradable polymer nanocomposites represent a transformative advancement in materials science, offering both environmental and technological benefits to meet the demands of a sustainable future.

REFERENCES

[1] A.K. Rana, Y.K. Mishra, V.K. Gupta, and V.K. Thakur, "Sustainable materials in the removal of pesticides from contaminated water: Perspective on macro to nanoscale cellulose", *Sci. Total Environ.*, vol. 797, p. 149129, 2021.

[2] I.Y. Jeon, and J.B. Baek, "Nanocomposites derived from polymers and inorganic nanoparticles", *Materials (Basel)*, vol. 3, no. 6, p. 3654, 2010.

[3] T. Narancic, F. Cerrone, N. Beagan, and K.E. O' Connor, "Recent advances in bioplastics", *Application and Biodegradation. Polym.*, vol. 12, p. 920, 2020.

[4] P. Rai, S. Mehrotra, S. Priya, E. Gnansounou, and S.K. Sharma, "Recent advances in the sustainable design and applications of biodegradable polymers", *Bioresour. Technol.*, vol. 325, p. 124739, 2021.

[5] J. Sun, "Nanofiller reinforced biodegradable PLA/PHA composites: current status and future trends", *Polym.*, vol. 10, p. 505, 2018.

[6] J. Yan, "Performance enhancements in poly(vinylidene fluoride)-based piezoelectric nanogenerators for efficient energy harvesting", *Nano Energy*, vol. 56, pp. 662-692, 2019.

[7] B. Jaleh, "Application of biowaste and nature-inspired (nano)materials in fuel cells", *J. Mater. Chem. A Mater. Energy Sustain.*, vol. 11, no. 17, pp. 9333-9382, 2023.

[8] A. Das, T. Ringu, S. Ghosh, and N. Pramanik, "A comprehensive review on recent advances in preparation, physicochemical characterization, and bioengineering applications of biopolymers", *Polym. Bull.*, vol. 80, no. 7, pp. 7247-7312, 2023.

[9] L. Averous, and N. Boquillon, "Biocomposites based on plasticized starch: thermal and mechanical behaviours", *Carbohydr. Polym.*, vol. 56, no. 2, pp. 111-122, 2004.

[10] C.I. Idumah, J.T. Nwabanne, and F.A. Tanjung, "Novel trends in poly (lactic) acid hybrid bionanocomposites", *Clean. Mater.*, vol. 2, p. 100022, 2021.

[11] W.S. Chow, W.L. Tham, and P.C. Seow, "Effects of maleated-PLA compatibilizer on the properties of poly(lactic acid)/halloysite clay composites. J Thermo", *Comp. Mater.*, vol. 26, no. 10, pp. 1349-1363, 2012.

[12] R.T.D. Silva, P. Pasbakhsh, K.L. Goh, S.P. Chai, and J. Chen, "Synthesis and characterisation of poly (lactic acid)/halloysite bionanocomposite films", *J. Compos. Mater.*, vol. 48, no. 30, pp. 3705-3717, 2013.

[13] W. Zhuang, J. Liu, J.H. Zhang, B.X. Hu, and J. Shen, Preparation, characterization, and properties of TiO_2/PLA nanocomposites by *in situ* polymerization, *Polym. Compos.*, vol. 30, no. 8, pp. 1074-1080, 2009.

[14] Y. Nekhamanurak, P. Patanathabutr, and N. Hongsriphan, "Mechanical properties of hydrophilicity modified CaCO3-poly (lactic acid)", *Nanocomposite. Int. J. Appl. Phys. Math.*, vol. 2, pp. 98-103, 2012.

[15] L. Basilissi, G. Di Silvestro, H. Farina, and M.A. Ortenzi, "Synthesis and characterization of PLA nanocomposites containing nanosilica modified with different organosilanes II: Effect of the organosilanes on the properties of nanocomposites: Thermal characterization", *J. Appl. Polym. Sci.*, vol. 128, no. 5, pp. 3057-3063, 2013.

[16] E. Fortunati, "Multifunctional bionanocomposite films of poly(lactic acid), cellulose nanocrystals and silver nanoparticles", *Carbohydr. Polym.*, vol. 87, no. 2, pp. 1596-1605, 2012.

[17] W. Zhuang, J. Liu, J.H. Zhang, B.X. Hu, and J. Shen, "Preparation, characterization, and properties of TiO2/PLA nanocomposites by in situ polymerization", *Polym. Compos.*, vol. 30, no. 8, pp. 1074-1080, 2009.

[18] M. Abrisham, "The role of polycaprolactone-triol (PCL-T) in biomedical applications: A state-of-t-e-art review", *Eur. Polym. J.*, vol. 131, p. 109701, 2020.

[19] M. Zhang, Z. Chang, X. Wang, and Q. Li, "Synthesis of poly(L-lactide-co-ε-caprolactone) copolymer: Structure, toughness, and elasticity", *Polymers (Basel)*, vol. 13, no. 8, p. 1270, 2021.

[20] H. Liu, C. Han, and L. Dong, "Preparation and characterization of poly(ε-caprolactone)/calcium carbonate nanocomposites and nanocomposite foams", *Polym. Compos.*, vol. 31, no. 9, pp. 1653-1661, 2010.

[21] K.J. Wu, C.S. Wu, and J.S. Chang, "Biodegradability and mechanical properties of polycaprolactone composites encapsulating phosphate-solubilizing bacterium Bacillus sp. PG01", *Process Biochem.*, vol. 42, no. 4, pp. 669-675, 2007.

[22] A. Muñoz-Bonilla, M.L. Cerrada, M. Fernández-García, A. Kubacka, M. Ferrer, and M. Fernández-García, "Biodegradable polycaprolactone-titania nanocomposites: preparation, characterization and antimicrobial properties", *Int. J. Mol. Sci.*, vol. 14, no. 5, pp. 9249-9266, 2013.

[23] H. Pan, J. Yu, and Z. Qiu, "Crystallization and morphology studies of biodegradable poly(ε-caprolactone)/polyhedral oligomeric silsesquioxanes nanocomposites", *Polym. Eng. Sci.*, vol. 51, no. 11, pp. 2159-2165, 2011.

[24] K.K. Yang, Y.H. Guo, Y.Z. Wang, X.L. Wang, and Q. Zhou, "AlEt3-H2O-H3PO4 catalyzed polymerizations of 1, 4-dioxan-2-one", *Polym. Bull.*, vol. 54, no. 3, pp. 187-193, 2005.

[25] H.X. Huang, K.K. Yang, Y.Z. Wang, X.L. Wang, and J. Li, "Synthesis, characterization, and thermal properties of a novel pentaerythritol-initiated star-shaped poly(p-dioxanone)", *J. Polym. Sci. A Polym. Chem.*, vol. 44, no. 3, pp. 1245-1251, 2006.

[26] H. Nishida, M. Yamashita, T. Endo, and Y. Tokiwa, "Equilibrium polymerization behavior of 1,4-dioxan-2-one in bulk", *Macromolecules*, vol. 33, no. 19, pp. 6982-6986, 2000.

[27] L.M. Esteves, L. Márquez, and A.J. Müller, "Optimization of the coordination–insertion ring-opening polymerization of poly(p-dioxanone) by programmed decreasing reaction temperatures", *J. Appl. Polym. Sci.*, vol. 97, no. 2, pp. 659-665, 2005.

[28] J.M. Raquez, P. Degée, R. Narayan, and P. Dubois, "Synthesis of melt-stable and semi-crystalline poly(1,4-dioxan-2-one) by ring-opening (co)polymerisation of 1,4-dioxan-2-one with different lactones", *Polym. Degrad. Stabil.*, vol. 86, no. 1, pp. 159-169, 2004.

[29] K.R. Yoon, "Surface-initiated, ring-opening polymerization of p-dioxanone from gold and silicon oxide surfaces", *J. Mater. Chem.*, vol. 13, no. 2, pp. 2910-2914, 2003.

[30] M.A. Sabino, S. González, L. Márquez, and J.L. Feijoo, "Study of the hydrolytic degradation of polydioxanone PPDX", *Polym. Degrad. Stabil.*, vol. 69, no. 2, pp. 209-216, 2000.

[31] Sabino, "Crystallisation and morphology of poly(p-dioxanone)", *Macromol Chem Phys*, 2000. Available from: https://onlinelibrary.wiley.com/doi/abs/10.1002/1521-

[32] A.P.T. Pezzin, G.O.R. Alberda van Ekenstein, and E A R. Duek, "Melt behaviour, crystallinity and morphology of poly(p-dioxanone)", *Polymer (Guildf.)*, vol. 42, no. 19, pp. 8303-8306, 2001.

[33] K.K. Yang, X.L. Wang, Y.Z. Wang, and H.X. Huang, "Effects of molecular weights of bioabsorbable poly(p-dioxanone) on its crystallization behaviors", *J. Appl. Polym. Sci.*, vol. 100, no. 3, pp. 2331-2335, 2006.

[34] M.A. Sabino, G. Ronca, and A.J. Müller, "Heterogeneous nucleation and self-nucleation of poly(p-dioxanone)", *J. Mater. Sci.*, vol. 35, no. 20, pp. 5071-5084, 2000.

[35] K. Yang, X. Wang, and Y-Z. Wang, "Progress in nanocomposite of biodegradable polymer", *J. Ind. Eng. Chem.*, vol. 13, no. 4, pp. 485-500, 2007.

[36] S.S. Ray, K. Okamoto, and M. Okamoto, "Structure−property relationship in biodegradable poly(butylene succinate)/layered silicate nanocomposites", *Macromolecules*, vol. 36, no. 7, pp. 2355-2367, 2003.

[37] S.S. Ray, S. Vaudreuil, A. Maazouz, and M. Bousmina, "Dispersion of multi-walled carbon nanotubes

in biodegradable poly(butylene succinate) matrix", *J. Nanosci. Nanotechnol.,* vol. 6, no. 7, pp. 2191-2195, 2006.

[38] Y. Someya, T. Nakazato, N. Teramoto, and M. Shibata, "Thermal and mechanical properties of poly(butylene succinate) nanocomposites with various organo-modified montmorillonites", *J. Appl. Polym. Sci.,* vol. 91, no. 3, pp. 1463-1475, 2004.

[39] G.X. Chen, E.S. Kim, and J.S. Yoon, "Poly(butylene succinate)/twice functionalized organoclay nanocomposites: Preparation, characterization, and properties", *J. Appl. Polym. Sci.,* vol. 98, no. 4, pp. 1727-1732, 2005.

[40] S.T. Lim, C.H. Lee, H.J. Choi, and M.S. Jhon, "Solidlike transition of melt-intercalated biodegradable polymer/clay nanocomposites. J. Polym. Sci. Part B Polym", *Phys.,* vol. 41, no. 17, pp. 2052-2061, 2003.

[41] S.T. Lim, Y.H. Hyun, H.J. Choi, and M.S. Jhon, "Synthetic biodegradable aliphatic polyester/montmorillonite nanocomposites", *Chem. Mater.,* vol. 14, no. 2, pp. 1839-1844, 2002.

[42] S. Wang, "Characteristics and biodegradation properties of poly(3-hydroxybutyrate-c--3-hydroxyvalerate)/organophilic montmorillonite (PHBV/OMMT) nanocomposite", *Polym. Degrad. Stabil.,* vol. 87, no. 1, pp. 69-76, 2005.

[43] W.M. Choi, T.W. Kim, O.O. Park, Y.K. Chang, and J.W. Lee, "Preparation and characterization of poly(hydroxybutyrate-co-hydroxyvalerate)–organoclay nanocomposites", *J. Appl. Polym. Sci.,* vol. 90, no. 2, pp. 525-529, 2003.

[44] D. Dubief, E. Samain, and A. Dufresne, "Polysaccharide microcrystals reinforced amorphous poly(β-hydroxyoctanoate) nanocomposite materials", *Macromolecules,* vol. 32, no. 18, pp. 5765-5771, 1999.

[45] C.Y. Tang, D.Z. Chen, C.P. Tsui, P.S. Uskokovic, P.H.F. Yu, and M.C.P. Leung, "Nonisothermal melt-crystallization kinetics of hydroxyapatite-filled poly(3-hydroxybutyrate) composites", *J. Appl. Polym. Sci.,* vol. 102, no. 6, pp. 5388-5395, 2006.

[46] I. Armentano, "Nanocomposites based on biodegradable polymers", *Materials (Basel),* vol. 11, no. 5, p. 795, 2018.

[47] M. Stegelmann, M. Müller, A. Winkler, A. Liebsch, and N. Modler, "Polymer analyses for an adapted process design of the pipe-extrusion of polyetherimide", *Mater. Sci. Appl.,* vol. 09, no. 07, pp. 614-624, 2018.

[48] I. Armentano, "Processing and characterization of plasticized PLA/PHB blends for biodegradable multiphase systems", *Express Polym. Lett.,* vol. 9, no. 7, pp. 583-596, 2015.

[49] I. Armentano, "Bio-based PLA_PHB plasticized blend films: Processing and structural characterization", *Lebensm. Wiss. Technol.,* vol. 64, no. 2, pp. 980-988, 2015.

[50] C. Argentati, "Surface hydrophilicity of poly(l-lactide) acid polymer film changes the human adult adipose stem cell architecture", *Polymers (Basel),* vol. 10, no. 2, p. 140, 2018.

[51] A. Ilaria, "Role of PLLA plasma surface modification in the interaction with human marrow stromal cells", *J. Appl. Polym. Sci.,* vol. 114, no. 6, pp. 3602-3611, 2009.

[52] S. Mattioli, J.M. Kenny, and I. Armentano, "Plasma surface modification of porous PLLA films: Analysis of surface properties and in vitro hydrolytic degradation", *J. Appl. Polym. Sci.,* vol. 125, no. S2, pp. E239-E247, 2012.

[53] E. Fortunati, "Keratins extracted from Merino wool and Brown Alpaca fibres: Thermal, mechanical and biological properties of PLLA based biocomposites", *Mater. Sci. Eng. C,* vol. 47, pp. 394-406, 2015.

[54] D. Puglia, "Effect of processing techniques on the 3D microstructure of poly (l-lactic acid) scaffolds reinforced with wool keratin from different sources", *J. Appl. Polym. Sci.,* vol. 132, no. 48, p. 42890, 2015.

[55] A. Aluigi, "Keratins extracted from Merino wool and Brown Alpaca fibres as potential fillers for PLLA-based biocomposites", *J. Mater. Sci.*, vol. 49, no. 18, pp. 6257-6269, 2014.

[56] N. Rescignano, "PVA bio-nanocomposites: A new take-off using cellulose nanocrystals and PLGA nanoparticles", *Carbohydr. Polym.*, vol. 99, pp. 47-58, 2014.

[57] G. Ciapetti, "Enhancing osteoconduction of PLLA-based nanocomposite scaffolds for bone regeneration using different biomimetic signals to MSCs", *Int. J. Mol. Sci.*, vol. 13, no. 2, pp. 2439-2458, 2012.

[58] I. Armentano, *Development of PLGA nanocomposite films and scaffolds for bone tissue engineering*, 2023. Available from: https://dspace.unitus.it/handle/2067/32237

[59] E. Lizundia, "Biocompatible poly(L-lactide)/MWCNT nanocomposites: morphological characterization, electrical properties, and stem cell interaction", *Macromol. Biosci.*, vol. 12, no. 7, pp. 870-881, 2012.

[60] E. Fortunati, F. D'Angelo, S. Martino, A. Orlacchio, J.M. Kenny, and I. Armentano, "Carbon nanotubes and silver nanoparticles for multifunctional conductive biopolymer composites", *Carbon N. Y.*, vol. 49, no. 7, pp. 2370-2379, 2011.

[61] C A D Rodriguez, and G Tremiliosi-Filho, "Electrochemical deposition", *Encycl Tribol*, pp. 918-922, 2013.

[62] S. Prakash, and J. Yeom, *Advanced Fabrication Methods and Techniques.* Nanofluidics Microfluid, 2014, pp. 87-170.

[63] S. Sinha Ray, K. Yamada, M. Okamoto, and K. Ueda, "New polylactide-layered silicate nanocomposites. 2. Concurrent improvements of material properties, biodegradability and melt rheology", *Polymer (Guildf.)*, vol. 44, no. 3, pp. 857-866, 2003.

[64] K. Fukushima, C. Abbate, D. Tabuani, M. Gennari, and G. Camino, "Biodegradation of poly(lactic acid) and its nanocomposites", *Polym. Degrad. Stabil.*, vol. 94, no. 10, pp. 1646-1655, 2009.

[65] M.P. Arrieta, M.A. Peltzer, J. López, and L. Peponi, "PLA-based nanocomposites reinforced with CNC for food packaging applications: From synthesis to biodegradation", In: *Appl. Renew. Biomass Prod. Past, Present Futur*Ind., 2017, pp. 265-300.

[66] D.P. Pabba, "MXene-based nanocomposites for piezoelectric and triboelectric energy harvesting applications", *Micromachines (Basel)*, vol. 14, no. 6, p. 1273, 2023.

[67] L.J. Bakhtar, H. Abdoos, and S. Rashidi, A review on fabrication and *in vivo* applications of piezoelectric nanocomposites for energy harvesting, *J. Taiwan Inst. Chem. Eng.*, vol. 148, p. 104651, 2023.

[68] M. Ali, M.J. Bathaei, E. Istif, S.N.H. Karimi, and L. Beker, "Biodegradable Piezoelectric Polymers: Recent Advancements in Materials and Applications", *Adv. Healthc. Mater.*, vol. 12, p. 2300318, 2023.

[69] J. Chang, "Large d33 and enhanced ferroelectric/dielectric properties of poly(vinylidene fluoride)-based composites filled with Pb(Zr0.52Ti0.48)O3 nanofibers", *RSC Advances*, vol. 5, no. 63, pp. 51302-51307, 2015.

[70] N.D. Bikiaris, "Recent advances in the investigation of poly(lactic acid) (PLA) nanocomposites: incorporation of various nanofillers and their properties and applications", *Polym.*, vol. 15, no. 5, p. 1196, 2023.

[71] W. Xu, "3D printing for polymer/particle-based processing: A review", *Compos., Part B Eng.*, vol. 223, p. 109102, 2021.

[72] S. Bairagi, "Shahid-ul-Islam, M. Shahadat, D. M. Mulvihill, and W. Ali, "Mechanical energy harvesting and self-powered electronic applications of textile-based piezoelectric nanogenerators: A systematic review,"", *Nano Energy*, vol. 111, p. 108414, 2023.

[73] U. Sundar, Z. Lao, and K. Cook-Chennault, "Investigation of piezoelectricity and resistivity of surface modified barium titanate nanocomposites", *Polym.*, vol. 11, no. 12, p. 2123, 2019.

[74] U. Sundar, Z. Lao, and K. Cook-Chennault, "Enhanced dielectric permittivity of optimized surface modified of barium titanate nanocomposites", *Polym.*, vol. 12, no. 4, p. 827, 2020.

[75] S.M. Purushothaman, "A review on electrospun PVDF-based nanocomposites: Recent trends and developments in energy harvesting and sensing applications", *Polymer (Guildf.)*, vol. 283, p. 126179, 2023.

[76] I. Pleşa, P.V. Noţingher, S. Schlögl, C. Sumereder, and M. Muhr, "Properties of polymer composites used in high-voltage applications", *Polym.*, vol. 8, no. 5, p. 173, 2016.

[77] F. Mokhtari, B. Azimi, M. Salehi, S. Hashemikia, and S. Danti, "Recent advances of polymer-based piezoelectric composites for biomedical applications", *J. Mech. Behav. Biomed. Mater.*, vol. 122, p. 104669, 2021.

[78] R. Nasseri, C.P. Deutschman, L. Han, M.A. Pope, and K.C. Tam, "Cellulose nanocrystals in smart and stimuli-responsive materials: a review", *Mater. Today Adv.*, vol. 5, p. 100055, 2020.

[79] A. Kausar, I. Ahmad, T. Zhao, M. Maaza, and P. Bocchetta, "Green nanocomposite electrodes/electrolytes for microbial fuel cells—cutting-edge technology", *J. Compos. Sci.*, vol. 7, no. 4, p. 166, 2023.

[80] T. Sathish, R. Sathyamurthy, S. Sandeep Kumar, G.B. Huynh, R. Saravanan, and M. Rajasimman, "Amplifying power generation in microbial fuel cells with cathode catalyst of graphite-based nanomaterials", *Int. J. Hydrogen Energy*, vol. 47, pp. 1-11, 2022. [http://dx.doi.org/10.1016/j.ijhydene.2022.12.077]

[81] N. Mahato, H. Jang, A. Dhyani, and S. Cho, "Recent progress in conducting polymers for hydrogen storage and fuel cell applications", *Polym.*, vol. 12, no. 11, p. 2480, 2020.

[82] T.W. Chen, "Recent developments in carbon-based nanocomposites for fuel cell applications: a review", *Mol.*, vol. 27, no. 3, p. 761, 2022.

[83] H. Liu, "Application of biodegradable and biocompatible nanocomposites in electronics: current status and future directions", *Nanomater.*, vol. 9, no. 7, p. 950, 2019.

[84] G.M. Rani, C.M. Wu, K.G. Motora, and R. Umapathi, "Waste-to-energy: Utilization of recycled waste materials to fabricate triboelectric nanogenerator for mechanical energy harvesting", *J. Clean. Prod.*, vol. 363, p. 132532, 2022.

[85] N. Sezer, and M. Koç, "A comprehensive review on the state-of-the-art of piezoelectric energy harvesting", *Nano Energy*, vol. 80, p. 105567, 2021.

[86] Y. Mi, "Biodegradable polymers in triboelectric nanogenerators", *Polym.*, vol. 15, no. 1, p. 222, 2023.

[87] J. Han, Y. Wang, Y. Ma, and C. Wang, "Enhanced energy harvesting performance of triboelectric nanogenerators via dielectric property regulation", *ACS Appl. Mater. Interfaces*, 2023.

[88] W. Hou, Y. Xiao, G. Han, and J.Y. Lin, "The applications of polymers in solar cells: a review", *Polym.*, vol. 11, no. 1, p. 143, 2019.

[89] M.M.H. Shuvo, T. Titirsha, N. Amin, and S.K. Islam, "Energy harvesting in implantable and wearable medical devices for enduring precision healthcare", *Energies*, vol. 15, no. 20, p. 7495, 2022.

CHAPTER 8

Advanced Materials with Carbon Nanostructure-Based Composites for Environmental Energy Harvesting

Nitin Kumar[1,*], **Nripesh Kumar**[1,2], **Manoj Kumar Prajapati**[3], **Sushil Kumar Verma**[4] and **Sonika**[5]

[1] *Department of Physics, National Institute of Technology Mizoram, Aizawl 796012, Mizoram, India*

[2] *Department of Physics, Bharat Institute of Engineering and Technology, Hyderabad 501510, Assam, India*

[3] *Department of Physics, SSSVS Government Post Graduate College Chunar Mirzapur, Mirzapur 231304, Uttar Pradesh, India*

[4] *Department of Chemical Engineering, CoE-Suspol, Indian Institute of Technology Guwahati, Guwahati 781039, Assam, India*

[5] *Department of Physics, Rajiv Gandhi University, Rono Hills, Doimukh, Itanagar 791112, Arunachal Pradesh, India*

Abstract: In recent years, carbon-based nanostructured materials have emerged as promising candidates for addressing the challenges associated with increasing energy demand and environmental degradation. These materials exhibit exceptional properties such as enhanced electrical conductivity, mechanical strength, and thermal stability, making them attractive for various applications in energy harvesting and environmental remediation. Carbon-based nanostructured materials have garnered significant attention from the scientific community due to their superior electrical, mechanical, and electronic/optoelectronic properties. In this chapter, we present a systematic overview of the advancements in carbon-based nanostructured materials for environmental and energy harvesting applications. Furthermore, we discuss their practical applications in environmental remediation and energy technologies. Specific focus is given to their roles in supercapacitors, solar cells, and batteries.

Keywords: Carbonn nanostructured specimen, Electronics devices, Environmental application, Energy harvesting applications.

---

[*] **Corresponding author Nitin Kumar:** Department of Physics, National Institute of Technology Mizoram, Aizawl 796012, Mizoram, India; E-mail: nitinphysicskushawaha@gmail.com

INTRODUCTION

A significant portion of the scientific community has recently raised concerns about the extraordinary challenges posed by the rise in energy demand. Rapid industrialization, technical advancement, and population increase are the primary factors contributing to the increasing energy demand. Currently, the world's per capita energy consumption has increased due to the increase in population density. Additionally, by 2035, energy demand is expected to increase by up to 40% from its current level. As a result, the manufacture of energy from sustainable sources is attracting a lot of interest in the context of global energy demand. In general, the term "carbon" refers to a unique element that can be found in both microscopic and macroscopic structures and compounds. The periodic table only has one element that can arrange four valence electrons under various hybridization conditions (sp, sp^2, sp^3), leading to both strong covalent and weak π-π bonds. Overall, about ninety-five percent of known chemical compounds can be categorized as carbon-based. Carbon has four valence electrons (namely 2s and 2p) that play an important role in bond formation (single, double, and triple). Further, it can react to form some stable substances by means of many electronegative and electropositive elements. The resultant massive variety of the specimen compounds and nanostructures is improved through a huge range of different physical, chemical, and biological features. These key features make carbon the most extensively researched specimen element in both research and materials science domains [1 - 6].

At the end of the 20th century, carbon technology and chemistry were marked by noteworthy technological and scientific progress. It led to the further establishment of novel kinds of nanosized carbon and emerging carbon specimens/nanostructures/nanomaterials.

Furthermore, Carbon Nanostructured Materials (referred to as CNMs) have received considerable attention because of their numerous uses in many domains (namely, environment, energy, biomedicine, water, and so on). The CNMs form various allotropes in different dimensions (namely zero-, one-, two-, and three-dimension) at the nanoscale, such as fullerene, carbon nanotubes (referred to as CNTs), graphene, and porous carbon/ diamond, respectively [7]. Moreover, carbon materials also have novel utilities due to their physical, chemical, optical, and electrical characteristics, which can be achieved through the architecture or assembly with different functional materials/nanomaterials [7]. The same material can be produced in several forms, such as oxide nanomaterials, carbon nano-coatings comprising functional metal elements, and graphene modified with carbon compounds.

To address new global concerns like environmental degradation, enormous energy production, and the need for improved agricultural and nutrition, novel technology solutions are needed. New advanced technologies must be developed to improve and automate operations that can independently monitor infrastructures, the environment, and process effectiveness in order to increase productivity while reducing the emission of pollutants [8 - 15]. The potential offered by sensing technologies has implications in this context.

NANOCOMPOSITES AND CARBON NANOSTRUCTURE MATERIALS

The demand for the commercial utilization of transformed carbon-based specimens is on the rise in today's technology-driven fields of agriculture, medicine, and the environment. Considering numerous characteristics exhibited by such carbon-based organisms, a group of researchers and businessmen has become highly concentrated, enabling the development of novel methods of manufacturing for significant industries [16]. One of the stimulating specimens with the ability to generate an extensive range of arrangements, typically with different qualities, is the element named "carbon" [17]. One or two of the most noteworthy allotropes of carbon are "soft" graphite and "hard" diamond [18].

In general, a composite material is a specimen that is formed from two/more constituent specimens. These element specimens have characteristics that are particularly distinct from the different constituents in terms of their physical or chemical composition. The various components become distinct and separate inside the completed structure, distinguishing composites from other interactions and solid solutions.

Multi-phase materials termed nanocomposites have at least one phase with a size in the nano range (10–100 nm) [19, 20]. Due to the high surface-to-volume ratio, contact between the matrix and reinforcement in nanocomposites is quite high. The characteristics of each component, the proportions between them, and the overall shape of the nanocomposites all affect how well they perform. In recent years, the scientific world has paid more attention to nanocomposite specimens. There are many phases in it, all of which have at least one, two, or three dimensions that are in the nanoscale range.

The innovative constituents of carbon elements are Carbon Nanotubes (CNTs), graphene, fullerenes, *etc*. They show excellent characteristics that serve as encouraging opportunities in rich application domains upon closer examination by the scientific industries because of their superior abilities. This element is therefore classified as an ensemble of "wonder materials" [21]. Fullerene, an allotropic form of carbon, is usually a chemical constituent of carbon. Sixty carbon atoms contribute to the highly symmetrical spherical specimen known as fullerene (abbreviated as C60).

The most well-known carbon-based nanomaterials are Carbon Nanotubes (CNTs). Graphene is a single-layer, two-dimensional (2D) allotrope of carbon. The variations of the carbon nanocomposites are depicted in Fig. (**1**). The carbon nanostructures shelter many low-dimensional allotropes of carbon, including Carbon Black (CB), Carbon Nanotubes (CNTs), Carbon Fiber (CF), fullerene (C60), and graphene [22].

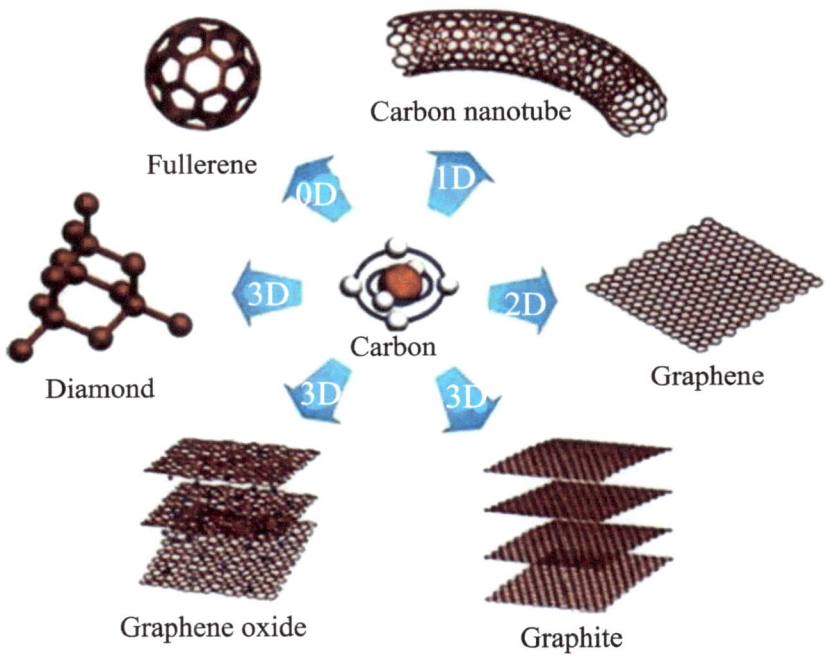

Fig. (1). The classification of carbon nanocomposite materials: 0-D (fullerene), 1-D (CNTs), 2-D (graphene), and 3-D (diamond, graphene oxide, graphite). Figure collected and reproduced from [22].

AREAS OF APPLICATION OF CARBON NANOSTRUCTURED COMPOSITE MATERIALS (ENVIRONMENTAL ENERGY HARVESTING)

Indeed, carbon-based materials have a wide range of properties that make them versatile and valuable in various fields. They are extensively used in various industries, such as metallurgy, medicine, optics, and environmental protection, due to their unique characteristics. However, as industries rapidly develop and new challenges arise, there is a need for more advanced materials with novel properties to meet future demands. Hybrid materials have emerged as a solution to address this need. Carbon Nanostructured (CN) specimens have garnered attention because of their capabilities in numerous fields like environment, energy, electronics, medicine, water, and so on. Carbon nanostructured specimens/materials form many allotropes in zero-dimensional (0-D), one-dimensional (1-D), two-dimensional (2-D), and three-dimensional (3-D) nanoscale schemes like fullerene (C-60), Carbon Nanotubes (CNTs), graphene, and diamond, respectively [7].

These specimens' samples display characteristics different from their bulk counterparts because of their size [23, 24]. The differences are determined by their pore dimension, surface functionality, distribution, conductivity, reactivity, optical functions, mechanical features, electrical features, tensile strength, and so on. A lot of benefits depend on the kinds of materials that feature these properties, which can be easily enhanced/modified at very small scales to achieve precise roles [25]. A group of different metals (like titanium, cobalt, nickel, copper, iron) and a group of nonmetallic specimens (in particular carbon-based specimens (like carbon, nitrogen, oxygen, chlorine, *etc.*) have already been utilized and reported for environmental energy harvesting. These specimens show the significantly enhanced substitute parameters to meet the world's cumulative demands for environmental conservation and energy needs. Carbon Nanostructured Composite (CNM) materials are used in many areas of our lives, such as the environment, medicine, energy harvesting devices, and materials science.

Environmental Applications

Environmental cleanup is a critical concern due to urbanization, industrialization, and changing lifestyles, which release harmful substances like poisonous gases, smoke, and fumes. Pollution encompasses chemical, biological, radiological, and physical contaminants, each posing unique risks to both the environment and living organisms. Chemical contaminants include toxic substances, heavy metals, and pesticides, with long-lasting effects on human and environmental health. Biological contaminants, such as pathogens, threaten human health and ecosystems. Radiological contaminants, like radioactive materials, have severe consequences, while physical contaminants, such as particulate matter, contribute to respiratory and health problems [26].

Nanosized materials have significant applications in environmental remediation. Nanotechnology offers opportunities to develop efficient and environmentally friendly solutions for treating air, water, manufacturing and domestic effluent, soil, and sediments. Numerous nanosized materials have been synthesized and show promise in decontaminating different environmental compartments. Nanosized materials used in environmental remediation can be categorized into three main types: nanoadsorbents, nanocatalysts, and nanomembranes.

i. Nanoadsorbents: Nanoadsorbents are materials designed to remove contaminants from water or air through adsorption processes. These materials have high surface areas and unique surface properties, allowing them to effectively adsorb pollutants onto their surfaces. Carbon-based nanomaterials like graphene, carbon nanotubes, and activated carbon are commonly used as nanoadsorbents due to their large surface area and adsorption capacity. Other nanomaterials, such as metal oxides and clay minerals, can also act as efficient nanoadsorbents. These materials have indeed demonstrated efficient removal of pollutants from wastewater and air [26]. A wide range of materials, including activated carbon, Carbon Nanotubes (CNTs), Carbon Nanofibers (CNFs), graphene, metals and their oxides, silica, and clay, have been widely used as adsorbents in environmental remediation applications. Activated carbon is a commonly employed nanoadsorbent due to its large surface area and high adsorption capacity. It effectively adsorbs a wide range of organic and inorganic pollutants from water and air, including dyes, heavy metals, and Volatile Organic Compounds (VOCs). Carbon Nanotubes (CNTs) and Carbon Nanofibers (CNFs) are nanomaterials with exceptional properties, such as high surface area, mechanical strength, and thermal stability. These properties make them effective adsorbents for various pollutants, including organic contaminants and heavy metals.

Graphene, a two-dimensional carbon material, has gained significant attention in recent years. Its unique structure and high surface area make it an excellent adsorbent for a wide range of pollutants. Graphene-based nanomaterials hold promise for water and air remediation due to their adsorption capacity and potential for functionalization to target specific contaminants. Metals and their oxides, such as iron oxide, titanium dioxide, and manganese oxide, are commonly used as nanoadsorbents. These materials exhibit adsorption properties and can also act as catalysts for pollutant degradation through processes like photocatalysis. Silica-based materials, such as mesoporous silica nanoparticles, have been explored for their adsorption capacity and surface functionalization capabilities. They can effectively absorb organic pollutants, heavy metals, and even radioactive materials. Clay minerals, including bentonite and montmorillonite, possess a layered structure with high cation exchange capacity. These materials have been widely used as nanoadsorbents for various pollutants, including heavy metals and organic contaminants.

i. Nanocatalysts: Nanocatalysts are materials used to accelerate chemical reactions and promote the degradation of pollutants. These materials possess catalytic properties, enabling them to break down contaminants into less harmful substances. Metal-based nanoparticles, such as iron nanoparticles, titanium dioxide nanoparticles, and palladium nanoparticles, are commonly used as nanocatalysts in environmental remediation processes. They can facilitate the removal of organic pollutants, detoxification of hazardous compounds, and conversion of harmful substances into harmless byproducts.

ii. Nanomembranes: Nanomembranes are thin, selective barriers that can separate contaminants from water or air. These membranes are engineered at the nanoscale to have specific properties, such as high permeability and selectivity. They can effectively filter out pollutants and allow the passage of clean water or air. Examples of nanomembranes include nanofiltration membranes, reverse osmosis membranes, and nanocomposite membranes. These membranes can be used in water treatment processes to remove contaminants, such as heavy metals, bacteria, viruses, and organic compounds

In addition, carbon-based nano-specimens, including Carbon Nanotubes (CNTs), Carbon Nanofibers (CNFs), graphene, and fullerenes, are widely used in the removal of contaminants from air and water, as noted by a study [27]. However, challenges such as minor particle size, poor dispersion, and exertion of parting have been identified in their applications. Researchers have addressed these concerns by modifying CNTs into Multiwalled CNTs (MWCNTs), which offer improved dispersion and ease of separation from wastewater. Magnetic MWCNTs, in particular, exhibit enhanced dispersion abilities and can be readily separated from water due to their magnetic properties. Studies have shown that MWCNTs effectively eliminate heavy metal ions such as Pb(II) and Mn(II) from water, as highlighted by Khulbe and Matsuura [28]. Surface modification of CNTs has also been found to enhance their sorption abilities compared to unmodified counterparts. This surface functionalization allows for tailored interactions with specific contaminants, improving their removal efficiency. Carbon Nanofibers (CNFs), produced using Chemical Vapor Deposition (CVD) techniques, are relatively new materials and have found applications in environmental, sensor, and biological fields, as mentioned by Sharma *et al.* [29]. CNFs can be produced on solid substrates like phenolic beads or activated carbon fibres. Metal nanoparticles can additionally exist in CNFs that have been developed on Anisotropic Conductive Films (ACFs) or phenolic resin beads. The elimination of several pollutants, such as arsenic (As), chromium (Cr), lead (Pb), fluoride (F), vitamin B12, and pharmaceutical effluents, can be enhanced through the dispersion of metal nanoparticles in CNFs [29]. In general, carbon-based nanomaterials have the ability to remove an array of pollutants from water and the atmosphere. Surface functionalization and the addition of metal nanoparticles can improve these materials' ability for sorption even further. Using carbon-based nanomaterials with metal nanoparticles scattered across them is a promising way to successfully remove an array of toxins from water and the environment. To better understand and exploit the potential of carbon-based nanomaterials in environmental remediation applications, more research and development in this field is vital.

Energy Harvesting Applications

In recent years, the discovery and development of novel materials have advanced various fields, particularly in environmental and energy applications. Researchers have highlighted their use in pollution control, water purification, waste treatment, and site remediation through adsorbents, membranes, catalysts, and sensors [30]. These materials offer more effective and sustainable environmental solutions. In the energy sector, material innovations have supported progress in renewable technologies, including solar cells, fuel cells, and energy storage systems, as discussed by Barber *et al.* [31].

New materials with improved properties, such as enhanced light absorption, efficient charge transport, and high catalytic activity, have contributed to the development of more efficient and cost-effective energy conversion as well as storage devices. Bio-based materials, as highlighted by Razavi, are another exciting area of material development [32]. These materials are derived from renewable sources, such as plants or biological organisms, and have found applications in various industries. Bio-based materials offer several advantages, including lower environmental impact, biodegradability, and the potential to replace conventional materials derived from fossil fuels. They are used in the production of consumer goods, packaging materials, medicine, electronics, construction materials, transportation, and other green technology-oriented applications. The advancements in material development have opened up new possibilities and opportunities in various sectors. By harnessing the unique properties and capabilities of novel materials, researchers and industries are working towards developing more sustainable, efficient, and environmentally friendly solutions. It is worth noting that the field of materials science and engineering is continuously evolving, with ongoing research efforts aimed at discovering new materials, understanding their properties, and exploring their potential applications. Such advancements hold great promise for addressing global challenges and improving the quality of life for individuals worldwide.

Supercapacitors

In situ polymerization was adopted by Gholivand *et al.* to produce a nanocomposite of polyaniline and copper oxide (CuO/PANI) [33]. Following that, cyclic voltammetry and electrochemical impedance spectroscopy were applied to characterize the resources. According to Gholivand *et al.*, the CuO/PANI nanocomposite exhibited improved capacitance properties compared to pure CuO nanoparticles. They reported a capacitance retention rate of 75% and a specific capacitance of 185 F/g for the CuO/PANI nanocomposite. In comparison, the pure CuO nanoparticles showed a capacitance retention rate of 30% and a specific capacitance of 76 F/g. Moreover, further study by Ates *et al.* reported the highest specific capacitance of 286.35 F/g for the same CuO/PANI material at a scanning rate of 20 mV/s [34]. This suggests that different experimental conditions and characterization techniques can lead to variations in the reported results. These findings indicate that the incorporation of polyaniline into the CuO nanocomposite can enhance its electrochemical properties, resulting in improved capacitance and capacitance retention. The specific capacitance values reported by Gholivand *et al.* and Ates *et al.* highlight the potential of the CuO/PANI nanocomposite for energy storage applications. It is worth observing that the specific capacitance can vary depending on factors such as the synthesis method, composition, morphology, and experimental conditions. Therefore, it is important to consider these factors when comparing the results from different studies. Overall, the research conducted by Gholivand *et al.* and Ates *et al.* demonstrates the potential of the CuO/PANI nanocomposite as an electrode material for energy storage devices [33, 34].

Solar Cells

Photovoltaic cells, also referred to as solar cells, are a cheap source of renewable energy. Specific wavelength photons can be transformed into electricity employing solar cells. Regan and Graetzel have been associated with the discovery of solar cells in 1991. Crystalline silicon and thin films are the two basic groups into which solar cells are frequently divided. The most common and widely used variety of solar cell is one made of crystalline silicon. They offer enormous efficiency in converting sunlight into power and are often produced from silicon that is highly pure. These cells are classified into the first-generation solar cell group. In first- and second-generation solar cells, crystalline silicon has been used in addition to other semiconductors. These semiconductors include cadmium telluride, copper indium selenide/sulfide, and III-V compounds (such as gallium arsenide). These materials have various characteristics and can be utilized to raise the effectiveness and performance of solar cells. These materials have been explored by authors such as Zulkifili *et al.* [35], Yun *et al.* [36], and Zhenqing *et al.* [37] for application in solar cell technologies. The second class of energy conversion technologies includes thin-film solar cells, which enable a simple and easy method of synthesis or manufacturing. Different semiconductor materials are often put in thin layers on an array of substrates to produce thin-film solar cells. Comparing these cells to solar cells comprised of crystalline silicon, they have a number of benefits, such as flexibility, lightweightness, and potentially lower cost. Technologies for thin-film solar cells are continually evolving and becoming better with the goal of increasing efficiency and reducing manufacturing expenses. It is essential to emphasize that research and development in solar cell technology are persistent. The efficiency, cost-effectiveness, and scalability of solar cells are constantly being enhanced by scientists and engineers as they experiment with novel materials, fabrication techniques, and designs. Solar cells are a potential technology for a sustainable and clean energy future and are essential to the development of renewable energy systems.

Furthermore, the remarkable features of carbon-based materials make them ideal substitutes for expensive substances like platinum (Pt) in a variety of applications, including Dye-Sensitized Solar Cells (DSSCs). Carbon-based materials are suitable for use in energy conversion and storage devices due to their high conductivity, favorable electrochemical activity, and low cost. Graphene-based materials received the most attention among the various kinds of carbon-based materials because of their distinctive characteristics and potential applications [38]. The two-dimensional sheet of graphene, which consists of carbon atoms organized in a hexagonal lattice, has an enormous surface area and noteworthy electrical conductivity. These characteristics make it a good choice for increasing solar energy conversion efficiency in DSSCs and improving electron transportation. Researchers prefer to enhance the photoelectrode's conductivity and encourage effective electron transport between the dye and the electrode by incorporating graphene-based materials into DSSCs. The efficiency of energy conversion as an entire process is improved by this enhanced electron transport, which also improves photocurrent. Other energy-associated applications, such as batteries and fuel cells, also utilize the use of the electron transportation capacity of graphene-based materials. The electrodes of batteries and fuel cells can be manufactured with materials based on graphene to increase electron mobility and improve overall device performance. High conductivity, an extensive surface area, and significant electrochemical activity are just some of the special qualities that make graphene-based materials ideal for an array of energy-related applications. The potential of graphene-based materials will be further explored, and their performance in solar cells, batteries, fuel cells, and other energy conversion and storage technologies will be enhanced through further research and development in this sector [5, 38].

Batteries

Energy is preserved and discharged chemically in Li-ion batteries, a common type of energy storage. They have higher energy and power density than capacitors, as well as many other benefits. Due to their excellent energy efficiency and potential to lower greenhouse gas emissions, Li-ion batteries have received a lot of attention in both research and industry [4]. Carbon-based compounds are believed to perform the best of all the materials used in Li-ion batteries. Several favorable characteristics of carbon materials include their low cost, light weight, flexible porosity, ease of manufacturing, and efficiency of chemical modification [39]. These characteristics make carbon-based materials, such as 2D graphene nanosheets, 1D Carbon Nanotubes (CNTs), and activated carbons, excellent possibilities for environmentally friendly energy storage technologies. Carbon materials' performance in Li-ion batteries is heavily affected by their structure. The structure of the carbon material affects factors such as the quantity of lithium that can be reversibly incorporated into the carbon lattice, the faradic losses during the initial charge-discharge cycle, and the voltage profile during charging and discharging [39]. Improved electrochemical capacitance performance can be achieved by increasing the specific surface area and pore size distribution of carbon structures, which increases Li-ion batteries' capacity to store energy as well as deliver power [40]. Energy densities of carbon-based Li-ion batteries can exceed 200 Wh/kg, which is higher than that of batteries constructed from other metal-based materials. This makes them highly desirable for various electronics applications and contributes to their widespread use in the industry.

CONCLUSION

In recent years, composite materials with carbon nanostructures have been shown to be essential in a variety of applications. Both the development of energy harvesting applications and environmental concerns depend on these materials. The incorporation of carbon nanostructured materials (CNMs) from distinct allotropes occurs in zero-, one-, two-, and three-dimensional (*i.e.*, nanoscale) nanostructures, such as fullerene, carbon nanotubes (CNTs), graphene, and porous carbon/diamond, respectively. The addition of CNTs and graphene nanofillers improves the novel characteristics of specific specimens, such as their ideal mechanical stability, optical features, physical properties, low water/gas permeability, and high thermal conductivity. Due to these much-enhanced properties, nanocomposite materials containing CNTs and graphene find widespread applications in various fields. For example, in energy harvesting and storage, they can be used in advanced batteries, supercapacitors, and solar cells to improve their performance and efficiency. In environmental concerns, these materials can be utilized in water purification membranes, gas separation membranes, and sensors to address pollution and environmental monitoring challenges. Overall, the integration of CNTs and graphene nanofillers into a matrix has revolutionized material science and opened up new possibilities for innovative applications, making them highly promising materials for the 21st century.

REFERENCE

[1] A. Hirsch, "The era of carbon allotropes", *Nat. Mater.*, vol. 9, pp. 868-871, 2010.

[2] S. Rajput, M. Averbukh, and A. Yahalom, "Electric power generation using a parallel-plate capacitor", *Int. J. Energy Res.*, vol. 43, pp. 3905-3913, 2019.

[3] P.J.F. Harris, "New perspectives on the structure of graphitic carbons", *Crit. Rev. Solid*, vol. 30, pp. 235-253, 2005.

[4] S. Rajput, M. Averbukh, and N. Rodriguez, "Energy Harvesting and Energy Storage Systems", *Electronics (Basel)*, vol. 11, no. 7, p. 984, 2022.

[5] S. Rajput, A. Sharma, V. Jately, M. Ram, Ed., *Recent advances in energy harvesting technologies.* 1st ed. River Publishers, 2023.

[6] P. Slepicka, N. Slepickova Kasalkova, J. Siegel, Z. Kolska, L. Bacakova, and V. Svorcik, "Nano-structured and functionalized surfaces for cytocompatibility improvement and bactericidal action", *Biotechnol. Adv.*, vol. 33, pp. 1120-1129, 2015.

[7] M-W. Moon, Ho-Y. Kim, A. Wang, A. Vaziri, "Nanostructured carbon materials", *J. Nanomater.*, vol. 2015, 2015.
[http://dx.doi.org/10.1155/2015/916834]

[8] A. Artunedo, R. Del Toro, and R.E. Haber, "Consensus-based cooperative control based on pollution sensing and traffic information for urban traffic networks", *Sensors (Basel)*, vol. 17, p. 953, 2017.

[9] G.K.A. Kolumban-Antal, V. Lasak, B. Razvan, and B. Groza, "A secure and portable multi-sensor module for distributed air pollution monitoring", *Sensors (Basel)*, vol. 20, p. 403, 2020.

[10] T. Eifert, K. Eisen, M. Maiwald, and C. Herwig, "Current and future requirements to industrial analytical infrastructure-part 2: Smart sensors", *Anal. Bioanal. Chem.*, vol. 412, pp. 2037-2045, 2020.

[11] G.P. Hancke, B. De Carvalho e Silva, and G.P. Hancke, "The role of advanced sensing in smart cities", *Sensors (Basel)*, vol. 13, pp. 393-425, 2013.

[12] R.P. Sishodia, R.L. Ray, and S.K. Singh, "Applications of remote sensing in precision agriculture: a review", *Remote Sens.*, vol. 12, p. 3136, 2020.

[13] G. Duffy, and F. Regan, "Recent developments in sensing methods for eutrophying nutrients with a focus on automation for environmental applications", *Analyst (Lond.)*, vol. 142, pp. 4355-4372, 2017.

[14] J. Lozano, C. Apetrei, M. Ghasemi-Varnamkhasti, D. Matatagui, and J.P. Santos, *Sensors and Systems for Environmental Monitoring and Control.* Hindawi, 2017, p. 6879748.

[15] H. Hayat, "Griffiths, Brennan D, Lewis R P, Barclay M, Weirman C, Philip B, Searle JR. The state-o--the-art of sensors and environmental monitoring technologies in buildings", *Sensors (Basel)*, vol. 19, p. 3648, 2019.

[16] N.M. Julkapli, and S. Bagheri, "Graphene supported heterogeneous catalysts: an overview", *Int. J. Hydrogen Energy,* vol. 40, no. 2, pp. 948-979, 2015.

[17] D. Deng, L. Xiao, I.M. Chung, I.S. Kim, and M. Gopiraman, "Industrial-quality graphene oxide switched highly efficient Metal- and solvent-free synthesis of β-ketoenamines under feasible conditions", *ACS Sustain. Chem.& Eng.*, vol. 5, pp. 1253-1259, 2017.

[18] O. Zaytseva, and G. Neumann, "Carbon nanomaterials: production, impact on plant development, agricultural and environmental applications", *Chem. Biol. Technol. Agric.,* vol. 3, p. 17, 2016.

[19] Sonika Verma SK, Samanta S, Srivastava AK, Biswas S, Alsharabi RM, Rajput S, "Conducting polymer nanocomposite for energy storage and energy harvesting systems", *Adv. Mater. Sci. Eng.,* pp. 1-23, 2022.

[20] B. Ates, S. Koytepe, A. Ulu, C. Gurses, and V.K. Thakur, "Chemistry, structures, and advanced applications of nanocomposites from biorenewable resources", *Chem. Rev.,* vol. 120, no. 17, pp. 9304-9362, 2020.

[21] P.O. Brien, R. Nuzzo, H. Kroto, and J. Rocha, *Hierarchical nanostructures for energy devices.* Royal Society of Chemistry, 2014.

[22] X. Hu, X. Bao, J. Wang, X. Zhou, H. Hu, L. Wang, and S. Rajput, "Enhanced energy harvester performance by a tension annealed carbon nanotube yarn at extreme temperatures", *Nanoscale,* vol. 14, pp. 16185-16192, 2022.

[23] S.P. Muduli, L. Lipsa, A. Choudhary, S. Rajput, and S. Parida, "Modulation of electrical characteristics of polymer-ceramic-graphene hybrid composite for piezoelectric energy harvesting", *ACS Appl. Electron. Mater.,* vol. 5, no. 6, pp. 3023-3037, 2023.

[24] S.P. Muduli, S. Parida, S.K. Behura, S. Rajput, S.K. Rout, and S. Sareen, "Synergistic effect of graphene on dielectric and piezoelectric characteristic of PVDF-(BZT-BCT) composite for energy harvesting applications", *Polym. Adv. Technol.,* vol. 33, pp. 3628-3642, 2022.

[25] I. Khan, K. Saeed, and I. Khan, "Nanoparticles: properties, applications and toxicities", *Arab. J. Chem.,* 2017.
[http://dx.doi.org/10.1016/j.arabjc.2017.05.011]

[26] M. Fang, S. Rajput, Z. Dai, Y. Ji, Y. Hao, S. Ren, and X. Ren, "Understanding the mechanism of thermal-stable high-performance piezoelectricity", *Acta Mater.,* vol. 169, pp. 155-161, 2019.

[27] K. Scida, P.W. Stege, G. Haby, G.A. Messina, and C.D. García, "Recent applications of carbon-based nanomaterials in analytical chemistry: critical review", *Anal. Chim. Acta,* vol. 691, no. 1–2, pp. 6-17, 2011.

[28] K. Khulbe, and T. Matsuura, "Removal of heavy metals and pollutants by membrane adsorption techniques", *Appl. Water Sci.,* vol. 8, p. 19, 2018.

[29] A. Sharma, N. Verma, A. Sharma, D. Deva, and N. Sankararamakrishnan, "Iron doped phenolic resin based activated carbon micro and nanoparticles by milling: synthesis, characterization and application in arsenic removal", *Chem. Eng. Sci.,* vol. 65, pp. 3591-3601, 2010.

[30] J. Theerthagiri, R. Senthil, B. Senthilkumar, A.R. Polu, J. Madhavan, and M. Ashokkumar, "Recent advances in MoS2 nanostructured materials for energy and environmental applications–a review", *J. Solid State Chem.,* vol. 252, pp. 43-71, 2017.

[31] S. Rajput, S. Parida, and A. Sharma, https://ieeexplore.ieee.org/servlet/opac?bknumber=10177830

[32] M. Razavi, 2- Bio-based nanostructured materials.*Nanobiomaterials.,* R. Narayan, Ed., Woodhead Publishing, 2018.

[33] M.B. Gholivand, H. Heydari, A. Abdolmaleki, and H. Hosseini, "Nanostructured CuO/PANI composite as supercapacitor electrode material", *Mater. Sci. Semicond. Process.,* vol. 30, pp. 157-161, 2015.

[34] M. Ates, M.A. Serin, I. Ekmen, and Y.N. Ertas, "Supercapacitor behaviors of polyaniline/CuO, polypyrrole/CuO and PEDOT/CuO nanocomposites", *Polym. Bull.,* vol. 72, pp. 2573-2589, 2015.

[35] A.N.B. Zulkifili, T. Kento, M. Daiki, and A. Fujiki, "The basic research on the dye-sensitized solar cells (DSSC)", *J Clean Energy Technol,* vol. 3, pp. 382-387, 2015.

[36] S. Yun, A. Hagfeldt, and T. Ma, "Pt-free counter electrode for dye-sensitized solar cells with high

efficiency", *Adv. Mater.*, vol. 26, no. 36, pp. 6210-6237, 2014.

[37] X. Zhenqing, A. Kumar, and A. Kumar, "Amperometric detection of glucose using a modified nitrogen-doped nanocrystalline diamond electrode", *J. Biomed. Nanotechnol.*, vol. 1, no. 4, pp. 1-5, 2005.

[38] S. Afreen, R.A. Omar, N. Talreja, D. Chauhan, and Md. Ashfaq, *Carbon-Based Nanostructured Materials for Energy and Environmental Remediation Applications*. Approaches in Bioremediation, 2018.

[39] C. Julien, A. Mauger, A. Vijh, and K. Zaghib, *Lithium batteries*. Springer: Cham, 2015.

[40] J. Wang, and S. Kaskel, "KOH activation of carbon-based materials for energy storage", *J. Mater. Chem.*, vol. 22, no. 45, pp. 23710-23725, 2012.

CHAPTER 9

Evolution of Advanced Polymer-Based Nanogenerators for Energy Harvesting Applications

Vijyendra Kumar[1], Sonika[2], Deo Karan Ram[3], Gamini Sahu[4], Nutan Kumar Sahu[1] and Sushil Kumar Verma[5,*]

[1] *Department of Chemical Engineering, Raipur Institute of Technology, Raipur 49001, Chhattisgarh, India*

[2] *Department of Physics, Rajiv Gandhi University, Rono Hills, Doimukh, Papumpare 791112, Arunachal Pradesh, India*

[3] *Department of Petroleum & Chemical Engg., NIMS University, Jaipur 303121, Rajasthan, India*

[4] *School of Life Science, Pt. Ravishankar Shukla University, Raipur 492001, Chhattisgarh, India*

[5] *Centre for Sustainable Polymer, Technology Complex, Indian Institutes of Technology, Guwahati 781039, Assam, India*

Abstract: The burgeoning demand for sustainable and renewable energy solutions has accelerated advancements in Nanogenerators (NGs) based on polymers. These innovative devices harness piezoelectric, triboelectric, and pyroelectric effects for energy harvesting, offering transformative solutions in wearable electronics, biomedical applications, and the Internet of Things (IoT). With their unique ability to convert mechanical and thermal energies into electrical power, polymer-based nanogenerators exemplify the synergy between nanotechnology and material science. This review delves into the evolution of these devices over the last two decades, highlighting key advancements in materials, fabrication techniques, and application domains. Additionally, challenges such as scalability, energy conversion efficiency, and environmental impact are discussed alongside strategies for addressing these limitations. By emphasizing the sustainable potential of polymer-based nanogenerators, this work aims to provide insights into their role in the future of energy technology and their application in diverse fields, from healthcare to smart cities.

Keywords: Energy harvesting, IoT applications, Piezoelectric effect, Polymer-based nanogenerators, Sustainable technology.

---

* **Corresponding author Sushil Kumar Verma:** Centre for Sustainable Polymer, Technology Complex, Indian Institutes of Technology, Guwahati 781039, Assam, India; E-mail: sushilnano@gmail.com

INTRODUCTION

The escalating global energy crisis and environmental concerns have prompted researchers to explore renewable energy harvesting technologies [1 - 3]. Traditional energy sources such as fossil fuels are rapidly depleting, and their continued use poses significant environmental challenges, including climate change, air pollution, and resource scarcity [4]. These issues have catalyzed the urgent need for alternative energy solutions that are not only sustainable but also adaptable to a wide range of applications [5].

Nanogenerators represent a groundbreaking approach in this domain, offering the capability to convert ambient mechanical, thermal, and other forms of energy into electrical power [6]. This ability to harvest energy from ubiquitous sources, such as human motion, environmental vibrations, and temperature gradients, positions nanogenerators as pivotal in addressing the growing energy demands of the modern world [7].

Polymer-based nanogenerators, in particular, have emerged as a promising subset of this technology due to their unique combination of properties [8]. Polymers are inherently lightweight, flexible, and cost-effective, making them suitable for applications in wearable electronics, biomedical devices, and the Internet of Things (IoT) [9]. Additionally, the ability to engineer polymers with tailored properties has enabled the development of multifunctional materials that combine energy harvesting with other functionalities, such as sensing and actuation [10].

Since their introduction in the early 2000s, polymer-based nanogenerators have undergone significant evolution, driven by advancements in material science, nanotechnology, and fabrication techniques [11]. Researchers have focused on enhancing the energy conversion efficiency, mechanical robustness, and environmental stability of these devices [12]. Innovative approaches, such as the incorporation of nanostructured fillers, hybrid composites, and advanced surface engineering, have propelled this field forward, enabling the realization of highly efficient and durable devices [13].

This review aims to provide a detailed examination of the evolution of polymer-based nanogenerators over the past two decades. It delves into the fundamental mechanisms underlying their operation, the innovative materials that have been developed, and the cutting-edge fabrication techniques that have enabled their deployment in diverse applications. Furthermore, this review highlights the challenges that remain in the field, such as scalability, energy output optimization, and environmental impact, while exploring future directions that can unlock the full potential of this transformative technology in addressing global energy challenges.

POLYMER-BASED NANOGENERATORS: MECHANISMS AND MATERIALS

Polymer-based nanogenerators operate primarily on piezoelectric, triboelectric, and pyroelectric principles. Each mechanism leverages specific material properties to convert mechanical or thermal energy into electrical energy [14].

Piezoelectric Nanogenerators (PENGs)

PENGs utilize piezoelectric polymers such as Polyvinylidene Fluoride (PVDF) and its copolymers. These materials exhibit high piezoelectric coefficients and mechanical flexibility [15]. Advanced fabrication techniques, including electrospinning and 3D printing, have enabled precise control over material morphology, enhancing energy conversion efficiency [16].

Recent innovations include the integration of nanoscale fillers such as carbon nanotubes, graphene, and metal oxides, which significantly enhance the piezoelectric response and mechanical durability of PVDF-based devices [17]. Additionally, the introduction of piezoelectric polymer blends and composite structures has improved the coupling efficiency, making these devices more efficient in converting mechanical stress into electrical energy (Fig. **1**) [18]. Research has also explored high-performance ferroelectric polymers to expand the range of applications in both flexible electronics and large-scale energy harvesting systems [19].

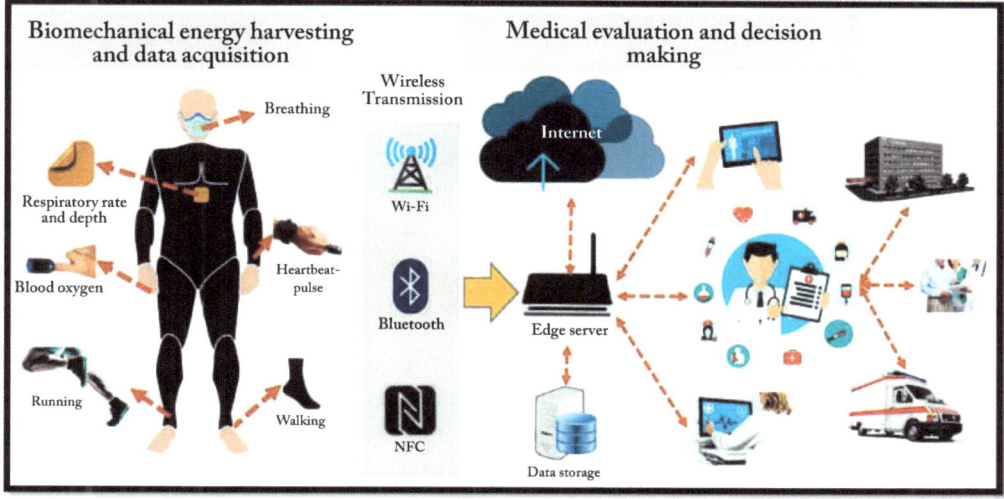

Fig. (1). Potential application of piezoelectric and triboelectric nanogenerators to change biomechanical energy into electrical energy [45].

Triboelectric Nanogenerators (TENGs)

TENGs rely on the triboelectric effect and electrostatic induction. Polymers such as Polytetrafluoroethylene (PTFE) and Polydimethylsiloxane (PDMS) are widely used due to their high triboelectric polarity [20].

The development of advanced surface engineering techniques has significantly enhanced the performance of TENGs [21]. Nanopatterning and chemical functionalization of polymer surfaces have increased surface roughness and charge density, leading to greater energy output [22]. Researchers have also explored multilayer configurations and hybrid systems that combine different triboelectric materials to maximize energy harvesting efficiency [23]. Furthermore, the incorporation of self-healing polymers and elastomers ensures longevity and reliability in wearable and deformable applications [24]. These advancements are enabling TENGs to be deployed in demanding environments, such as underwater energy harvesting and wearable electronics [25].

Pyroelectric Nanogenerators

Pyroelectric nanogenerators exploit temperature fluctuations to generate electricity. PVDF-based polymers are frequently employed, given their combined piezoelectric and pyroelectric properties [26].

Emerging research in this domain focuses on hybrid structures that integrate pyroelectric polymers with thermal insulation layers to optimize temperature gradients and improve energy harvesting efficiency [27]. Moreover, advanced pyroelectric composites, combining organic polymers with ceramic nanomaterials, have demonstrated significant improvements in energy output [28]. These devices are increasingly being used in applications where both mechanical and thermal energy sources coexist, providing a dual-mode energy harvesting approach [29].

Emerging Multifunctional Materials

Multifunctional polymers that exhibit combined piezoelectric, triboelectric, and pyroelectric properties are being actively explored to develop versatile energy harvesting systems [30]. These materials are designed to capture energy from diverse sources, including vibrations, body movements, and thermal gradients, within a single device architecture [31]. Additionally, bio-inspired polymers and eco-friendly biodegradable materials are gaining traction to align with sustainability goals [32].

Recent developments include polymers embedded with nanostructures that simultaneously exhibit piezoelectric and triboelectric properties, enabling them to efficiently harvest energy from complex environmental conditions [33]. For example, materials with tunable surface functionalities have demonstrated the ability to optimize energy harvesting across varying mechanical and thermal stimuli [34]. Moreover, the integration of self-assembling biomimetic structures has opened avenues for multifunctional devices that align with the principles of green energy [35]. Flexible energy-harvesting devices identified as paper-based nanogenerators (PENGs and TENGs) use paper as an active material or substrate. Table 1 compares the main types of paper-based nano generators [33 - 35].

Table 1. Comparison of key paper-based nanogenerator types [33 - 35].

Type	Mechanism	Common Polymers	Energy Output (mW/m^2)	Key Applications	Advantages	References
Piezoelectric (PENG)	Converts mechanical stress to electrical energy	PVDF, PVDF-TrFE, ZnO	5	Wearable sensors, biomedical implants	High flexibility, compatibility with polymers	[1] [15] [17]
Triboelectric (TENG)	Generates electricity *via* friction and electrostatics	PTFE, PDMS, Kapton	100	IoT devices, environmental monitoring	High output, variety of material choices	[1] [20] [22]
Pyroelectric (PyNG)	Utilizes temperature fluctuations	PVDF, ceramics (*e.g.*, PZT)	10	Waste heat recovery, thermal sensing	Harvests energy from waste heat, dual-mode	[1] [26] [28]
Hybrid	Combines piezoelectric, triboelectric, or pyroelectric effects	Composite structures (*e.g.*, PVDF + CNTs)	50	Multifunctional systems, wearable devices	Versatile, higher efficiency, multimodal	[1] [18] [34]

Types of Nanogenerators and Their Mechanisms

The Nanogenerators (NGs) discussed in the document are primarily based on **piezoelectric**, **triboelectric**, and **pyroelectric** effects. Each type harnesses different principles of energy conversion and has distinct advantages and limitations depending on the application.

Piezoelectric Nanogenerators (PENGs)

Mechanism: PENGs operate by converting mechanical stress or pressure into electrical energy *via* the piezoelectric effect in materials such as **PVDF (Polyvinylidene Fluoride)** and **ZnO (Zinc Oxide)**.

Advantages:

- **High Flexibility and Adaptability**: PENGs made from polymer-based materials like PVDF are highly flexible and lightweight, which makes them ideal for applications in wearable electronics and biomedical devices.
- **High Efficiency in Small-Scale Applications**: Piezoelectric materials have high energy conversion efficiency when subjected to mechanical deformations such as vibrations or movements. This is particularly useful for health monitoring and biomedical implants where energy is harvested from body movements.
- **Durability and Stability**: The materials used in PENGs exhibit good mechanical properties, ensuring longevity and stability, which is critical in implantable devices.

Disadvantages:

- **Low Energy Output**: PENGs typically generate low energy output, which may limit their use in large-scale energy harvesting systems.
- **Sensitivity to Mechanical Stress**: PENGs are highly sensitive to mechanical stress, which limits their effectiveness in environments where mechanical forces are not consistent.
- **Material Degradation**: Despite being flexible, piezoelectric materials such as PVDF are prone to degradation over time, which affects their performance in long-term applications.

Application Scenarios:

- **Wearable Electronics**: They are suitable for energy harvesting from body movements, powering devices like health monitoring systems.
- **Biomedical Applications**: They can power implantable devices using energy generated from physiological motions.

Triboelectric Nanogenerators (TENGs)

Mechanism: TENGs generate electricity by exploiting the triboelectric effect, where certain materials, like **PTFE (Polytetrafluoroethylene)** and **PDMS (Polydimethylsiloxane)**, exchange charges when rubbed together.

Advantages:

- **High Output Power**: TENGs are capable of generating significantly higher energy output compared to PENGs, especially useful in applications that require high power output.
- **Material Versatility**: A wide range of materials can be used in TENGs, providing flexibility for various applications.
- **Self-Healing and Durability**: The integration of self-healing polymers and elastomers ensures longevity and reliability, making TENGs suitable for wearable and deformable applications.

Disadvantages:

- **Lower Efficiency in Harvesting Low-frequency Energy**: TENGs are less efficient at low frequencies compared to PENGs, limiting their use in applications requiring consistent low-frequency energy harvesting.
- **Wear and Tear**: While TENGs have a high energy output, they can be prone to wear and degradation, especially in rough or harsh environments.
- **Environmental Sensitivity**: TENGs can be sensitive to environmental factors such as humidity and temperature, which can affect their performance.

Application Scenarios:

- **Environmental Monitoring**: They are ideal for energy harvesting in environments with frequent mechanical movement, such as vibration harvesting from industrial machines.
- **Wearable Electronics**: They are used in self-powered systems where mechanical energy from body movement is converted into electrical power.

Pyroelectric Nanogenerators (PyNGs)

Mechanism: PyNGs utilize temperature changes to generate electricity, primarily using materials like **PVDF** that exhibit both piezoelectric and pyroelectric properties.

Advantages:

- **Dual-Mode Energy Harvesting**: PyNGs can simultaneously harvest energy from both mechanical and thermal sources, increasing their overall efficiency in environments where both forms of energy are available.
- **Effective in Energy Harvesting from Waste Heat**: They are particularly suitable for applications that involve waste heat recovery.

- **Integration with Thermal Systems**: PyNGs can be integrated with thermal sensors to simultaneously harvest energy and monitor temperature, ideal for smart systems.

Disadvantages:

- **Low Power Density**: PyNGs generally generate lower power densities compared to PENGs and TENGs, which may limit their applications in large-scale systems.
- **Temperature Dependency**: Their performance is highly dependent on ambient temperature fluctuations, which can be inconsistent.
- **Material Complexity**: Fabrication of pyroelectric materials, particularly composites, can be more complex and expensive compared to simpler piezoelectric or triboelectric systems.

Application Scenarios:

- **Waste Heat Recovery**: PyNGs are ideal for recovering energy from waste heat in industrial and thermal management systems.
- **Environmental Monitoring**: They can be used in temperature-sensing applications where temperature gradients are present.

Hybrid Nanogenerators

Hybrid nanogenerators combine the mechanisms of piezoelectric, triboelectric, and pyroelectric effects to harvest energy from multiple sources simultaneously.

Advantages:

- **Multifunctionality**: Hybrid systems are capable of harvesting energy from multiple sources (vibrations, thermal gradients, *etc.*), offering greater versatility and higher efficiency. By combining different energy harvesting mechanisms, hybrid devices can achieve a higher energy output compared to single-effect devices.
- **Flexible and Durable**: Many hybrid systems use flexible substrates, which are essential for wearable devices and integration into complex environments.

Disadvantages:

- **Complexity in Design**: The integration of multiple mechanisms can lead to more complex designs, requiring advanced fabrication techniques and precise material selection.

- **Expensive:** CID systems tend to be more expensive due to the combination of different materials and fabrication processes. In some cases, the combination of mechanisms may lead to reduced efficiency compared to individual systems when not optimized properly.

Application Scenarios:

- **Wearable Systems**: They are ideal for health monitoring and personalized healthcare systems that require harvesting energy from body motion and temperature fluctuations.
- **Environmental and Industrial Monitoring**: They are used in systems that require energy from multiple environmental factors, such as IoT devices and smart grids.

Polymer-based nanogenerators, including PENGs, TENGs, and PyNGs, offer promising energy solutions for a variety of applications, from wearable devices to environmental monitoring. However, the choice of nanogenerator type depends heavily on the specific energy harvesting needs, including power output, environmental conditions, and durability requirements. Moving forward, hybrid systems that integrate multiple mechanisms are expected to play a crucial role in maximizing energy output while improving efficiency, especially in wearable electronics and smart systems [52 - 56].

Emerging Techniques

Recent developments include inkjet printing and roll-to-roll manufacturing, which allow for large-scale and cost-effective production of nanogenerators [36, 1]. Nano-structuring processes, such as lithography and soft molding, have enabled the creation of micro- and nanoscale features that enhance energy conversion efficiency [3]. Additionally, self-assembly techniques have been explored for fabricating hierarchical structures that mimic biological systems, providing insights into bio-inspired energy harvesting devices [4].

Hybrid Nanogenerator Fabrication Advancements

Hybrid Nanogenerators (HNGs) combine multiple energy harvesting mechanisms, such as piezoelectric, triboelectric, and pyroelectric effects, into a single device (Fig. **2**). Fabrication advancements for HNGs have been crucial in optimizing their performance and expanding their application scope [2].

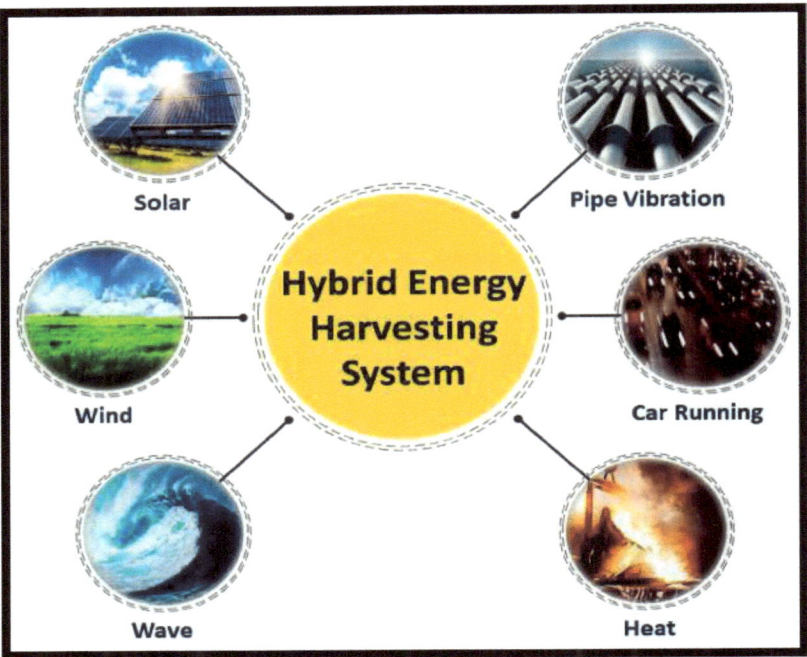

Fig. (2). Applications of hybrid nanogenerators [43].

Examples of Recent Advancements

1. Layered Architectures: Hybrid NGs often involve stacking multiple functional layers, each dedicated to a specific energy conversion mechanism. For instance, a TENG-PENG hybrid device integrates triboelectric and piezoelectric layers to simultaneously harvest mechanical and vibrational energy, enhancing overall efficiency [5].
2. Nanostructured Interfaces: Advanced techniques such as Atomic Layer Deposition (ALD) have been used to create highly conductive and durable interfaces between different functional layers in hybrid NGs. This improves charge transfer and reduces energy loss [6].
3. 3D Hybrid Devices: Utilizing 3D printing and microfabrication, researchers have developed HNGs with intricate geometries that enhance surface area and enable energy harvesting from multiple directions, making them ideal for complex environments like wearable systems [7].
4. Functional Composite Materials: Hybrid NGs often utilize composite materials that combine triboelectric and piezoelectric nanomaterials, such as PDMS mixed with $BaTiO_3$ nanoparticles, to achieve dual energy conversion capabilities. These composites improve material flexibility and device robustness [9].

5. Integration with Flexible Substrates: Flexible substrates, such as graphene or bio-inspired polymers, have been incorporated into hybrid NGs to enhance their adaptability for wearable and implantable devices [8].

By leveraging these advancements, hybrid nanogenerators have demonstrated enhanced energy conversion efficiency and broader applicability, from wearable electronics to environmental monitoring systems [10].

APPLICATIONS

Polymer-based NGs have diverse applications across various domains. In wearable electronics, these devices enable self-powered sensors and health-monitoring systems [36]. Biomedical devices, such as implantable NGs, harness energy from physiological movements to power medical implants [41]. Furthermore, the integration of NGs in IoT devices has paved the way for energy-autonomous smart systems [42].

Environmental and Industrial Applications

Beyond personal electronics and biomedical uses, NGs are increasingly being utilized in environmental monitoring systems, capturing energy from wind, water, and ambient vibrations [46]. In industrial settings, NGs are used for powering remote sensors in hazardous or hard-to-reach areas. These applications demonstrate the versatility and potential of NGs in addressing diverse energy challenges [43].

For instance:

- Smart Grids: NGs integrated with smart grids harness environmental vibrations for decentralized power systems, aiding in efficient energy distribution [1].
- Agriculture: Triboelectric nanogenerators are deployed to monitor soil conditions, such as moisture and nutrient levels, thereby enhancing agricultural productivity [3].
- Marine Systems: NGs designed for underwater use capture energy from waves and currents, powering underwater communication systems and monitoring sensors [4].
- Industrial IoT: Triboelectric and piezoelectric NGs are increasingly used in harsh industrial environments to power sensors that track machinery performance and structural health [9].

Schematic Representation of Applications

Below is a schematic representation of the diverse applications of polymer-based nanogenerators, illustrating their use in wearable devices, biomedical applications, IoT devices, environmental monitoring, industrial sensors, and smart grids (Fig. 3):

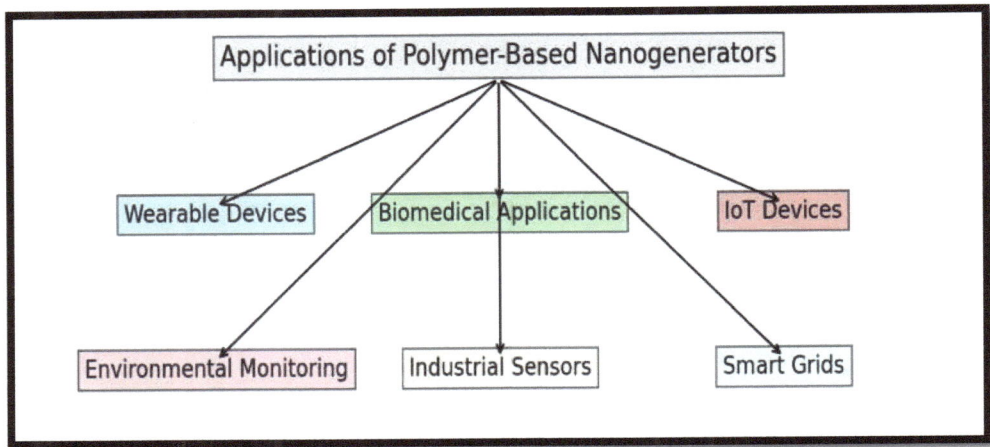

Fig. (3). Schematic representation of the application of a polymer-based nanogenerator.

CHALLENGES AND FUTURE DIRECTIONS

Despite significant progress, challenges such as low energy conversion efficiency, material degradation, and scalability remain [8]. Future research should focus on the development of multifunctional polymers, hybrid materials, and innovative device architectures to overcome these limitations. Additionally, exploring bio-based polymers and eco-friendly fabrication processes will align NG development with sustainability goals [10].

Enhancing Stability and Longevity

The stability and longevity of polymer-based NGs under cyclic loading and environmental stress are the critical challenges. Developing advanced encapsulation techniques and robust materials can mitigate these issues [12]. Furthermore, the integration of self-healing polymers offers a promising approach to prolong device lifespan and maintain performance over extended periods [11].

Self-healing materials that can repair microcracks and recover electrical functionality after mechanical damage are particularly promising [43, 47]. Encapsulation with hydrophobic coatings can protect NGs from moisture and environmental pollutants, ensuring reliable operation over long durations. Additionally, the development of wear-resistant and fatigue-resistant composites will further enhance the durability of these devices in practical applications [48].

Advancing Multifunctionality

Future NGs are expected to incorporate multifunctional capabilities, such as energy harvesting combined with sensing or data storage. This can be achieved through the design of smart materials and hybrid structures that enable seamless integration of multiple functionalities in a single device [44, 49].

For example:

- Energy and Memory Devices: NGs embedded with memory elements can simultaneously harvest energy and store it for later use [44].
- Integrated Sensing: Hybrid NGs with chemical or biological sensing capabilities are under development for use in healthcare and environmental monitoring [44, 45, 50].
- Wearable Multifunctional Systems: These devices combine energy harvesting with motion tracking, vital sign monitoring, and wireless data transmission, opening new possibilities for personalized healthcare [45, 51].

Efficiency

- **Root Cause**: Efficiency in nanogenerators is primarily affected by factors like energy loss during the conversion process, material properties (such as dielectric constant and conductivity), and the mismatch between the mechanical energy source and the energy harvesting mechanism [46].
- **Current Solutions and Limitations**: While advancements like nanostructured fillers (*e.g.*, carbon nanotubes and graphene) have improved the energy output, these improvements are often offset by high material costs and processing complexities. Additionally, current materials may not be able to fully harness energy from low-frequency or low-intensity mechanical sources, limiting the overall efficiency [48].

Stability

- **Root Cause**: Stability issues in polymer-based nanogenerators are linked to the inherent mechanical and environmental stress experienced by the materials. Polymers, particularly those used in triboelectric and piezoelectric nanogenerators, may degrade under repetitive mechanical deformation, exposure to UV radiation, moisture, and temperature fluctuations.

- **Current Solutions and Limitations**: Solutions like encapsulation techniques and the development of self-healing polymers have shown promise, but they often come with trade-offs in terms of added complexity, cost, and the potential for reduced flexibility. Additionally, while self-healing polymers offer the ability to repair microcracks, their long-term effectiveness and scalability in real-world applications are still under evaluation [53].

Lifetime

- **Root Cause**: The lifetime of nanogenerators is often limited by the mechanical fatigue of materials and the wear and tear that occurs with cyclic loading, which can lead to the breakdown of the material structure over time. Furthermore, environmental factors such as humidity, temperature, and exposure to pollutants also accelerate material degradation [55].
- **Current Solutions and Limitations**: Research into durable composite materials, hybrid structures, and coatings has led to improvements, but achieving long-term stability and reliability remains challenging, particularly for high-performance applications such as **wearable electronics** and **implantable medical devices**. Additionally, maintaining high energy output while extending the device's operational life remains an unresolved issue in some cases [41.

Scalability

- **Root Cause**: Scalability challenges stem from the difficulties in mass-producing nanogenerators with uniform material properties and consistent performance. The fabrication processes, such as electrospinning, 3D printing, and nano-patterning, often face issues related to cost-effectiveness and complexity when applied on a large scale. Moreover, integrating nanogenerators into flexible substrates for large-area applications poses additional hurdles in terms of both performance and manufacturing costs.
- **Current Solutions and Limitations**: While techniques such as roll-to-roll processing and inkjet printing have been explored for large-scale production, the high cost of high-quality materials and the need for precision in device fabrication hinder the widespread commercialization of polymer-based nanogenerators [54]. High-performance materials designed to exhibit better mechanical, thermal, chemical, and functional properties are called advanced polymers. In Table **2**, advanced polymers and performance metrics, including biodegradable, self-healing, shape-memory, and conductive polymers, are discussed.

Table 2. Advanced Polymers and Performance Metrics [19, 53 - 56].

Polymer/Composite Material	Permittivity	Durability (Years)	Flexibility Rating	Energy Efficiency (%)	References
PVDF-TrFE	12–14	10+	High	40–50	[1] [5] [7]
PTFE	2.1	15+	Moderate	30–35	[3] [8] [12]
PDMS	2.65	12+	High	25–30	[4] [6] [13]
PVDF + BaTiO3	35–40	8–10	Moderate	50–60	[10] [14] [15]
Graphene Composite	60–70	5–7	Low	70–80	[11] [16] [18]

CONCLUSION

The evolution of advanced polymer-based nanogenerators over the past two decades underscores their potential as a sustainable energy harvesting technology. Continued advancements in materials, fabrication techniques, and device integration are expected to unlock new possibilities, driving their adoption in emerging applications.

While challenges remain, such as scalability, efficiency, and durability, the trajectory of research in this field highlights a promising future. Polymer-based nanogenerators, with their inherent flexibility and adaptability, are poised to play a significant role in powering the next generation of wearable electronics, IoT devices, and environmental monitoring systems. As the demand for renewable and sustainable energy solutions grows, these innovative devices are set to become a cornerstone of energy technologies in the 21st century.

REFERENCES

[1] Saha, A., 2024. Polymer nanocomposites: a review on recent advances in the field of green polymer nanocomposites. Current Nanoscience, 20(6), pp.706-716.

[2] Z. Lin, Z. Chen, and Z.L. Wang, "Eco-friendly materials in polymer-based nanogenerators", *ACS Sustain. Chem.& Eng.,* vol. 8, no. 5, pp. 1234-1242, 2020.

[3] Kakim, A., Nurkesh, A., Sarsembayev, B., Dauletiya, D., Balapan, A., Bakenov, Z., Yeshmukhametov, A. and Kalimuldina, G., 2024. Incorporating MIL-125 Metal-Organic Framework for Flexible Triboelectric Nanogenerators and Self-Powered Sensors for Robotic Grippers. Advanced Sensor Research, 3(8), p.2300163.

[4] Farzin, M.A., Naghib, S.M. and Rabiee, N., 2024. Advancements in bio-inspired self-powered wireless sensors: Materials, mechanisms, and biomedical applications. ACS Biomaterials Science & Engineering, 10(3), pp.1262-1301.

[5] Yadegari, A., Akbarzadeh, M., Kargaran, F., Mirzaee, R., Salahshoori, I., Nobre, M.A. and Khonakdar, H.A., 2024. Recent advancements in bio-based dielectric and piezoelectric polymers and their biomedical applications. Journal of Materials Chemistry B, 12(22), pp.5272-5298.

[6] Gołąbek, J. and Strankowski, M., 2024. A review of recent advances in human-motion energy harvesting nanogenerators, self-powering smart sensors and self-charging electronics. Sensors, 24(4),

p.1069.

[7] S. Kumar, S. Verma, and A. Singh, "Self-healing polymer nanogenerators for extended device durability", *ACS Appl. Mater. Interfaces,* vol. 12, no. 12, pp. 1489-1498, 2020.

[8] Liu, Y., Chen, L., Li, W., Pu, J., Wang, Z., He, B., Yuan, S., Xin, J., Huang, L., Luo, Z. and Xu, J., 2024. Scalable production of functional fibers with nanoscale features for smart textiles. ACS nano, 18(43), pp.29394-29420.

[9] Rahul, T.P. and Sreekanth, P.R., 2025. Synergies in materials and manufacturing: a review of composites and 3D printing for triboelectric energy harvesting. Journal of Composites Science, 9(8), p.386.

[10] Zhi, C., Shi, S., Wu, H., Si, Y., Zhang, S., Lei, L. and Hu, J., 2024. Emerging trends of nanofibrous piezoelectric and triboelectric applications: Mechanisms, electroactive materials, and designed architectures. Advanced Materials, 36(26), p.2401264.

[11] Afshar, H., Kamran, F. and Shahi, F., 2025. Recent progress in energy harvesting technologies for self-powered wearable devices: the significance of polymers. Polymers for Advanced Technologies, 36(4), p.e70187.

[12] Liu, Z., Chen, X. and Wang, Z.L., 2025. Biopolymer and biomimetic techniques for triboelectric nanogenerators (TENGs). Advanced Materials, 37(22), p.2409440.

[13] L. Sun, M. Chen, and P. Zhao, "Advanced composites for high-performance nanogenerators", *Compos. Sci. Technol.,* vol. 178, pp. 42-50, 2020.

[14] Singh, A.K., Singh, V. and Anand, P.I., 2025. An innovative laser decal transfer of ZnO ceramic in LIG for advanced hybrid nanogenerator applications. Ceramics International, 51(4), pp.4957-4970.

[15] Z. Fang, H. Chen, and Y. Wang, "Integrated sensing and energy harvesting with hybrid nanogenerators", *Sens. Actuators A Phys.,* vol. 310, p. 112043, 2020.

[16] Chen, S., Fan, S., Qiao, Z., Wu, Z., Lin, B., Li, Z., Riegler, M.A., Wong, M.Y.H., Opheim, A., Korostynska, O. and Nielsen, K.M., 2025. Transforming healthcare: Intelligent wearable sensors empowered by smart materials and artificial intelligence. Advanced Materials, p.2500412.

[17] S. Xu, Y. Liu, and H. Hu, "Nanostructured surfaces for enhanced energy harvesting in polymer-based devices", *Nano Today,* vol. 35, p. 100998, 2020.

[18] Thilakarathne, N.N., Nanomaterials on the internet of things and smart technologies

[19] Liu, H., Deng, Y., Li, J., Liu, Q., Mu, L., Zhang, R., Sun, C.L., He, J. and Qu, M., 2025. Stretchable, anti-freezing, self-healing, and degradable high-performance conductive hydrogel-based triboelectric nanogenerator for energy harvesting and human activity recognition. Materials Today Energy, p.102013.

[20] Cui, X., Wu, L., Zhang, C. and Li, Z., 2025. Implantable self-powered systems for electrical stimulation medical devices. Advanced Science, 12(24), p.2412044.

[21] Sengupta, J. and Hussain, C.M., 2025. Graphene-Enhanced Piezoelectric Nanogenerators for Efficient Energy Harvesting. C, 11(1), p.3.

[22] Wang, F., Wang, S., Wang, H., Liu, Y., Ouyang, S., Zhang, S., Hu, Y., Zhao, J., Ma, S., Wu, Z. and Wang, L., 2025. Enhancing the output of cellulose-based triboelectric nanogenerators *via* dual interfacial polarization effects. Carbohydrate Polymers, p.124093.

[23] Li, A., Qiliang, Z., Mi, Y., Ur Rehman, H., Shoaib, M., Cao, X. and Wang, N., 2025. Triboelectric nanogenerator drives electrochemical water splitting for hydrogen production: Fundamentals, progress, and challenges. Small, 21(1), p.2407043.

[24] Wang, C., Chai, H., Li, G., Wang, W., Tian, R., Wen, G.L., Wang, C.H. and Lai, S.K., 2024. Boosting biomechanical and wave energy harvesting efficiency through a novel triple hybridization of piezoelectric, electromagnetic, and triboelectric generators. Applied Energy, 374, p.123876.

[25] Sagar, P. and Kumar, B., 2025. PANI-reinforced-ZnS/PDMS-based flexible hybrid piezo-triboelectric nanogenerator for self-powered wearable electronics and sensing. Materials Research Bulletin, p.113482.

[26] Krishnan, S. and Giwa, A., 2025. Advances in real-time water quality monitoring using triboelectric nanosensors. Journal of Materials Chemistry A.

[27] J. Wang, X. Liu, and S. Zhang, "Piezoelectric polymers for energy harvesting applications", *ACS Appl. Energy Mater.*, vol. 3, no. 10, pp. 9756-9765, 2020.

[28] Hou, X., Ren, P., Guo, Z., Tian, W., Wang, Y., Fan, B., Chen, H., Chen, Z. and Jin, Y., 2025. Triboelectric nanogenerators and supercapacitors based on biowaste-derived porous carbon for efficient energy harvesting and storage. Journal of Power Sources, 649, p.237422.

[29] Wang, C., Niu, H., Shen, G. and Li, Y., 2025. Self-healing hydrogel-based triboelectric nanogenerator in smart glove system for integrated drone safety protection and motion Control. Advanced Functional Materials, 35(17), p.2419809.

[30] Yu, Y., Zhou, Z., Ruan, H. and Li, Y., 2025. High conductivity, low-hysteresis, flexible PVA hydrogel multi-functional sensors: Wireless wearable sensor for health monitoring. Chemical Engineering Journal, 505, p.158877.

[31] Ye, Z., Liu, T., Du, G., Shao, Y., Wei, Z., Zhang, S., Chi, M., Wang, J., Wang, S. and Nie, S., 2025. Bioinspired superhydrophobic triboelectric materials for energy harvesting. Advanced Functional Materials, 35(2), p.2412545.

[32] Li, X., Li, J., Li, K., Zhang, S., Yang, Z., Zhang, C., Zhang, J., Li, Y., Zhang, D., Liu, Y. and Hu, X., 2025. From Fiber to Power: Recent Advances Toward Electrospun-Based Nanogenerators. Advanced Functional Materials, 35(13), p.2418066.

[33] Safaei, B., Peiravian, M. and Siamaki, M., 2025. Eco-Friendly IoT: Leveraging Energy Harvesting for a Sustainable Future. IEEE Sensors Reviews.

[34] Qian, H., Zhou, Y., Cao, Z., Tang, T., Deng, J., Huo, X., Zhou, H., Wang, L. and Wu, Z., 2025. A Hybrid Nanogenerator Based on Rotational-Swinging Mechanism for Energy Harvesting and Environmental Monitoring in Intelligent Agriculture. Sensors, 25(16), p.5041.

[35] Rosca, C.M. and Stancu, A., 2025. Integration of AI in Self-Powered IoT Sensor Systems. Applied Sciences, 15(13), p.7008.

[36] D.H. Kim, J.H. Kim, and T.H. Kwon, "Enhancing triboelectric performance through nanostructured hybrid composites", *Nano Energy,* vol. 62, pp. 400-410, 2019.

[37] J. Wang, Y. He, and T. Zhao, "Durability of polymer-based nanogenerators in cyclic environments", *J. Mater. Chem. A Mater. Energy Sustain.*, vol. 9, no. 32, pp. 18765-18776, 2021.

[38] F.-R. Fan, Z.-Q. Tian, and Z. L. Wang, "Flexible triboelectric nanogenerators for self-powered electronics," *Nature Communications*, vol. 8, 15938, 2017

[39] Cho, W., Kim, S., Lee, H., Han, N., Kim, H., Lee, M., Han, T.H. and Wie, J.J., 2024. High-performance yet sustainable triboelectric nanogenerator based on sulfur-rich polymer composite with MXene segregated structure. Advanced Materials, 36(44), p.2404163.

[40] Peng, W., Ni, Q., Zhu, R., Fu, X., Zhu, X., Zhang, C. and Liao, L., 2024. Triboelectric-electromagnetic hybrid wind energy harvesting and multifunctional sensing device for self-powered smart agricultural monitoring. Nano Energy, 131, p.110272.

[41] Wang, F., Zhu, X. and Du, X., 2025. Intelligent Poly (vinylidene fluoride)-Based Materials for Biomedical Applications. Advanced Functional Materials, p.2500685.

[42] Hajra, S., Ali, A., Panda, S., Song, H., Rajaitha, P.M., Dubal, D., Borras, A., In-Na, P., Vittayakorn, N., Vivekananthan, V. and Kim, H.J., 2024. Synergistic integration of nanogenerators and solar cells: advanced hybrid structures and applications. Advanced Energy Materials, 14(21), p.2400025.

[43] Kwon, Y.H., Meng, X., Xiao, X., Suh, I.Y., Kim, D., Lee, J. and Kim, S.W., 2024. Triboelectric energy harvesting technology for self-powered personal health management. International Journal of Extreme Manufacturing, 7(2), p.022005.

[44] E. Delgado-Alvarado, E.A. Elvira-Hernández, J. Hernández-Hernández, J. Huerta-Chua, H. Vázquez-Leal, J. Martínez-Castillo, P.J. García-Ramírez, and A.L. Herrera-May, "Recent Progress of Nanogenerators for Green Energy Harvesting: Performance, Applications, and Challenges", *Nanomaterials (Basel),* vol. 12, p. 2549, 2022.

[45] E. Delgado-Alvarado, J. Martínez-Castillo, L. Zamora-Peredo, J.A. Gonzalez-Calderon, R. López-Esparza, M.W. Ashraf, S. Tayyaba, and A.L. Herrera-May, "Triboelectric and Piezoelectric Nanogenerators for Self-Powered Healthcare Monitoring Devices: Operating Principles, Challenges, and Perspectives", *Nanomaterials (Basel),* vol. 12, p. 4403, 2022.

[46] X. Tao, X. Chen, and Z.L. Wang, "Design and synthesis of triboelectric polymers for high performance triboelectric nanogenerators", *Energy Environ. Sci.,* vol. 16, no. 7, pp. 3654-3678, 2023.

[47] Y. Zhang, J. Li, and Z. Wang, "Porous polymer materials in triboelectric nanogenerators: A review", *Polymers (Basel),* vol. 15, no. 22, p. 4383, 2023.

[48] J. Liu, H. Zhang, and Y. Wang, "A review of polymer-based environment-induced nanogenerators", *Polymers (Basel),* vol. 16, no. 4, p. 555, 2023.

[49] L. Zhang, M. Chen, and Y. Li, "Recent advances in stretchable hydrogel-based triboelectric nanogenerators", *Mater. Horiz.,* vol. 11, pp. 784-803, 2024.

[50] J. Wang, X. Zhang, and Y. Liu, "Advances in conducting polymer-based nanogenerators for biomechanical energy harvesting", *Micromachines (Basel),* vol. 12, no. 11, p. 1308, 2024.

[51] H. Zhang, Y. Li, and Z. Wang, "A novel polymer composite from polyhexamethylene guanidine hydrochloride for high performance triboelectric nanogenerators", *RSC Advances,* vol. 10, no. 1, pp. 7768-7776, 2025.

[52] Du, T., Chen, Z., Dong, F., Cai, H., Zou, Y., Zhang, Y., Sun, P. and Xu, M., 2024. Advances in green triboelectric nanogenerators. Advanced Functional Materials, 34(24), p.2313794.

[53] Gupta, A., Gajula, P., Yoon, J.U., Lee, S.H., Kim, H., Adusumalli, V.N., Bae, J.W. and Park, Y.I., 2024. Revolutionizing energy harvesting: Eco-friendly and adaptable triboelectric sensors with recycled cloth and metallo-dielectric loaded Ecoflex hybrid films. Nano Energy, 122, p.109346.

[54] Kakim, A., Nurkesh, A., Sarsembayev, B., Dauletiya, D., Balapan, A., Bakenov, Z., Yeshmukhametov, A. and Kalimuldina, G., 2024. Incorporating MIL-125 Metal-Organic Framework for Flexible Triboelectric Nanogenerators and Self-Powered Sensors for Robotic Grippers. Advanced Sensor Research, 3(8), p.2300163.

[55] Padha, B., Verma, S., Ahmed, A., Patole, S.P. and Arya, S., 2024. Plastic turned into MXene–based pyro-piezoelectric hybrid nanogenerator-driven self-powered wearable symmetric supercapacitor. Applied Energy, 356, p.122402.

[56] Yadegari, A., Akbarzadeh, M., Kargaran, F., Mirzaee, R., Salahshoori, I., Nobre, M.A. and Khonakdar, H.A., 2024. Recent advancements in bio-based dielectric and piezoelectric polymers and their biomedical applications. Journal of Materials Chemistry B, 12(22), pp.5272-5298.

CHAPTER 10

Polymeric Energy Materials: Development, Challenges, and Future Benefits for Industrialization

Debasish Banerjee[1], Sumit Bhowmik[1], Arbind Prasad[2,*] and Sudipto Datt[3,*]

[1] Mechanical Engineering Department, Omdayal Group of Institutions, Howrah 711316, West Bengal, India

[2] Mechanical Engineering Department, Katihar Engineering College (Under Department of Science, Technology and Technical Education, Government of Bihar), Katihar 854109, Bihar, India

[3] Department of Materials Engineering, Indian Institute of Science, Bangalore 560012, Karnataka, India

Abstract: Polymeric active materials have emerged as a promising class of materials that will likely lead the way in energy conversion and storage. Materials such as conductive polymers, polymer composites, and polymer electrolytes are being processed to enhance energy-delivery devices like batteries, supercapacitors, and photovoltaic cells. The challenges in polymeric energy materials are related to their stability, scalability, and behavior under extreme operating conditions. It calls for extraordinary solutions in designing materials, manipulating methods of processing the material, and enhancing properties like electrical conductivity, mechanical properties, and flexibility. The prospects for industrializing polymeric energy materials offer advantages because they can be engineered to be lighter, more efficient, and eco-friendly for energy delivery. These new materials, if incorporated successfully into commercially viable products, can lead to the fourth generation of energy materials needed for further development of sustainable energy and non-renewable resources such as fossil fuels.

Keywords: Conductive polymers, Energy conversion, Energy storage, Industrialization, Material development, Polymer composites, Polymeric energy materials, Solar cells, Supercapacitors, Sustainability.

---

* **Corresponding author Arbind Prasad:** Mechanical Engineering Department, Katihar Engineering College (Under Department of Science, Technology and Technical Education, Government of Bihar), Katihar 854109, Bihar, India; E-mail: sudiptodatta1990@gmail.com

INTRODUCTION

Polymer energy materials can now be recognized as one of the emerging themes in the search for adequate energy supplies. These materials, including conductive polymers, polymer composites, and polymer electrolytes, are considered more promising substitutes for traditional inorganic materials in energy storage and conversion equipment [1]. The use of polymeric materials in energy technologies, such as batteries, supercapacitors, and solar cells, has attracted a lot of interest because of their inherent advantages of flexibility, lightweight nature, and potential for cost-effective manufacturing processes [2]. Namely, conductive polymers possess certain distinctive features, including high electrical conductivity, ease of processing, and ability to be tuned, all of which have contributed to the increasing use of conductive polymers as materials for next-generation energy devices [3].

Nevertheless, the application of polymeric energy materials in the field of development and industrialization has various problems. Some of the main concerns are the ability of these materials to perform stably in practical applications, the lack of a wide range of manufacturing approaches, and the high demand for efficiency in storage devices, such as batteries and supercapacitors [4]. The problem of optimizing the relationship between electrical conductivity and mechanical properties, as well as some questions related to environmental issues, remainsrelevant for investigators [5]. However, despite these challenges, the constant improvement of the material and its processing methods is gradually eradicating these barriers [6].

In emerging areas, polymeric energy materials have massive potential for industrial applications. As recent trends in nanotechnology, polymer blends, and hybrid systems dictate, such materials are expected to play a major role in the generation of further improved, cleaner, and more cost-effective energy technologies [7]. Specifically, it has been pointed out that there is an opportunity for developing novel energy storage devices that can pave the way for change in industries, starting with electronics and moving to transportation, through the use of easy-to-fold, lightweight, and flexible wearable electronics, mobile power sources, and skin-like solar cells [8]. While the problems are being investigated today, polymeric energy materials are likely to become the cornerstone for advancing the future of energy storage and conversion, providing new opportunities for industrialization and environmental sustainability [9].

POLYMERIC ENERGY MATERIALS: APPLICATIONS IN MODERN INDUSTRY

Thermoplastic energy materials have been widely incorporated into contemporary manufacturing industries due to their versatility, lightweight nature, and potential for cost-effective conversion [10]. These are polymers, polymer composites, and polymer electrolytes that are critical to different technologies involving theconversion, storage, and management of energy [11]. They are used in almost all industries, including electronics, the automotive industry, renewable energies, healthcare, and many more. In this section, we also cover key industrial applications of polymeric energy materials such as energy storage, energy conversion, flexible electronics, *etc*.

Energy Storage Systems

Polymeric energy materials have emerged as critical components for new-age energy solutions, especially in the design of batteries, supercapacitors, and fuel cells. Copolymer and polymer composites are widely used in lithium-ion batteries, supercapacitors, and other energy storage technologies because of their high electrical conductivity, flexibility, and ease of processing. Conductive polymers, such as PEDOT, PANI, and polyacetylene, are among the most researched polymers owing to their conductive and charge/discharge properties, which can improve energy storage systems [12]. To illustrate this, conductive polymer composites can be blended with battery electrodes to enhance cyclability, discharge capacity, and energy density.

Flexible and lightweight energy storage devices can be built without the need for solid-state electrolytes in supercapacitors by incorporating polymeric materials. Supercapacitors that can be bent should be utilized in wearable electronic devices and portable gadgets because size and flexibility play a vital role in these products [13]. Conductive polymers combined with carbon-based nanomaterials exhibit good charge/discharge abilities, long cycling stability, and energy density, which primarily enhance the supercapacitor's performance [14]. The use of such materials in large-scale energy storage applications for grid-scale storage is also being investigated, which is beneficial for the use of renewable energy storage systems [15]. This chapter also explores theinterconnections among polymeric composites, bioeconomy, and Amazonian residues. With sections representing important components, including chemical interactions, sustainable resource management, material reuse, and the production of novel materials, the study illustrates the role of human activity in sustainable development, as studied in the literature [16].

Solar Cells and Energy Conversion

Polymeric materials are critically involved in the renewable energy industry, with a particular emphasis on the use of Organic Photovoltaic (OPV) cells. Based on an organic polymer as the active material, OPVs have attracted considerable attention due to their features, such as low cost, flexibility, and lightweight nature, which make them suitable for photovoltaic solar panels [17, 18]. These materials serve as substitutes for silicon materials in solar cells, making them again suitable for large-scale production and implementation in regions where conventional solar power may be uneconomical [19]. Polymer systems, such as P3HT and PPV, have been widely examined for these roles in OPVs because of several reasons, including the ease of synthesizing these polymers, good charge transport abilities, and good light absorption properties [20]. Despite the opportunities for OPV development, some problems exist at present related to the efficiency and stability of OPVs [21]. New developments in polymer blends, like the use of fullerene derivatives, have boosted the PCE of the devices in recent years [22]. Current activities include fine-tuning the molecular weight of conductive polymers and optimizing the overall efficiency of the active layer/ electrode interfaces [23]. Work is also being done on increasing the resistance of the devices to outdoor conditions [24]. Consequently, until they become an economic reality, polymeric materials are expected to present viable long-term solutions for utilizing solar power at both the utility-scale and for distributed applications [25].

Flexible and Wearable Electronics

In the modern industry, flexible and wearable electronics are one of the most significant breakthrough areas in which polymeric energy materials are applied [26]. The excellent mechanical flexibility and electrical conductivity of conductive polymers have made them ideal for use in wearable devices, sensors, and health-tracking devices, as well as modern medical imaging technology [27, 28]. For instance, Polypyrrole (PPy) and Poly(3,4-ethylenedioxythiophene) (PEDOT)-based materials are used in electrodes for wearable sensors for monitoring physiological signals, including heart rate, sweating rate, and skin temperature [29]. They can be incorporated into clothing or wearable skins, which makes such gadgets thin and non-intrusive for user comfort while giving real-time health information [30].

Polymeric materials are also used in energy conversion devices, such as Triboelectric Nanogenerators (TENGs), which convert mechanical energy into electrical energy [31]. Polymer-based TENGs have been applied to power wearable electronics and small devices, thereby enabling self-powered sensors and wearable health monitoring systems [32]. The concept of harvesting energy from the motion of the human body has created new opportunities to generate clean energy. Portable and lightweight electricity transformer applications can be found in the medical and smart fiber sectors.

Automotive Industry: Lightweight Energy Materials

The automotive sector has also benefited from the use of polymeric energy materials, especially in the fabrication of light, energy-friendly cars. Polymers and composites are integrated into electrodes and electrolyte components of Electric Vehicle (EV) batteries to enhance the performance characteristics and energy density of battery systems [26]. This paper also found that employing lightweight polymer composites in vehicle manufacturing has helped lower the weight of the vehicle, thus reducing fuel consumption by ensuring energy conservation [33].

Furthermore, several polymeric energy materials are under consideration for applications in fuel cells, which are regarded as the future power plant for emission-less automobiles. As electrolytes in Proton Exchange Membrane Fuel Cells (PEMFCs), conductive polymer membranes such as Nafion allow the passage of protons but not electrons [27]. Current work emphasizes the optimization of the ionic conductivity of these polymer materials, as well as their overall stability, to enhance the technical and economic feasibility of fuel cell technology for automobiles.

Energy-Efficient Building Materials

Polymeric energy materials have also been incorporated into the construction industry to improve the energy performance of construction materials. Polymeric coatings and composites are being applied in solar thermal collectors used for the conversion of sunlight into useful heat. These coatings consist of polymers, such as polycarbonate and polyurethane, which have good thermal stability and resistance to UV [34]. Furthermore, polymeric materials find their use in thermal insulation, where thickness, manpower, and trapped air space lose or insulate heat, thereby saving energy.

Polymeric materials are also beginning to appear in smart windows that can control light and heat transfer in buildings. These windows utilize electrochromic polymers that change color in response to voltage; this approach makes it easier to regulate the indoor temperature, thereby minimizing the use of artificial heating/cooling [35]. The role of polymers in building applications will increase progressively due to the demand for energy-efficient buildings.

POLYMERIC ENERGY MATERIALS IN HEALTHCARE APPLICATIONS

In recent years, interest in polymeric energy materials has quickly grown in the healthcare sector because of their malleable, biocompatible, and electrochemical properties. These materials are slowly finding applications in different fields of healthcare, such as the fabrication of medical instruments, sensors, and drug delivery systems, as well as wearable health monitoring systems. The ability to incorporate energy storage and conversion into polymer boundaries has expanded new opportunities in healthcare. This section focuses on polymeric energy materials for healthcare and the role they have played in advancing healthcare technology.

Wearable Health Monitoring Devices

Smart clothing systems for health monitoring include devices that constantly monitor various physiological parameters, such as heart rate, blood sugar, skin temperature, and so on, making them an indispensable part of modern healthcare. Many of these systems demand methods for supplying energy that are portable, adaptable, and capable of delivering power over long intervals of time. Conductive polymers and polymer composites are particularly suitable for these applications because they are flexible electrochemical materials [36].

Flexible and conducting materials such as PEDOT and PPy are mainly used in the electrodes of wearable health sensors owing to their high conductivity and compatibility with flexible substrates [35]. Such materials can be incorporated into fabric or skin patches, allowing users to have continuous and non-intrusive monitoring. In addition, energy scavenging devices, especially Triboelectric Nanogenerators (TENGs) based on polymeric materials, are used in the development of wearable devices that do not require additional power supply [30]. These milestones allow for constant monitoring of health information, thus enhancing the effectiveness of individualized or personalized medicine.

Drug Delivery Systems

Polymers have traditionally enjoyed great popularity in the formulation of drug delivery systems, most notably in the development of controlled and sustained release forms. Liposomes, polymeric nanoparticles, hydrogels, and polymeric micelles have been reported as suitable carriers for drug delivery systems as they are designed to target specific sites and reduce side effects. The concepts of incorporating energy-responsive polymeric materials into drug delivery systems have recently been reported as a strategy with the potential to provide the specific release of therapeutic agents [37].

Many applications of polymeric nanocomposites involve loading polymers with magnetic or conductive nanoparticles in a slow-release process. For instance, magnetic polymer composites can be used in drug delivery where drug release is initiated by a magnetic field [38]. This process of on-demand drug delivery means that certain therapeutic agents can be delivered to a particular site, which is ideal for treating cancerous cells and tissues, as drug concentrations can be specifically aimed at these locations with minimal impact on healthy body tissues and organs [39]. Moreover, conducting polymers, such as PEDOT, are also incorporated into the drug delivery system, which is responsive to electric stimulus and serves as an added advantage with proper timing and distinct doses of drug release [40].

Energy Harvesting for Implantable Medical Devices

Pacemakers, neural stimulators, and cochlear implants are examples of implantable medical devices that need power and must, therefore, last for the duration of the implant. Conventionally, these devices are battery-operated, and batteries must be replaced periodically, which an expensive and sometimes invasive procedure. Polymeric energy materials represent another favorable type owing to the possibility of energy accumulation and the device's autonomy.

Piezoelectric and triboelectric materials based on polymeric materials can convert energy from physiological movements or functions, including heartbeat, respiration, and joint movement, into electricity [26]. For instance, piezoelectric polymeric materials, such as Polyvinylidene Fluoride (PVDF), can convert mechanical energy from body movement to electrical energy, which is optimal for the powering of small implantable devices. The positive aspect of generating energy from natural body movement also helps in avoiding periodic external power source requirements, which lowers implant durability and patient comfort.

Biosensors for Disease Diagnosis

Polymeric materials are also being gradually applied to biosensors for the diagnosis of diseases. These sensors are designed to detect particular biomolecules in body fluids, including whole blood, urine, saliva, and tissues, to offer real-time information about a patient's condition. Polymeric energy materials are crucial to the design and synthesis of biosensors, as they can store and release energy and also respond to biomolecular signals.

For example, based on the application of electrochemical biosensors, conductive polymers are widely applied for electrode materials. These materials allow for the identification of biomolecules through electrical signals, such as current or voltage changes, that can be linked to biomarker presence [51]. Additionally, the application of polymeric energy materials can also help develop sensors with lower energy consumption for portable diagnostic instruments used in point-of-care diagnosis. The combination of these flexible and biocompatible polymeric materials also allows for patient comfort while providing accurate diagnostic data on time [41 - 45].

Flexible and Stretchable Electronics

In the context of healthcare, flexible and stretchable electronics have provided novel utility for diagnostics, therapeutic tools, and monitoring systems. As these applications require flexible and stretchable materials, polymeric materials are especially suitable for these uses [46 - 50]. The materials that have been employed in stretchable electronics include conductive polymers such as PEDOT, polypyrrole, and Poly (ethylene oxide) (PEO) because of their conductivity even when deformed [52]. Both wearable health monitoring systems and implantable systems are two areas of application for these flexible electronics. For example, wearable Electrocardiogram (ECG) sensors, which track cardiopulmonary activity, encompass conductive polymer electrodes that can perform when the fabric, on which the device is draped, is stretched or folded. In the same way, polymeric materials are employed in flexible TE devices that can measure temperature and are used in monitoring patients' health in real-time without the need for invasive procedures.

Tissue Engineering and Regenerative Medicine

Polymeric energy materials are also being used in tissue engineering and regenerative medical applications due to their appropriateness in supporting cell growth and providing an electricity stimulus. For instance, conductive polymers are employed in fabricating substrates for cellular growth, and the electrical signals are capable of influencing the activities of the cell, including differentiation, migration, and proliferation. These materials can also be used in bioelectronic interfaces to enhance the compatibility of manipulating prosthetic devices with the body [22].

In regenerative medicine, polymeric scaffolds are generally developed to afford a three-dimensional structure that resembles the natural ECM to support tissue restoration. These scaffolds, containing or coated with conductive polymers or materials exhibiting piezoelectricity, can actively stimulate the healing of damaged tissues, as electrical stimulation has been demonstrated to enhance the rate of reparative tissue regeneration [52].

CHALLENGES AND FUTURE DIRECTIONS

Polymer-type energy materials have great potential in enhancing healthcare tools because of certain characteristics, like biocompatibility, flexibility, and energy storage. Nevertheless, as mentioned above, there are some difficulties associated with the formation and utilization of these materials in healthcare. This section covers the issues addressed and presents the directions for the further developement of polymeric energy materials in medicine.

CHALLENGES

a. Biocompatibility and Toxicity

Some of the issues peculiar to the healthcare application of polymeric energy materials include concerns about the biocompatibility and potential toxicity of such materials. Some polymers usually exhibit biocompatibility issues due to their inherent electrical characteristics, the storage of energy, or the incorporation of conducting agents or other components, such as metal nanoparticles or carbon-based materials in the polymer matrix, which may bring toxicity effects [53]. For example, methods of using nanoparticles in drug delivery systems, biosensors, or energy storage devices can cause problems such as immune reactions or cytotoxicity. All these materials must possess good electrochemical properties to work safely in medical applications, although biocompatibility also plays a key role.

Future Direction: Materials are composite, biocompatible, and non-toxic, but they must retain the ability to maintain energy efficiency; therefore, researchers need to continue to strive for these qualities. Replacing traditional non-degradable polymers with fresh biodegradable polymers or natural polymers like PLA or chitosan might offer a safer atmosphere for healthcare applications [32]

b. Stability and Durability

Flexible electronics and energy harvesting devices require flexible polymeric materials that can degrade over time owing to moisture, heat, or physiological conditions. This degradation can impact the performance of wearable health monitoring technology or implantable medical technology, shortening their lifespan and compromising performance conditions. Additionally, since most polymer-based drug delivery systems require sustained release, the stability of these systems is paramount to providing treatment efficacy for extended periods.

Future Direction: Future work in the development and use of polymeric materials for tissue engineering should be aimed at increasing the chemical and physical stability of these materials when exposed to biological settings. Horizons can also be developed in the area of nanostructuring and coatings, which may lead to improved protection of polymeric materials against environmental degradation and increased service life. Furthermore, novel multifaceted polymeric systems that can heal themselves or remain stable in physiological environments for a prolonged timeframe can enhance healthcare applications.

c. Manufacturing Scalability

One challenging issue is the fabrication of polymeric energy materials specifically for the applications of flexible electronics, energy harvesting, and drug delivery systems. Although there was a laboratory-scale synthesis that had positive outcomes, it is quite challenging to produce these materials at a commercial level for use in healthcare applications due to high cost, complex manufacturing processes, and uneven material properties, as pointed out by Barua *et al.* (2024) [38]. However, using these materials to create medical devices requires close control over the production processes to get consistent quality in production batches.

Future Direction: The future direction should cover the establishment of efficient and cheap synthesis methods for polymeric energy materials. Breakthroughs in 3D printing, roll-to-roll processing, and other high-throughput fabrication processes can therefore hold the key to the large-volume production of flexible, biocompatible, and energy-efficient polymeric systems. Moreover, if the processes of manufacturing these technologies were environmentally friendly, consumed little energy, and produced less waste, this would go a long way in encouraging the application of these technologies in healthcare.

d. Energy Efficiency and Power Output

Energy storage and conversion of electrolytes in polymeric materials is one of the challenges that has not yet been addressed. It has been recognized that although conductive polymers and composites exhibit satisfactory electrochemical performance, their energy density and output power density are usually lower than those of metals or inorganic semiconductors [35]. This limitation can hinder the application of polymeric energy materials in demanding applications, such as implantable medical devices or wearable electronics, where the overall power density and operation time are critical.

Future Direction: Future investigations should be aimed at increasing the energy density, power, and efficiency of polymeric energy materials by identifying new polymers, polymer nanocomposites, and hybrid materials [28]. These limitations can perhaps be addressed by integrating energy-harvesting materials with high-performance polymers or by fabricating polymeric supercapacitors/batteries [53]. Furthermore, if the lifecycles of healthcare devices were to be extended and maintained without constant recharging and power source replacement, searching for energy-efficient and self-charging materials that can harness energy from various sources, such as light, heat, human motion, *etc.*, would be essential.

e. Integration with Biological Systems

It is quite a challenge to interface polymeric energy materials with biological systems. The adhesion between an implantable synthetic material and human tissue results in complications, such as inflammation, rejection, or tissue irritation. For example, conducting polymers incorporated into neural interfaces or medical pacemakers may not be fully biocompatible, which can affect their operation [54]. Moreover, polymeric materials used in energy harvesting in medical devices should possess characteristics of elasticity, stretchability, and conformability to the human body.

Future Direction: The understanding of the optimal electrical and mechanical properties of bioelectronic materials, which enables the integration of polymeric materials with the human body, is therefore exceptional. This field needs improvements in the interface characteristics between the biological tissues and the polymeric materials used for integration. This may also involve synthesizing bioactive coatings or surface modifications to prevent or inactivate the probability of rejection or irritation.

f. Regulatory and Standardization Issues

Two key challenges are time and regulatory approval, as well as the standardization of polymeric energy materials for use in medical applications. The healthcare industry is thoroughly regulated, and the use of new materials in the construction of medical devices must meet the primary tests and certification to ensure the safety of applications. The absence of proper testing milestones in polymeric energy materials, especially those used in implantable devices or drug delivery systems, will inevitably slow down their commercialization [34].

Future Direction: Advancements in the use of polymeric energy materials in the health sector will thus require fixed legal boundaries as well as standard measures of assessment for such materials [28]. There will be a major emphasis in the coming years on the coordination of material scientists, doctors, and other related regulatory agencies to set up appropriate guidelines necessary for the proper use of suitable materials in medical applications. Moreover, the introduction of reliable testing methods, such as using advanced testing methods and test tube models, in the approval process of these technologies might help speed up the development of these technologies.

Despite the great opportunities that polymeric energy materials can provide in the development of healthcare technologies, specific difficulties must be overcome to achieve successful application. Compatibility with biological tissues, dimensional stability, techniques for large-scale production, energy costs of fabrication, and the potential to interface with biological structures are five challenges that are still the subject of investigation. Thus, prospective future research in the area of polymeric energy materials and their applications in healthcare should focus on the aforementioned challenges and try to develop new material formulations, fabrication methods, and regulatory approaches. This field has the potential to develop more inventions due to technological advancements in wearable devices, implantable medical devices, and drug delivery systems, among others, which will enhance the delivery of patient care.

CONCLUSION

Polymeric energy materials are steadily gaining importance in modern industrial applications that apply energy storage, energy converting systems, and sustainability. The use of nanomaterials in batteries, solar panels, flexible electronics, automobiles, and construction materials demonstrates that their applications are diverse. Despite these issues, current and ongoing improvements in performance, manufacturing, and long-term durability of these materials enable the industrialization of advanced materials for market applications. Polymeric energy materials represent a major driving force for further progress in sustainable energy technologies and are a critical enabler of future industrial ecosystems. Polymeric energy materials are now revolutionizing the healthcare industry as they aim at providing innovative technologies for energy-related issues, specifically in drug delivery systems, implantable devices, biosensors, and tissue engineering systems. These materials can be easily shaped, are biocompatible, and can store and transform energy, making them highly suitable for medical uses. As research continues in this area, polymeric solar cells and other polymeric energy materials have significant potential to enhance the performance, efficiency, and sustainability of healthcare technology, ultimately benefiting patients and providing cheaper solutions for the healthcare industry.

REFERENCES

[1]   Y. Chen, L. Zhang, and J. Liu, "The role of conductive polymer composites in energy storage devices," Journal of Energy Materials, vol. 34, no. 7, pp. 1345-1360", *Online (Bergh.),* 2023. [http://dx.doi.org/10.1016/j.jenmat.2023.03.012]

[2]   L. Gao, X. Li, and T. Zhang, *Advances in polymer electrolytes for lithium-ion batteries: Challenges*

and future perspectives, 2023.

[3] R. Barua, A. Sarkar, and S. Datta, Emerging advancement of 3D bioprinting technology in modern medical science and vascular tissue engineering education.

[4] S. Ghosh, and H. Lee, *Conductive polymers and their applications in energy storage: A review*, 2022.

[5] R. Barua, and S. Datta, "Revolutionizing nerve repair", *Advances in Medical Diagnosis, Treatment, and Care (AMDTC) Book Series*, pp. 275-300, 2024.
[http://dx.doi.org/10.4018/979-8-3693-3065-4.ch010]

[6] Y. EL-Ghoul, F. M. Alminderej, F. M. Alsubaie, R. Alrasheed, and N. H. Almousa, "Recent advances in functional polymer materials for energy, water, and biomedical applications: A review", *Polymers*, vol. 13, no. 24, p. 4327, 2021.

[7] S. Datta, R. Barua, R. Sarkar, A. Barui, A.R. Chowdhury, and P. Datta, "Design and development of alginate: Poly-l-lysine scaffolds by 3D bio printing and studying their mechanical, structural and cell viability properties", *IOP Conf. Series Mater. Sci. Eng.*, vol. 402, p. 012113, 2018.
[http://dx.doi.org/10.1088/1757-899x/402/1/012113]

[8] R. Barua, and S. Datta, The ongoing advancements in surgical robotics

[9] R. Barua, Advanced biomimetic compound continuum robot for minimally invasive surgical applications.*Modeling, Simulation, and Control of AI Robotics and Autonomous Systems*. IGI Global, 2024, pp. 213-231.

[10] S. Chowdhury, N. Jain, Y.O. Waidi, R. Barua, S. Das, A. Prasad, and S. Datta, Biological smart biomaterials: Materials for biomedical applications.*Applications of Biotribology in Biomedical Systems*. Springer Nature Switzerland: Cham, 2024, pp. 313-325.

[11] N. Jain, Y.O. Waidi, S. Chowdhury, R. Barua, S. Das, A. Prasad, and S. Datta, Medical devices tribology.*Applications of Biotribology in Biomedical Systems*. Springer Nature Switzerland: Cham, 2024, pp. 235-250.

[12] R. Barua, The emerging role of artificial intelligence in organ-on-a-chip (OOAC) biomedical devices.*Reshaping Healthcare with Cutting-Edge Biomedical Advancements*. IGI Global, 2024, pp. 369-381.

[13] Y. Wang, Y. Xu, and Z. Sun, "Challenges and strategies in the development of polymer-based energy storage materials," Journal of Power Sources, vol. 530, p. 230974", *Online (Bergh.)*, 2023.
[http://dx.doi.org/10.1016/j.jpowsour.2023.230974]

[14] F. Zhang, and D. Yang, *High-performance polymeric materials for energy conversion and storage: Progress and challenges*. vol. Vol. 39. Energy Storage Materials, 2022, pp. 1-19. [Online]

[15] R. Barua, An investigation of AI techniques for detecting kidney stones in CT scan images through advanced image processing.*Enhancing Medical Imaging with Emerging Technologies*. IGI Global, 2024, pp. 133-150.

[16] O. Leite-Barbosa, C.C.O. Pinto, J.M. Leite-da-Silva, E.M.M.M. de Aguiar, and V.F. Veiga-Junior, "Polymer composites reinforced with residues from Amazonian agro-extractivism and timber industries: A sustainable approach to enhancing material properties and promoting bioeconomy", *Polymers (Basel)*, vol. 16, no. 23, p. 3282, 2024.

[17] H. Chen, X. Xu, and W. Wang, "Flexible triboelectric nanogenerators based on conductive polymers for wearable electronics," Nano Energy, vol. 85, p. 106025", *Online (Bergh.)*, 2022.

[18] R. Barua, Exploring artificial intelligence in evolving healthcare environments: A comprehensive analysis.*Advances in Computational Intelligence for the Healthcare Industry 4.0*. IGI Global, 2024, pp. 123-138.

[19] Y. Chen, and J. Liu, "Organic photovoltaics based on polymeric materials: Challenges and opportunities," Renewable and Sustainable Energy Reviews, vol. 152, p. 111682", *Online (Bergh.)*,

2023.
[http://dx.doi.org/10.1016/j.rser.2021.111682]

[20] S. Kim, J. Lee, and J. Park, "Recent advancements in polymer blends for high-performance organic photovoltaics," Journal of Polymer Science, vol. 61, no. 8, pp. 1033-1045", *Online (Bergh.)*, 2023.
[http://dx.doi.org/10.1002/pol.20230123]

[21] H. Li, X. Zhang, and Z. Zhao, "Conductive polymer composites in lithium-ion batteries: A review of recent progress," Journal of Materials Science, vol. 57, no. 1, pp. 14-30", *Online (Bergh.)*, 2022.
[http://dx.doi.org/10.1007/s10853-021-06543-6]

[22] R. Barua, and S. Datta, Emerging applications of nanomaterials with less toxicity and more efficacy in modern biomedical engineering

[23] R. Barua, H. Giria, S. Datta, A.R. Chowdhury, and P. Datta, "Force modeling to develop a novel method for fabrication of hollow channels inside a gel structure", *Proc. Inst. Mech. Eng. H,* vol. 234, no. 2, pp. 223-231, 2019.
[http://dx.doi.org/10.1177/0954411919891654]

[24] S. Das, S. Datta, A. Barman, and R. Barua, Smart Biodegradable and Bio-Based polymeric biomaterials for biomedical applications

[25] L. Zhang, X. Zhu, and Y. Wang, "Nanostructured conductive polymers for efficient energy storage: A review", *Adv. Funct. Mater.,* vol. 33, no. 11, pp. 2234-2249, 2022.

[26] R. Barua, Unleashing drones for medical advancements

[27] R. Barua, and J. Mondal, Study of the current trends of CAD (Computer-Aided Detection) in modern medical Imaging

[28] Y. Zhang, S. Li, and H. Ma, "Polymer-based conductive materials for wearable bioelectronics", *Adv. Mater. Interfaces,* vol. 10, no. 5, p. 230235, 2023.

[29] R. Barua, and S. Datta, An extensive evaluation of new federated learning approaches for COVID-19 identification

[30] R. Barua, S. Das, and J. Mondal, Emerging applications of artificial intelligence (AI) and machine learning (ML) in modern urology

[31] H. Li, X. Xu, and J. Wang, "High-performance polymer nanocomposites for energy applications", *Nano Today,* vol. 48, p. 101923, 2023.

[32] R. Barua, and S. Datta, "A. RoyChowdhury, and P. Datta, "Study of the surgical needle and biological soft tissue interaction phenomenon during insertion process for medical application: A Survey", *Proc. Inst. Mech. Eng. H,* vol. 236, no. 10, pp. 1465-1477, 2022.
[http://dx.doi.org/10.1177/09544119221122024]

[33] J. Zhang, Y. Chen, and X. Li, "Conductive polymer-based nanomaterials for flexible electronics and energy harvesting", *Journal of Energy Chemistry,* vol. 75, pp. 34-55, 2023.

[34] R. Barua, S. Das, S. Datta, P. Datta, and A.R. Chowdhury, "Study and experimental investigation of insertion force modeling and tissue deformation phenomenon during surgical needle-soft tissue interaction", *Proc. Inst. Mech. Eng., C J. Mech. Eng. Sci.,* vol. 237, no. 5, pp. 1007-1014, 2022.
[http://dx.doi.org/10.1177/09544062221126628]

[35] X. Lin, Y. Wang, and Z. Zhou, "Recent progress in polymeric nanomaterials for advanced energy storage systems", *Materials Today Advances,* vol. 17, p. 100265, 2022.

[36] S. Datta, and R. Barua, "Fluorescent nanomaterials and its application in biomedical engineering", *Adv. Digit. Crime Forensics Cyber Terrorism Book Ser.,* pp. 164-186, 2023.

[37] R. Barua, D. Das, and N. Biswas, "Revolutionizing drug discovery with artificial intelligence", *Adv. Med. Technol. Clin. Pract. Book Ser.,* pp. 62-85, 2024.
[http://dx.doi.org/10.4018/979-8-3693-2238-3.ch004]

[38] R. Barua, "The emerging potential of 21st century bio-inspired swarm robotics in modern medical surgery", *Adv. Comput. Intell. Robot. Book Ser.,* pp. 28-45, 2024.
[http://dx.doi.org/10.4018/979-8-3693-1277-3.ch003]

[39] S. Datta, N. Jain, Y. O. Waidi, and R. Barua, "Processing and applications of shape memory alloys", *Adv. Chem. Mater. Eng. Book Ser.,* pp. 151-165, 2024.
[http://dx.doi.org/10.4018/978-1-6684-9385-4.ch006]

[40] Y. O. Waidi, R. Barua, and S. Datta, "Metals, polymers, ceramics, composites biomaterials used in additive manufacturing for biomedical applications", *Adv. Chem. Mater. Eng. Book Ser.,* pp. 165-184, 2023.
[http://dx.doi.org/10.4018/978-1-6684-9224-6.ch008]

[41] R. Barua, D. Banerjee, and S. Bhowmik, "Study of the potential impact of microplastics and additives on human health", *Adv. Hum. Serv. Public Health Book Ser.,* pp. 128-147, 2022.
[http://dx.doi.org/10.4018/978-1-7998-9723-1.ch007]

[42] S. K. Verma, A. Prasad, and V. Katiyar, "State of art review on sustainable biodegradable polymers with a market overview for sustainability packaging", *Mater. Today Sustain.,* p. 100776, 2024.

[43] R. Barua, "An in-depth exploration of AI and humanoid robotics' role in contemporary healthcare", *Adv. Med. Technol. Clin. Pract.,* pp. 42-61, 2024.
[http://dx.doi.org/10.4018/979-8-3693-2238-3.ch003]

[44] G. Chakraborty, V. Pandey, A. Prasad, and A. Kumar, Introduction to Sustainable Manufacturing for Industries 4.0.*Sustainable Smart Manufacturing Processes in Industry 4.0.* CRC Press, 2023, pp. 1-17.

[45] A. Prasad, S.M. Bhasney, V. Prasannavenkadesan, M.R. Sankar, and V. Katiyar, "Polylactic acid reinforced with nano-hydroxyapatite bioabsorbable cortical screws for bone fracture treatment", *J. Polym. Res.,* vol. 30, no. 5, p. 177, 2023.

[46] A. Prasad, S.M. Bhasney, V. Prasannavenkadesan, M.R. Sankar, and V. Katiyar, "Nano☐hydroxyapatite reinforced polylactic acid bioabsorbable cancellous screws for bone fracture fixations", *J. Appl. Polym. Sci.,* vol. 140, no. 43, p. e54577, 2023.

[47] G. Chakraborty, A. Prasad, and A. Kumar, *Processing of biodegradable composites. In Biodegradable Composites for Packaging Applications.* CRC Press, 2022, pp. 33-48.

[48] A. Prasad, A. Kumar, K.K. Gajrani, Ed., *Biodegradable Composites for Packaging Applications.* CRC Press, 2022.

[49] A. Prasad, S. Datta, S. De, P. Singh, and B. Mahto, Bioresorbable Composite for Orthopedics and Drug Delivery Applications.*Applications of Biotribology in Biomedical Systems.* Springer Nature Switzerland: Cham, 2024, pp. 327-344.

[50] A. Banerjee, and R. Barua, A comprehensive overview of Carbon-Based nanofluids and related progress for heat transfer uses

[51] T.R. Nayak, A. Singh, and S. Kumar, "Biodegradable and bioactive polymeric materials for healthcare applications", *Biomacromolecules,* vol. 24, no. 1, pp. 1-14, 2023.

[52] X. Xie, T. Zhou, and Z. Zhang, "Toxicological considerations for the application of polymer-based nanocomposites in biomedical devices", *J. Nanobiotechnology,* vol. 19, no. 1, pp. 1-13, 2021.

[53] Z. Zhao, J. Zhang, and Y. Wang, "Enhancing the integration of polymeric materials with biological systems: Current challenges and future directions", *Biomater. Sci.,* vol. 9, no. 5, pp. 1267-1281, 2021.

CHAPTER 11

Enhanced Energy Harvester Performance by a Tension-Annealed Carbon Nanotube Yarn at Extreme Temperatures

Vikas Kashyap[1,*], Shivanshu Sharma[1], Chandra Kumar[2], Anand Kumar[3] and Kapil Saxena[4]

[1] *Department of Physics, Panjab University, Chandigarh 16001, Punjab, India*

[2] *Escuela de Ingeniería, Facultad de Ciencias, Ingeniería y Tecnología, Universidad Mayor, Santiago 7500994, Chile*

[3] *Department of Applied Sciences and Humanities, Invertis University, Bareilly, 243001, Uttar Pradesh, India*

[4] *Department of Applied Sciences, Kamla Nehru Institute of Technology, Sultanpur 228118, Uttar Pradesh, India*

Abstract: Carbon Nanotubes (CNTs) are nanoscale cylindrical structures composed of carbon atoms with diverse applications in supercapacitors, artificial muscles, and intelligent textiles. An Incandescent Tension Annealing Process (ITAP) can significantly enhance their mechanical and electrochemical properties, improving energy harvesting capabilities. The experimental observations reveal that during the 1 Hz sinusoidal stretching cycle, the peak-to-peak Open Circuit Voltage (OCV) and Short Circuit Current (SCC) generated by ITAP yarn were more than 1.5 times that of pristine CNT. The densified surface of ITAP yarn results in about a 20% reduction in capacitance when it is stretched to about 30% strain. A noticeable negative shift in the values of the potential of ITAP yarn suggests greater charge injection when immersed in the electrolyte. Thus, the factors—namely, the pronounced changes in capacitance and the increased initial charge injection—are key contributors to the enhanced energy harvesting performance of ITAP yarn compared to untreated CNTs.

Keywords: Carbon nanotubes, Incandescent tension annealing process, Open circuit voltage, Short circuit current.

INTRODUCTION

CNTs have good electrical and thermal conductance with exceptional mechanical properties. Its nanoscale-based properties make it an ideal candidate for manufact-

---

[*] **Corresponding author Vikas Kashyap**: Department of Physics, Panjab University, Chandigarh, 160014, India; E-mail: vikaskashyap78647@gmail.com

Sushil Kumar Verma, Sonika & Arbind Prasad (Eds.)
All rights reserved-© 2026 Bentham Science Publishers

uring lightweight and flexible fibers, potentially overcoming the issues of bulkiness, heaviness, and low-density power output associated with conventional electromagnetic devices [1]. CNT yarns are currently produced in different continuous processes using a variety of important technologies, such as liquid-state and dry-state techniques [2 - 5]. Yarns spun from spinnable nanotube forests have excellent mechanical strength, low impurities, a strong nanotube line-up, and novel structural versatility. Twisting these yarns increases their strength, making them behave as high-strain conductors and high-performance artificial muscles for torsion and tension [6].

The ITAP process involves heating a vertically suspended nanotube yarn from a two-point suspension through a current of about 20,000 A cm^{-2} in a vacuum. This treatment solves several problems existing in single-coil CNTs due to the weak interfacial connection between adjacent nanotubes present in yarns [7].

1. Except under tension, coiled and twisted strings tangle.
2. To avoid irreversible untwisting, CNT yarns should be torsionally tethered as actuatable muscles. Some non-actuating parts are needed for torquing back; these non-actuating parts give spring action.
3. The strength of twisted nanotube yarns is still some orders of magnitude below that of the individual nanotubes.

Hence, many practical applications of CNT yarns are opened up by ITAP [8 - 10]. Therefore, it becomes essential to introduce more mechanical bonding inside the structure of the yarn because it offers both twist retention and mechanical strength [11, 12]. We are focused on precisely controlling the parameters of the Incandescent Tension Annealing Process (ITAP) itself to ensure reproducible performance for yarns treated under specific conditions. We describe the controlled conditions applied during the process, which are fundamental to achieving consistent results. The yarns were heated to a specific, extreme temperature of about 2000 °C. This was typically achieved by applying an electrical current (electrothermally). Temperature measurement methods were used, including spectroscopic measurements calibrated by resistance for smaller yarns, highlighting efforts to accurately control this parameter. The duration of the high-temperature treatment was controlled. Major property improvements were observed within seconds, and a standard annealing time of 2 minutes was typically employed for optimal results, as longer times could degrade properties. However, in this chapter, we have described the experimental methods and characterization techniques used for the specific samples studied; it does not detail quality control measures or data analysis methods used to verify consistency, which have yet to be explored.

An understanding of the working mechanisms for converting mechanical energy into electrical energy becomes a prerequisite for improved performance of CNT-based twist-based harvesters [13 - 16]. CNTs act as electrodes upon immersion in an electrolyte and cause the ionic species to adsorb onto their accessible surface area naturally through self-charge injection, driven by the chemical potential difference between the carbon atoms of the polymer and the ionic species within the electrolytes [3]. Mechanical deformation, such as twisting or coiling, decreases the available surface area, causing self-charge redistribution and alteration of capacitance [17 - 20]. According to the relation:

$$\Delta V = \frac{Q}{\Delta C}$$

where ΔV is the change in potential, Q is the total injected charge, and ΔC is the change in capacitance. This means that the variation of capacitance directly impacts the potential with fixed Q [21]. Strategies to modify the electrochemical surface area, such as infiltration with polymers or irradiating carbon double-walled nanotubes, often compromise its conductivity and restrict power output. High-temperature annealing corrects structural flaws to enhance conductivity and strength, though its effect on electrochemical capacitance has not yet been ascertained [22 - 25].

The ITAP significantly enhances the mechanical properties. Neat coiled yarns tend to untwist, whereas ITAP-treated yarns remain stable and straight and show minimal untwisting, demonstrating enhanced structural integrity. The process is described as solving issues of weak interfacial connection and improving twist retention as well as mechanical strength. The ITAP was developed to stabilize twisted and coiled CNT yarns, preventing irreversible untwisting, which was a problem for pristine yarns used in applications like artificial muscles. ITAP-produced inter-nanotube connections act as internal springs, allowing for fast, reversible, torsional, and tensile actuation without needing external springs or torsional tethering. The long-term stability of ITAP-treated yarns under repeated mechanical or electrochemical stress is discussed in terms of their observed reversibility and performance during cyclic testing. The ITAP-treated yarns exhibit significant stability under repeated mechanical and electrochemical stress. Reversible torsional and tensile actuation is achieved with guest-free ITAP yarns in response to vapor absorption and desorption cycles. Unlike pristine yarns, which untwist irreversibly, the ITAP-produced inter-nanotube connections act as internal springs, enabling these yarns to retwist after untwisting, thereby solving the problem of unwanted untwisting for torsional actuators. For example, when subjected to a load causing untwisting, an ITAP-30 yarn is fully retwisted upon load removal, whereas a pristine yarn is only partially retwisted. Highly reversible torsional actuation is observed, with resonant operation enhancing stroke and speed during cyclic actuation tests driven by vapor absorption/desorption. In the context of energy harvesting, ITAP yarns demonstrated robust functionality during sinusoidal stretching cycles used for characterization across a frequency range of up to 25 Hz [26]. This enhanced performance under repeated mechanical stress is attributed to large changes in capacitance and increased initial charge injection resulting from the ITAP treatment. These results highlight the capability of ITAP yarns for repeated, reversible action and robust functionality under cyclic stress within the testing parameters.

There is a significant difference between ITAP and traditional annealing. The significant tensile stress is applied during the high-temperature annealing process, which is how ITAP differs from a conventional annealing approach. Thermally annealing twisted CNT yarns, usually in a vacuum at high temperatures (about 2000 K or 2000 °C) for several hours, is known as traditional annealing (without substantial stress). Importantly, this procedure is executed without a great deal of applied stress, in contrast to ITAP, which entails applying large tensile loads and heating twisted CNT yarns to extremely high temperatures (roughly 2000 °C), frequently by applying an electrical current (electrothermally). Up to 40% of the precursor yarns' room-temperature tensile strength may be applied as tensile stress. The main property changes are noticeable in a matter of seconds, and for best results, an annealing period of two minutes is usually used; longer intervals may cause properties to deteriorate. It has been demonstrated that traditional annealing, which does not involve a lot of stress, increases Young's modulus while somewhat reducing yarn strength. For a particular yarn type, annealing at 2000 °C without stress results in a 27% rise in modulus and a 10% drop in strength. The mechanical characteristics in the ITAP (under high stress) show notable improvements. It leads to significant increases in modulus and yarn strength. Yarn strength and modulus were enhanced by ITAP by up to 2.6 and 12 times, respectively. Since tension helps align nanotubes and bring them closer together for stronger connections, the combination of stress and high temperature is crucial for enhancing mechanical characteristics. The temperature is iteratively changed using the ITAP, a contemporary adaptive annealing technique, in response to convergence or performance behavior. ITAP's primary objective is to effectively adjust the "temperature" in order to maximize convergence speed and quality. ITAP has a high degree of reactivity and can adjust temperature in response to gradient changes or solution quality. Because of its flexibility, the ITAP procedure is frequently quicker. Because of adaptive logic, this method's implementation is moderately to very difficult. Inspired by physical annealing procedures, traditional annealing is a traditional method that follows a set or predetermined cooling schedule. The primary objective of traditional annealing is to use gradual cooling to obtain global optima and escape local minima. Regardless of the system condition, traditional annealing has low reactivity and declines rigidly in accordance with a schedule. Conventional annealing is frequently slower, particularly when conservative schedules are used to guarantee convergence. The implementation of this method is straightforward and easily understood. The ITAP was created to prevent unintended, irreversible untwisting of CNT yarns that are coiled or twisted. The application of significant tensile stress during the ITAP is essential for achieving simultaneous increases in strength and modulus, providing unprecedented structural stability (preventing untwisting and snarling), and facilitating the formation of internal springs through inter-nanotube cross-links. In summary, both ITAP and traditional high-temperature annealing (without stress) involve heat treatment of CNT yarns. Compared to pristine or simply annealed yarns, ITAP-treated yarns are far more resilient and appropriate for applications needing stable, reversible mechanical motion and improved environmental resistance due to these advantages [27].

In this study, the ITAP technique assisted in improving the energy-harvesting performance of CNT yarns. The electrochemical test confirmed the enhancement of open-circuit voltage, short-circuit current, and peak power density, attributed to enhanced capacitance changes and the initial level of injected charge. Pre-coiling thermal annealing also improved stretchability, which enhanced the energy output and expanded the applications for flexible devices. Here, we addressed the structural integrity and properties of the Carbon Nanotubes (CNTs) after the Incandescent Tension Annealing Process (ITAP) performed at temperatures around 2000 °C. We have used several characterisation techniques to investigate the changes occurring in the yarns and the nanotubes themselves. The ITAP was indeed performed at a high temperature, approximately 2000°C [28, 29].

Experimental Setups

1. Making a neat CNT and an ITAP-treated CNT yarn:

By using floating-catalyst chemical vapor deposition, spinnable CNT forests were generated as multi-walled carbon nanotubes with diameters varying between 8.2 nm and 9.8 nm. Through spinning, the stacked MWCNT sheets 40 mm wide were mechanically extracted from those spinnable CNT forests. Then, the sheets were twisted using a cone-spun technique under mechanical stress ranging from 7 to 12 MPa. A fully-coiled CNT yarn was created by over-twisting the same chirality of CNT yarn (Fig. **1**).

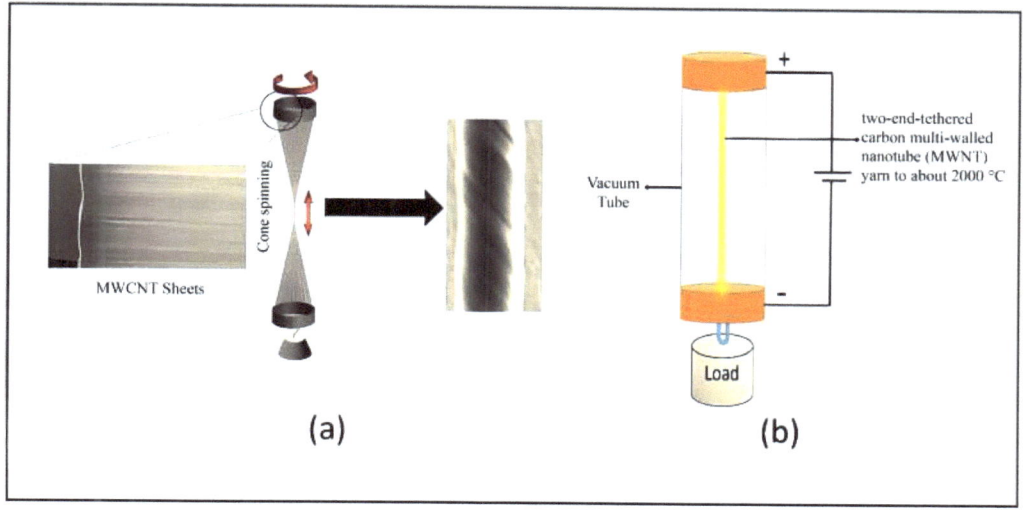

Fig. (1). (a) A coiled neat CNT yarn was made by inserting an extreme twist into forest-drawn MWCNT sheets until it forms a fully coiled structure [5]. (b) Configuration for MWNT yarn ITAP application.

The incandescent tension annealing process (Fig. **2**) was carried out in a vertical cylindrical glass tube. The air pressure in the tube was increased to 5.5×10^{-6} bar. A steel hook was used to hang fully-twisted CNT yarn within a glass tube. The yarn was electro-thermally heated to approximately 2000°C for 2 minutes using 50 Hz, 63 W mg^{-1} alternating electrical power [30, 31].

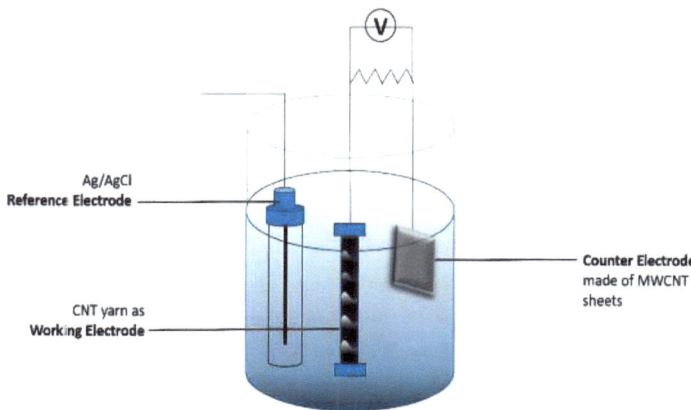

Fig. (2). Electrochemical Cell.

During high-temperature annealing, mechanical tensile stress was applied to the CNT yarn by hanging various weights through the bottom steel hook. The applied tensile stress during the ITAP ranged from 0 to 30 MPa.

Methods of characterization

A field emission scanning electron microscope (model: JSM-7800F) was used to characterize the microstructures of an ITAP-treated CNT yarn and a neat CNT yarn [19, 20]. A commercial step motor was used for the application of sinusoidal wave deformation in the CNT twistron harvester. Gamry Reference 3000 Electrochemical Analyzer (USA) was used for electrochemical characterization. X-ray diffraction studies were carried out using a Siemens D5000 diffractometer with Cu-Ka radiation (λ=1.5418 Å).

Setup for Electrochemical Analysis

An electrochemical cell with three electrodes was used (Fig. 2). An Ag/AgCl reference electrode, a high-surface-area counter electrode (a Pt mesh with layers of MWCNT sheets), and a coiled CNT yarn working electrode were all included in the cell. For the cell capacitance to be primarily governed by the twist-on electrode's capacitance, the MWCNT sheets employed in the counter electrode were used to ensure the counter electrode had a sufficiently high capacitance in comparison to the twistron electrode [23]. In this process, a 0.1 M HCl aqueous electrolyte was utilized.

For electrochemical analysis, a three-electrode electrochemical cell wasused (see Figure 2). The peak-to-peak Open Circuit Voltage (OCV) and Short Circuit Current (SCC) were monitored while carrying out a 1Hz sinusoidal stretch.

Morphological Analysis

For comparison, both neat CNT and ITAP yarns are studied. Electron microscope images reveal that before ITAP treatment, coiled pristine nanotube yarns either untwist into loose spring structures under load or snarl without tethering [5]. In contrast, ITAP-treated yarns remained stable and straight and showed minimal untwisting even after tethering was released (Fig. 3a). This stabilization effect was consistent for both twisted and coiled yarns, demonstrating that ITAP enhances the structural integrity of CNT yarns.

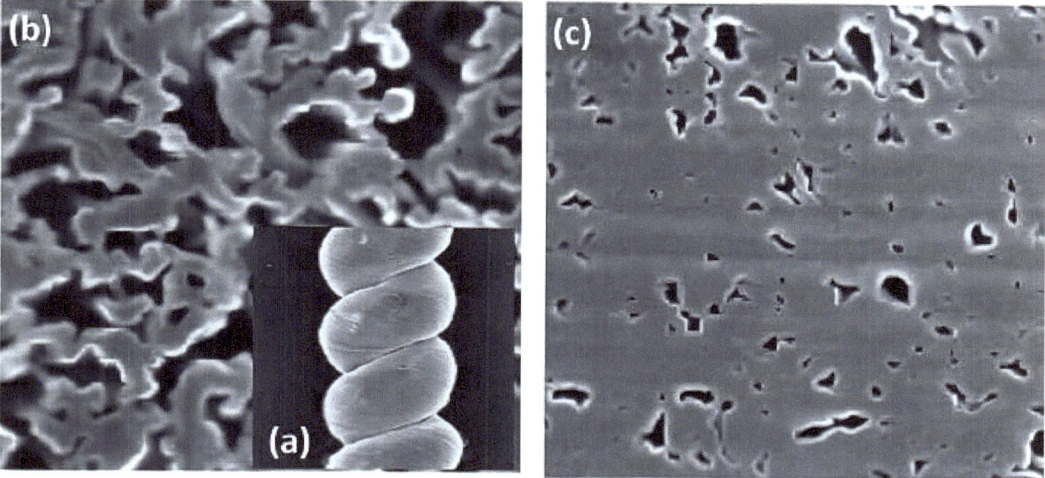

Fig. (3). (a) The ITAP stabilizes the coiled yarn by preventing untwisting (b) and (c) Top view of SEM images (after ITAP), showing the density increase caused by the ITAP [5].

Secondly, ITAP reduces the yarn bias angle (α) and diameter (d). Finally, cross-sectional SEM images show a twofold reduction in yarn porosity, leading to an increase in yarn density after ITAP treatment (Fig. **3b** and **3c**). Enhanced nanotube alignment and increased bundle size are also observed.

The open circuit voltage was measured as a peak-to-peak value relative to the Ag/AgCl reference electrode. The SCC was obtained using a two-electrode system without using the reference electrode. Results revealed that the ITAP-treated CNT yarn generated a peak-to-peak OCV that was about 1.5 times as high as that generated by the neat CNT yarn.

Experimental Results

The strain dependence of the peak-to-peak Open-Circuit Voltage (OCV) (Fig. **4a**) generated by ITAP-treated CNT yarns exhibits a linear relationship with the applied strain, highlighting their potential as effective strain sensors. This improvement can be attributed to the increased electrical conductivity imparted by the ITAP process, enabling ITAP yarns to generate higher Short-Circuit Currents (SCC) compared to pristine CNT yarns across all strain levels (Fig. **4b**).

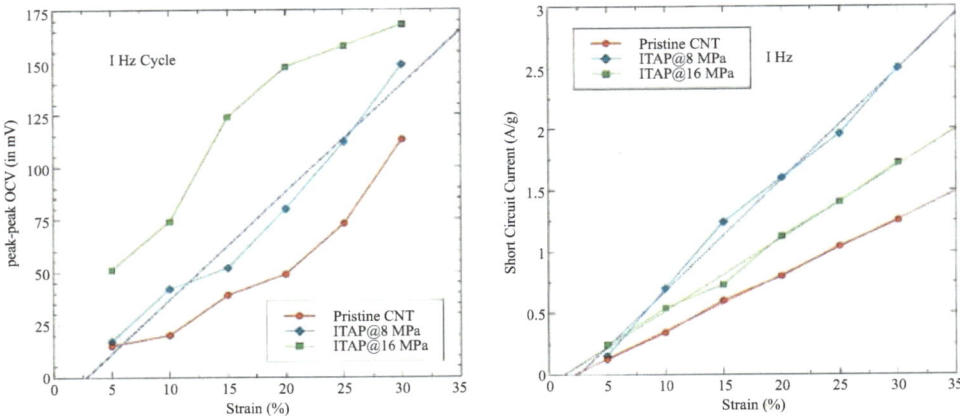

Fig. (4). (a) The peak-to-peak open-circuit voltage and (b) SCC as a function of the applied strain for the pure CNT yarn and the ITAP-treated CNT yarn subjected to mechanical loads of 8 MPa and 16 MPa, respectively.

Moreover, the ITAP process addresses the issue of twist retention in coiled CNT yarns, which typically untwist or snarl when torsional tethering is removed. Thermal annealing under mechanical loads stabilizes the inserted twist for a short period, about 10 seconds. ITAP-treated CNT yarns retain about the same density as the initial twist density, as opposed to more than 35% loss in pristine CNT yarns. This improvement in twist stability broadens the applicability of ITAP-treated yarns in stretchable electrical conductors and self-powered sensing technologies.

Capacitance decreaseslinearly with increasing load stress during the ITAP process in an ITAP yarn (Fig. **5a**), with a bias angle of 22 degrees and annealing for 90 seconds. This behavior is due to structural changes motivated by the applied stress, whichleads to better alignment of the nanotubes and improvement in crystallinity. In turn, these changes result in a lower accessible surface area for ion adsorption.

An increase in load stress compresses the yarn, which in turn decreases porosity and void spaces between nanotube bundles, making further interactions with the electrochemical surface less favorable. All these changes then restrain the movement and availability of ions within the space owing to an overall decrease in double-layer capacitance.

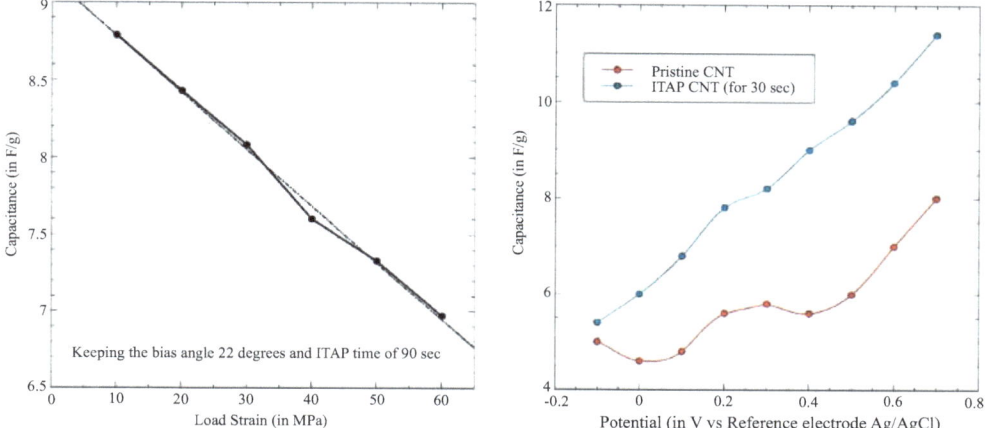

Fig. (5). (**a**) Dependence of capacitance on the stress applied during ITAP to a twisted CNT yarn. (**b**) The capacitance variation with potential applied to a pristine CNT and an ITAP-treated CNT yarn.

In a way, the observed decrease in capacitance reflects the interaction between mechanical stress and the structural densification of the yarn during ITAP, as explained herein. Fig. (**5b**) demonstrates that the capacitance measured by Electrochemical Impedance Spectroscopy (EIS) varies with the applied potential for different coiled twistron yarns. For the pristine CNT yarn, the capacitance reaches a minimum near 0 V, indicating the Potential of Zero Charge (PZC). In contrast, the ITAP-treated CNT yarn shows a monotonic increase in capacitance from 5.40 F/g (Farad per gram) at -0.1 V to 11.47 F/g at 0.7 V, with the minimum capacitance shifting to approximately more than 0.1 V. This shift in the PZC reflects the impact of ITAP treatment on the yarn's electrochemical properties. Additionally, the relationship between output power and resistance, governed by

$$P = \frac{(\Delta V)^2}{R}$$

means that reduced yarn resistance enhances output power for a given voltage change (Fig. **6a**). To further optimize performance, a platinum wire can be wrapped around the ITAP yarn, significantly decreasing internal resistance. For high-frequency applications, the performance of the harvester (Fig. **6b**) is characterized across a frequency range of up to 25 Hz.

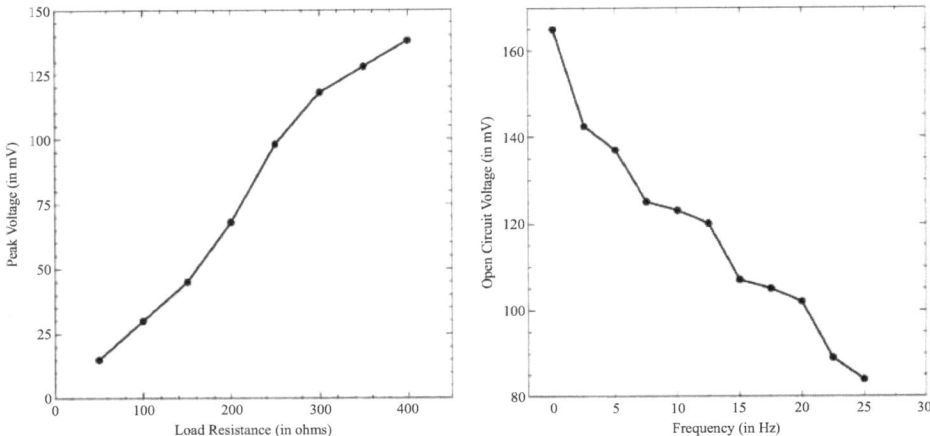

Fig. (6). **(a)** Peak voltage dependence on load resistance and **(b)** frequency dependence of OWC (20% strain is applied on ITAP yarn during a 1 Hz cycle).

The ITAP-treated CNT yarn demonstrated robust functionality, though the peak-to-peak Open-Circuit Voltage (OCV) decreased by 49% as the frequency increased from 1 Hz to 25 Hz. This decline was attributed to the inability of ion injection to keep pace with rapid mechanical deformation.

CONCLUSION

This study shows that the Incandescent Tension Annealing method (ITAP) is a changing approach capable of significantly enhancing both the mechanical and electrochemical properties of CNTs, leading to a marked increase in energy harvesting efficiency.

The coiled ITAP-treated CNT yarn attains more than a two-fold increase in peak electrical power density and about a 1.5-fold increase in output voltage compared to that of the untreated CNT yarn. The electrochemical analysis attributes these enhancements to significant changes in capacitance resulting from increased initial charge injection. CNT yarns can function as high-strain conductors and artificial muscles when infiltrated with electrolyte. Long-term stability is observed under repeated electrochemical stress, although it pertains to a different type of CNT yarn muscle. Electrochemical muscles utilize an ionic-liquid-in-nanofiber sheath. These yarns are characterized as highly robust, capable of reversibly contracting stably under various conditions, including long-term cycling and storage in air. They exhibit very stable contractile strokes during 1000 working cycles when driven electrochemically. This stability is credited to the ionic-liquid-in-nanofiber sheath electrolyte system. Therefore, ITAP yarns exhibit enhanced chemical stability and stability/reversibility under repeated mechanical stress. These findings point toward the ITAP-treated CNT yarns as advanced materials for next-generation energy systems.

REFERENCES

[1] J. Terrones, A.H. Windle, and J.A. Elliott, "The electro-structural behaviour of yarn-like carbon nanotube fibres immersed in organic liquids", *Sci. Technol. Adv. Mater.,* vol. 15, no. 5, 2014.
[http://dx.doi.org/10.1088/1468-6996/15/5/055008]

[2] C.D. Tran, W. Humphries, S.M. Smith, C. Huynh, and S. Lucas, "Improving the tensile strength of carbon nanotube spun yarns using a modified spinning process", *Carbon N. Y.,* vol. 47, no. 11, pp. 2662-2670, 2009.
[http://dx.doi.org/10.1016/j.carbon.2009.05.020]

[3] Y.C. Zhang, X. Chen, and X. Wang, "Effects of temperature on mechanical properties of multi-walled carbon nanotubes", *Compos. Sci. Technol.,* vol. 68, no. 2, pp. 572-581, 2008.
[http://dx.doi.org/10.1016/j.compscitech.2007.03.012]

[4] Y. Li, "Recent progress in self-powered wireless sensors and systems based on TENG", *Sensors (Basel),* vol. 23, no. 3, 2023.
[http://dx.doi.org/10.3390/s23031329]

[5] X. Hu, "Enhanced energy harvester performance by a tension annealed carbon nanotube yarn at extreme temperatures", *Nanoscale,* vol. 13, pp. 16185-16192, 2022.
[http://dx.doi.org/10.1039/d2nr05303a]

[6] J. Charoenpakdee, A. Hutem, and S. Boonchui, "Curvature-induced electrical properties of two-dimensional electrons on carbon nanotube springs", *Symmetry (Basel),* vol. 17, no. 3, pp. 1-18, 2025.
[http://dx.doi.org/10.3390/sym17030316]

[7] X. Hu, "Erratum: Enhanced energy harvester performance by a tension annealed carbon nanotube yarn at extreme temperatures", *Nanoscale,* vol. 14, no. 46, p. 17466, 2022.
[http://dx.doi.org/10.1039/d2nr90224a]

[8] J. Si, R. Duan, M. Zhang, and X. Liu, "Recent progress regarding materials and structures of triboelectric nanogenerators for AR and VR", *Nanomaterials (Basel),* vol. 12, no. 8, 2022.
[http://dx.doi.org/10.3390/nano12081385]

[9] L. Yang, Z. Ma, Y. Tian, B. Meng, and Z. Peng, "Progress on self-powered wearable and implantable systems driven by nanogenerators", *Micromachines (Basel),* vol. 12, no. 6, 2021.
[http://dx.doi.org/10.3390/mi12060666]

[10] H. Zhao, "Underwater wireless communication via TENG-generated Maxwell's displacement current", *Nat. Commun.,* vol. 13, no. 1, pp. 1-10, 2022.
[http://dx.doi.org/10.1038/s41467-022-31042-8]

[11] L. Li, Y. Yao, Z. Lin, Y. Liu, and C.P. Wong, "Low-cost micrometer-scale Silicon Vias (SVs) fabrication by Metal-assisted Chemical Etching (MaCE) and carbon nanotubes (CNTs) filling", *Electron. Components Technol. Conf.,* 2013pp. 581-585

[12] S. Oh, H.J. Kim, S. Lee, K.J. Kim, and S.H. Kim, "Carbon nanotube sheets/elastomer bilayer harvesting electrode with biaxially generated electrical energy", *Polymers (Basel),* vol. 16, no. 17, pp. 1-9, 2024.
[http://dx.doi.org/10.3390/polym16172477]

[13] Y. Zhu, H. Yue, M.J. Aslam, Y. Bai, Z. Zhu, and F. Wei, "Controllable preparation and strengthening strategies towards high-strength carbon nanotube fibers", *Nanomaterials (Basel),* vol. 12, no. 19, pp. 1-24, 2022.

[http://dx.doi.org/10.3390/nano12193478]

[14] U. Vainio, T.I.W. Schnoor, S. Koyiloth Vayalil, K. Schulte, M. Müller, and E.T. Lilleodden, "Orientation distribution of vertically aligned multiwalled carbon nanotubes", *J. Phys. Chem. C,* vol. 118, no. 18, pp. 9507-9513, 2014.
[http://dx.doi.org/10.1021/jp501060s]

[15] X. Han, "A stretchable, self-healable triboelectric nanogenerator as electronic skin for energy harvesting and tactile sensing", *Materials (Basel),* vol. 14, no. 7, pp. 1-11, 2021.
[http://dx.doi.org/10.3390/ma14071689]

[16] D. Qian, W.K. Liu, and R.S. Ruoff, "Load transfer mechanism in carbon nanotube ropes", *Compos. Sci. Technol.,* vol. 63, no. 11, pp. 1561-1569, 2003.
[http://dx.doi.org/10.1016/S0266-3538(03)00064-2]

[17] N. Matsumoto, A. Oshima, M. Yumura, D.N. Futaba, and K. Hata, "Current treatment of bulk single walled carbon nanotubes to heal defects without structural change for increased electrical and thermal conductivities", *Nanoscale,* vol. 7, no. 19, pp. 8707-8714, 2015.
[http://dx.doi.org/10.1039/c5nr00170f]

[18] B. Ni, *Artificial muscle based on coiled CNT yarns and biofriendly ionogels*, 2024.

[19] N. Sheng, Y. Peng, F. Sun, and J. Hu, "High-performance fasciated yarn artificial muscles prepared by hierarchical structuring and sheath–core coupling for versatile textile actuators", *Adv. Fiber Mater.,* vol. 5, no. 4, pp. 1534-1547, 2023.
[http://dx.doi.org/10.1007/s42765-023-00301-8]

[20] Wang, Z. (2021). Multifunctional Carbon Nanotube Yarns for Artificial Muscles and Energy Harvesters. The University of Texas at Dallas.

[21] A. Mayeen, L.K. Shaji, A.K. Nair, and N. Kalarikkal, *Morphological characterization of nanomaterials.* Elsevier Ltd., 2018.

[22] Y. Suzuki, D. Miki, M. Edamoto, and M. Honzumi, "A MEMS electret generator with electrostatic levitation for vibration-driven energy-harvesting applications", *J. Micromech. Microeng.,* vol. 20, no. 10, 2010.
[http://dx.doi.org/10.1088/0960-1317/20/10/104002]

[23] Z. Wang, "More powerful twistron carbon nanotube yarn mechanical energy harvesters", *Adv. Mater.,* vol. 34, no. 27, 2022.
[http://dx.doi.org/10.1002/adma.202201826]

[24] P. Poulin, B. Vigolo, and P. Launois, "Films and fibers of oriented single wall nanotubes", *Carbon N. Y.,* vol. 40, no. 10, pp. 1741-1749, 2002.
[http://dx.doi.org/10.1016/S0008-6223(02)00042-8]

[25] C. Wei, and X. Jing, *A comprehensive review on vibration energy harvesting: Modelling and realization,* 2017.

[26] M. Ren, "Strong and robust electrochemical artificial muscles by ionic-liquid-in-nanofiber-sheathed carbon nanotube yarns", *Small,* vol. 17, no. 5, 2021.
[http://dx.doi.org/10.1002/smll.202006181]

[27] J. Di, "Strong, Twist-Stable Carbon Nanotube Yarns and Muscles by Tension Annealing at Extreme Temperatures", *Adv. Mater.,* vol. 28, no. 31, pp. 6598-6605, 2016.
[http://dx.doi.org/10.1002/adma.201600628]

[28] R. Article, *Recent Advances in Twisted and Coiled Artificial Muscles and Their Applications.,* 2025, pp. 1-20.

[29] F. Invernizzi, S. Dulio, M. Patrini, G. Guizzetti, and P. Mustarelli, "Energy harvesting from human motion: Materials and techniques", *Chem. Soc. Rev.,* vol. 45, no. 20, pp. 5455-5473, 2016.
[http://dx.doi.org/10.1039/c5cs00812c]

[30]   X. Hu, "Harvesting continuous rotational mechanical energy using coiled sheath-core carbon nanotube yarn", *Carbon N. Y.*, vol. 229, no. August, p. 119541, 2024.
[http://dx.doi.org/10.1016/j.carbon.2024.119541]

[31]   A. Muzammil, "Emerging transition metal and carbon nanomaterial hybrids as electrocatalysts for water splitting: a brief review", *Mater. Horiz.*, vol. 10, no. 8, pp. 2764-2799, 2023.
[http://dx.doi.org/10.1039/d3mh00335c]

CHAPTER 12

Energy Harvesting: Innovating Advanced Technologies Through Polymer Nanocomposites

Akash Ranjan[1,*], **Jimli Sarma**[2] and **Sonika**[3]

[1] *Faculty of Education, Banaras Hindu University, Kamachha, Varanasi, Uttar Pradesh, India*

[2] *Centre for Multidisciplinary Research, Tezpur University, Tezpur 784028, Assam, India*

[3] *Department of Physics, Rajiv Gandhi University, Rono Hills, Doimukh, Papumpare 791112, Arunachal Pradesh, India*

Abstract: The demand for renewable and sustainable energy sources has made energy harvesting a crucial field of study. This study examines cutting-edge energy harvesting technologies with an emphasis on the application of polymer nanocomposites. Combining polymers with nanoscale components to create polymer nanocomposites offers special benefits such as improved thermal stability, electrical conductivity, and mechanical qualities. Because of these qualities, they are good candidates fordeveloping energy harvesting devices that effectively transform ambient energy sources—such as mechanical, thermal, and solar energy—into electrical energy that can be used. The review examines many processes for creating polymer nanocomposites and how they are used in energy harvesting devices. The usefulness of methods like electrospinning, solution casting, and melt blending in improving the efficiency of energy harvesting devices is highlighted. The electrical characteristics of polymer matrices are greatly enhanced by the use of nanoparticles, such as metal oxides, graphene, and carbon nanotubes, which raises energy conversion efficiency. The study also looks at certain energy harvesting uses for polymer nanocomposites, such as solar cells, thermoelectric devices, and piezoelectric generators. These materials are positioned as competitive substitutes for conventional energy harvesting technologies due to their ability to provide flexible, lightweight, and long-lasting energy harvesting devices. It is observed from the review study that by demonstrating the revolutionary potential of polymer nanocomposites in energy harvesting applications, the ongoing developments in materials science and nanotechnology open the door to the creation of novel solutions that can make a substantial contribution to the worldwide effort to achieve sustainable energy, tackling the problems brought on by environmental concerns and energy demand. Future research should concentrate on improving these technologies' scalability and performance to allow for broad adoption across a range of industries.

---

[*] **Corresponding author Akash Ranjan:** Faculty of Education, Banaras Hindu University, Kamachha, Varanasi, Uttar Pradesh, India; E-mail: 1aranjanmedmphil@gmail.com

Sushil Kumar Verma, Sonika & Arbind Prasad (Eds.)
All rights reserved-© 2026 Bentham Science Publishers

Keywords: Energy harvesting devices, Nanocomposites, Polymer, Sustainable energy.

INTRODUCTION

The need for energy to sustain transportation, industrial development, and quality of life increases with an increase in the human population. Currently, the main sources of energy for the ecosphere are natural resources like coal, oil, and petroleum gas. Vestige fuels, on the other hand, have disadvantages like short supply, rapid depletion, and environmental issues, such as air or water pollution [1]. Low-cost, clean, and renewable energy sources are therefore highly desired. Notable decisions are being taken to develop sophisticated techniques for transforming solar, wind, or thermal hydropower energy into affordable electricity [2, 3]. One of these technologies that has received a lot of interest is organic Photovoltaics (PVs), which directly convert sunlight into power. Every year, the sun's energy illuminates the planet with approximately 3×10^{24} J, which is 10^4 times more than the world's entire energy consumption. The entire planet can be powered with just 0.10% of the Earth's surface covered with solar cells that are 10% productive. PV-generated electricity only makes up 0.1% of the world's energy production, despite its huge potential [4 - 6]. The Power Conversion Influence (PCE) has significantly improved as a result of advancements in modern light-emitting materials, creative device design, the emergence of applied science or nanotechnology, and breakthroughs in material blends (in the form of composites or copolymers) or the design of Organic-Based Photovoltaic (OPV) devices. Research on energy generation is a global concern that has a significant impact on both industrialization and the everyday quality of live in contemporary civilization. Finding sustainable power sources with lower carbon emissions and renewable energy technologies is crucial for green economics and the healthy advancement of human civilization in light of the current environmental degradation and global energy constraints. Furthermore, as the world enters the era of fifth-generation wireless networks (5G) and the Internet of Things (IoT), a variety of self-powered electronics are thought to be the fundamental components of the upcoming industrial revolution that will lead to a smarter world. A sustainable power supply solution is required to meet the demands of the future intelligent world, where traditional batteries are struggling because of their short lifespan, environmental impact, and frequent replacement. This is due to the advanced characteristics of electronic devices, which include miniaturization, light weight, and portability. The only way to overcome these problems is to create gadgets that can harvest energy from their environment. Mechanical-to-electrical energy transduction is a viable method for powering tiny devices, as mechanical energy is one of the most abundant and pervasive forms of energy in the environment. Numerous energy harvesting technologies, such as Electromagnetic (EM) induction, piezoelectric effect, and electrostatic effect, have previously been developed. In order to overcome these drawbacks, Zhong Lin Wang created the Triboelectric Nanogenerator (TENG) in 2012. Since then, it has garnered international attention and experienced substantial advancements, emerging as a promising technology for energy harvesting, particularly in the field of Nanoenergy and Nanosystems (NENS) [7].

The most versatile materials available today are polymers, finding applications in countless areas due to their unique properties. Among the essential benefits are flexibility, cost-effectiveness, processability, environmental stability, lightweight design, and tailorability. As early as the fifth century, polymers were discovered to be electrically insulating materials. Among other things, polymers have been used as switches, insulating gloves, electrical wire insulating covers, and a shielding layer on an electronic circuit pane [8]. These saturated polymer systems are isolating by nature due to the long covalent carbonchain. To move electrons from the valence to the conduction band in these kinds of polymers, the molecular orbital bandgap must be more than 10eV [9]. Consequently, the surface electric resistance of insulating polymers is often more than 10^{12} Ω-cm. Since conducting performance is contained within the appearance of a polymer, the elasticity of polymer materials is increased. Two primary methods canalso be used to create the physical phenomenon of conductivity in polymers [10], which is characterized by π-electron coupling in polymers and polymer composites of conducting nanoparticles (carbon nanotube, graphite, metal and inorganic hybrid salt, *etc.*) with insulating polymers. The primary one is known as an "intrinsically conducting polymer", and the other is known as an "ion-conducting polymer". Composites of polyethylene oxide compounds [11], polyethylene adipate [12], polyethylene succinate [13] with lithium salts, as well as polyacetylene, polyaniline, polypyrrole, *etc.*, are examples of ion-conducting polymers [14]. Intrinsically conducting polymers are recognized as "synmet" or "synthetic metal" because they incorporate some particular metallic properties, such as conductivity and resistivity [15]. Since three scientists (Prof. Alan G. MacDiarmid, Prof. Alan J. Heeger, and Prof. Hideki Shirakawa) were awarded the Nobel Prize for the creation and enhancement of electrically conducting polymers in 2000, interest in this area has significantly expanded [16]. Conjugated double-bond polymers are electrically conductive. Lightly bound π-electrons in the conjugated polymer have the ability to delocalize along the polymer chain [17]. As a result, the conjugated polymers have a lower bandgap.

There are several covalent bonds in a chain of π-electron conjugated polymers. Every carbon atom or heteroatom undergoes sp^2 hybridization when a conjugated polymer chain is formed. It shows that all other p-orbitals continue to exist parallel to one another, whereas one unhybridized p-orbital per atom stays vertical to the polymer chain. Consequently, the p-orbitals are delocalized throughout the polymer chain once one p-orbital laterally overlaps with either of the next p-orbitals to generate a π-orbital. Two π-electrons in the Nuclear Orbital (NO) are located inside the bonding π-Molecular Orbital (MO) and antibonding π-orbital remnants that are accessible in the isolated covalent bond (Fig. 1). Also, π-MO energy representations are available for both localized and delocalized systems, such as ethylene. Delocalization causes two bonding π-orbitals with equivalent energies and two antibonding π-orbitals with similar energies to be transformed into two bonding and two antibonding π-orbitals in conjugated polymers with a double bond, such as butadiene [18, 19].

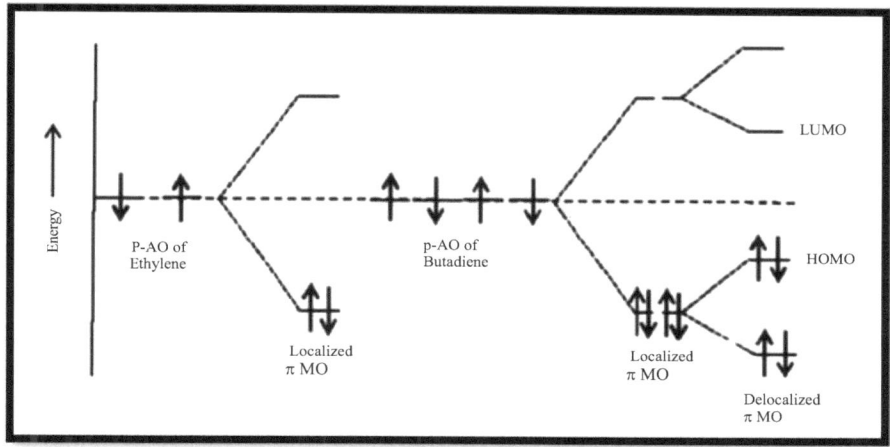

Fig. (1). π-MO energy representations for both localized and delocalized systems [18]

This results in a decrease in the energy of the Lower Unoccupied Molecular Orbital (LUMO) and an increase in the energy of the delocalised Highest Occupied Molecular Orbital (HOMO). Consequently, there is a reduction in the energy cavity with respect to the occupied bonding orbital and the unoccupied antibonding orbital. While electrons occupy the maximum energy band in metals, they only partially fill the highest energy band in semiconductors. In semiconductor polymers, on the other hand, electrons load the HOMO while leaving the LUMO vacant. Conduction requires energy to move an electron from the valence band to the conduction band. This is referred to as "bandgap energy." Fig. (2) represents of band gap structure of insulator, conductor, and semiconductor [19].

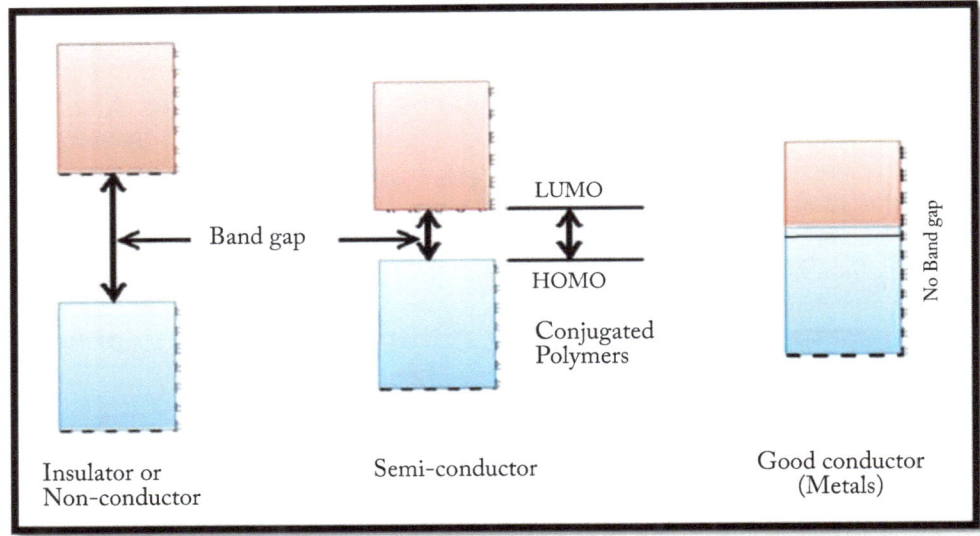

Fig. (2). Semiconductor, conductor, and insulator band gap structures [18].

These polymers' conductivity is increased by doping [8], either by donating electrons to the π conjugated system or by extracting electrons from it. The conductivity of conducting metal salts, including Cu, Ag, Au, and Sr, is still much lower than that of doped conducting polymers, even if it has surpassed that of non-conducting polymers [19, 20]. Additionally, conductivity in semiconductive areas is defined by the majority of doped conducting polymers and their derivatives. For possible use as an additive for inorganic semiconductor material, this conducting polymer is ideal. In order to produce distinctive, reliable solutions for applications such as sensors, energy harvesting devices, solar cells, and supercapacitors, the conjugated polymer industry is now concentrating on nanoscience and nanotechnology [21, 22]. Since they combine the characteristics of two or more different materials, conjugated polymer composites with a greater diversity of nanomaterials are particularly concerning, as they may offer unique mechanical, electrical, or chemical properties. Consequently, the conducting nanocomposites address a number of issues associated with intrinsic conducting polymers in addition to being composed of nanoparticles and macromolecules. Novel features of nanocomposites can be produced by carefully mixing the attributes of several elements into a single substance. Nanocomposites comprising conjugated polymers with nanoparticles have attracted attention because of their cooperative and hybrid properties produced from a wide range of components (Table **1**) [18].

Table 1. Energy harvesting applications for some of the most important conducting polymers [18].

Polymer	Possible applications
Polyacetylene	Solar cells, rechargeable battery, organic photovoltaics, chemical and gas sensors, radiation detectors, Schottky diode, anti-electrostatic, encapsulation, biotechnology, optoelectronics, *etc.*
Polypyrrole	Schottky diode, rechargeable battery, condenser, printed electronic circuit boards, chemical sensors, electroplating, electroacoustic device, fillers, adhesive, transparent coating, electromagnetic shielding, electro-photochemical cells, field effect transistor, photocatalysts, physiological implantations, optoelectronics, conductive textiles, and yarn
Polythiophene	Optoelectronics, rechargeable battery, display device, fillers, field effect transistor, Schottky diode, gas and chemical sensor, photocatalysts, and other devices
Polyaniline	Chemical and biosensor, rechargeable battery, electrochromic devices, indicator devices, textiles, *etc.*
Poly (p-phenylene)	Photocatalysts, rechargeable battery, fillers

PROPERTIES AND BENEFITS OF POLYMER NANOCOMPOSITES

Polymer substrates with trace quantities (*e.g.*, a few weight percent of the polymer matrix) of nanometer-sized fillers make up polymer nanocomposites. The goal of creating polymer nanocomposites is to enhance the mechanical, thermal, and electrical characteristics of polymers. The primary difference between polymer nanocomposites and conventional polymer composites, such as carbon fiber-reinforced polymers, is the size of the interfacial region between polymer matrices and nanometer-sized fillers. Moreover, in terms of cost, flexibility, and scalability, polymer nanocomposites present a strong substitute for conventional energy harvesting materials, such as metal-based composites and piezoelectric ceramics (like PZT). Despite having a higher energy conversion efficiency, PZT ceramics are frequently brittle and difficult to incorporate into flexible devices. Although metal-based composites have good thermal stability, they can be less flexible and more costly than polymer nanocomposites. In contrast to PZT ceramics, which call for specific materials and processing methods, polymer nanocomposites are frequently preferred due to their easier manufacturing procedures and lower cost. Because of the difficulties in synthesizing and processing them, inorganic semiconductors can be difficult to produce on a large scale. Since polymer materials are easy to manufacture and process, polymer nanocomposites typically offer better scalability than certain metal composites, which can be challenging to produce in large quantities. Inorganic semiconductors are generally less flexible and need rigid structures, which restricts their application in flexible settings. Also, although some metal composites are flexible, their form factor and adaptability may still be constrained by their rigidity. Applications requiring flexible energy harvesting solutions, such as wearable technology, flexible electronics, and other cutting-edge technologies, can benefit from polymer nanocomposites' exceptional flexibility and form factor adaptability. Despite having a high energy conversion rate, semiconductors may not be as flexible or low-power for low-power applications as polymer nanocomposites. Although metal-based composites have potential for energy harvesting applications, their overall efficiency may be constrained by their lack of adjustable properties and restricted flexibility. The flexibility, processing simplicity, and energy conversion efficiency of polymer nanocomposites are all well-balanced. According to the Department of Science & Technology (DST), the piezoelectric properties of nanoparticles and the polymer matrix can be optimized for a range of energy harvesting applications by carefully choosing them.

The performance properties of polymer nanocomposites may surpass those of conventional polymer composites [22 - 24]. The basic polymer's mechanical, electrical, and thermal characteristics are greatly improved by the addition of these nanoparticles. Among the main benefits are:

Better Electrical Conductivity: By creating channels for the movement of charges, nanoparticles enhance the electrical characteristics necessary for energy conversion.

Higher Mechanical Strength: The composites' endurance and structural integrity are improved by the use of nanoscale fillers.

Thermal Stability: Nanocomposites are appropriate for high-temperature applications because of their greater resilience to heat deterioration.

High conformability and flexibility: Polymer nanocomposites readily adapt to complicated surfaces and may be molded into a variety of forms.

Lightweight: Polymer nanocomposites are lighter than conventional piezoelectric materials, which makes them perfect for portable electronics.

Customizable properties: The electrical and mechanical characteristics of the nanocomposite can be tailored for particular purposes by varying the kind and concentration of nanoparticles.

Because of these properties, polymer nanocomposites are positioned as better materials than traditional alternatives for energy harvesting systems.

Types of nanofillers used in nanocomposites made of polymers: There are three forms of nanofillers:

i. Nanoplatelets, which have one nanoscale dimension,
ii. nanofibers, which have two nanoscale dimensions, and
iii. nanoparticulate, which have three nanoscale dimensions.

Nanofibers, for instance, are nanofillers with two nanoscale dimensions since their length is often in the micron range. Depending on the needs of the application, nanofillers with suitable dimensions and physical characteristics can be chosen. Since their discovery in 1991 (Iijima, 1991), Carbon Nanotubes (CNTs) have been widely employed due to their remarkable qualities and are regarded as new among the numerous nanofillers. They have superior mechanical, thermal, and electrical characteristics that were not present in earlier materials (24,25). Depending on how many curled cylinder-shaped graphene sheets they include, carbon nanotubes can be classified as either single-walled (SWCNTs) or multiwalled (MWCNTs). In general, SWCNTs have superior physical qualities (26,37). CNTs are used in a variety of technological fields, including gas adsorbents, actuators, composite reinforcements, catalyst support, and chemical sensors, in the automotive, aerospace, energy, and medical industries. MWCNTs are regarded as an economically viable substitute for SWCNTs of different uses due to the latter's 100-fold reduction in cost.

SYNTHESIS TECHNIQUES FOR POLYMER NANOCOMPOSITES

Electrospinning: It produces high-surface-area nanofiber mats that improve energy conversion efficiency.

Solution Casting: It is a straightforward but efficient method for incorporating nanoparticles into polymer matrices.

Melt Blending: It enables massive production of nanocomposites with uniform dispersion of fillers.

Each of these processes offers special benefits, and choosing the type of method depends on the intended usage and desired properties of the final product.

In situ polymerization, melt blending, and solution blending are common techniques for creating polymer nanocomposites [36]. The solution blending approach creates a composite by mixing nanofillers and polymers in an appropriate solvent, followed by solvent evaporation. Ultrasonication and magnetic stirring often improve mixing. For polymer chain disentanglement, a suitable solvent must be used. In order to avoid the high processing temperatures necessary for efficient polymer flow in the melt blending process, the solution blending approach is recommended, especially for specific high-performance polymers [34, 37,38]. The solution blending approach has many disadvantages despite its benefits. The performance of the product may be impacted by residual solvent, necessitating an additional solvent extraction step. This additional solvent extraction process might take a lot of time and effort. Furthermore, the solution blending approach may not be suitable for industrial use due to economic and environmental considerations as a significant amount of solvent must be utilized. By using a single or twin-screw extruder to provide shear force at temperatures higher than the glass transition temperature (Tg) for amorphous polymers and the melting temperature for semicrystalline polymers, thermoplastic polymers and nanofillers are combined in the melt blending method. The simplicity and compatibility of thismethod with modern industrial processes are its key benefits. This method's limitation to only processable thermoplastic polymers is a disadvantage. Some high-performance polymers need processing temperatures that standard industrial melt mixing equipment may not be able to reach. Furthermore, when nanofillers are present and the shear rate is significant, unanticipated polymer breakdown may occur. For insoluble and thermally unstable polymers that cannot be handled using solution blending or melt blending procedures, the *in situ* polymerization approach is particularly crucial. To obtain effective filler dispersion, nanofillers are combined with a low-viscosity monomer solution and mechanically treated with ultrasonication. This method has the benefit of allowing high diffusivity monomers to penetrate the filler agglomerates and improve their dispersion. Almost every type of polymer composite may be processed using this technique [26]. However, the synthesis procedure becomes more difficult due to the chemical polymerization step [24]. When nanoparticles are produced or mixed with a conducting polymer that has already been synthesized as a matrix, this is known as an *ex situ* technique. First, oxidative polymerization is used to fit the ICPs *in situ* using an electrochemical or organic method using a consistent monomer. Nanoparticles are incorporated into the polymer matrix over time using chemical or electrochemical techniques. Powdery or dusty nanomaterials are produced by chemical procedures and can be converted to polymer nanocomposites by employing an *ex situ* (Fig. 3) and *in situ* (Fig. 4) strategy during the production of a chemically pre-synthesized conjugated conducting polymer. The most widely used technique for cycling nanoparticles in the structure of a cluster polymer is intermingling a suitable solution phase. As an alternative, solution desertion can be used to chemically alter the different nanoparticles during the conjugated polymer film casting process. For instance, a single-step method for producing Silver (Ag) nanoparticles involves the simple thermal decomposition of $AgNO_3$ ammonia complex during the film casting of the derivative of the conducting polymer PANI as poly(m-aminophenol). Another embedding method for producing conjugated polymer nanocomposite films is electrochemical separation. One successful method for adding metal nanoparticles to pre-deposited polymers is the electrochemical *ex situ* technique. There are two steps involved in the *ex-situ* electrochemical blending of metal nanocomposites with conjugated polymers. In the first stage, an ICP that recognizes polyaniline, polypyrrole, or poly(methyl thiophene) is placed on a probe by electro-oxidizing an appropriate monomer. The conducting polymer encloses Au, Pt, Ag, Al, or Cu nanoclusters when polymer sheets are dipped in a reaction including metal salts of Pt^{6+}, silver (Ag^+), or copper (Cu^{2+}) prior to electrochemical reduction [27 - 31]. When inorganic metal particles are applied to pre-deposited polymers *via* metal ion reduction [32], the resulting structure often involves the metal nanoparticles being embedded within or on the surface of the polymer matrix. Inorganic nanostructures may be distributed precisely and consistently within the polymeric matrix using *ex situ* techniques, and their ordering ability can be fine-tuned. In appropriate polymer host systems, the nanocomposite is extensively distributed in relation to the fake nanoparticles. Conversely, *ex situ* methods enable the effective clustering of inorganic nanomaterials with well-defined polymer characteristics and the direct transfer of their primary size-dependent properties into the mass matrix. The application of a common solvent in *ex situ* development is often reinforced by the high surface energy of nanomaterials. This leads to aggregation, which results in the destructive failure of the thermal, mechanical, electrical, and optical properties of the critical nanocomposite materials [29, 30, 18]

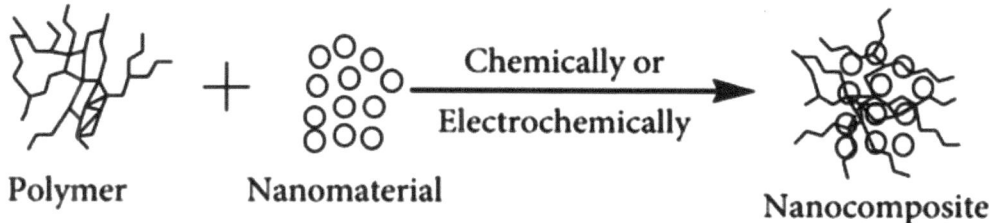

Fig. (3). Method of *ex situ* nanocomposite production [18].

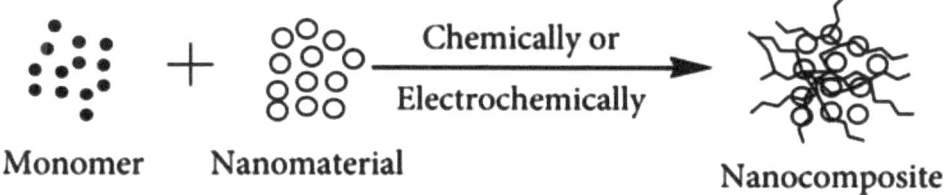

Fig. (4). Method for creating nanocomposite materials *in situ* [18].

USES IN ENERGY HARVESTING DEVICES

Polymer nanocomposites are used in a variety of energy harvesting devices, such as solar cells, where they improve photovoltaic cell efficiency through improved light absorption and charge transport.

Piezoelectric sensors: The piezoelectric effect may be greatly enhanced by adding nanoparticles to polymer matrices, opening the door to the creation of extremely sensitive pressure sensors for uses such as tactile sensing, human motion tracking, and structural health monitoring. In partnership with scientists from the National Chemical Laboratory (CSIR-NCL), Pune, researchers at the Centre for Nano and Soft Matter Sciences (CeNS), an independent institute of the Department of Science and Technology, have created a security warning system based on a piezoelectric polymer nanocomposite, as shown in Fig. (**5**). This system is based on a new material created by using two zirconia-based nanomaterials. The discovery that metal oxide nanoparticles with suitable surface characteristics and crystal structure significantly improve the piezoelectric response when added as fillers to polymer composites served as the basis for this advancement.

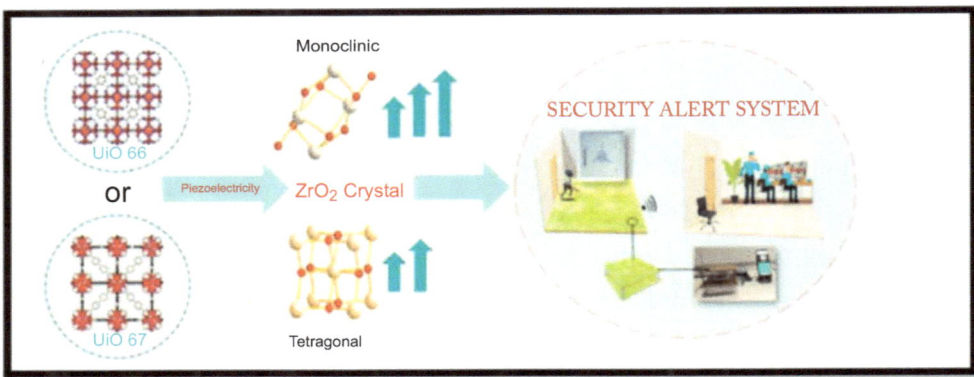

Fig. (5). Two zirconia-based metal-organic frameworks (UiO-66 and UiO-67) [34].

They were formed by the researchers and transformed into zirconia nanoparticles with remarkable control over their monoclinic and tetragonal crystallographic phases. These nanoparticles with various crystal forms were then added to Poly (Vinylidene Difluoride), a well-known piezoelectric polymer, to create polymer nanocomposite films (PVDF). The researchers' team assessed how different zirconia nanoparticle crystal structures affected a piezoelectric energy-generating zirconia-PVDF composite and found that the surface features and crystal structure of the nanofillers significantly affect the polymer material's piezoelectric qualities.

The monoclinic zirconia nanoparticle-based polymer nanocomposite made from UiO-66 performed better than other derivatives and had a higher piezoelectric output than pure polymer. Additionally, the manufactured prototype was used as an energy-generating and security alert pavement unit in a laboratory-scale demonstration of a wireless, Bluetooth-based security alarm system backed by an Android application. In a chamber, the security pavement prototype was put in place. Every time an unauthorized entrance occurred, the piezoelectric pavement produced voltage as a result of footfall (a mechanical energy conversion to electrical energy) [33-34].

The security system was engaged as a result, and the Bluetooth module wirelessly communicated with the relevant screen. This was shown in the system using an app that ran on an Android phone. The prototype may be utilized to generate electrical energy from mechanical energy input in addition to a touch sensor. PVDF-monoclinic ZrO_2 nanoparticle nanocomposites will be a great value addition for flexible, long-lasting energy production and pressure-sensing applications, according to this study. The American Chemical Society journal ACS Applied Nano Materials has recently published this paper. This work is a component of the Department of Science and Technology's ongoing "Materials for self-powered energy-generating and pressure-sensing devices" project, which is supported by the Inspire faculty fellowship program. The systematic study and crystal structure changes will surely pave the way for a deeper comprehension of the mechanism behind the PVDF-based polymer nanocomposites' piezoelectric properties [34]

Vibration energy harvesting: Wearable electronics, implanted medical devices, and mechanical energy harvesting may all benefit from the effective conversion of vibrational energy into electrical power that polymer nanocomposites can provide.

Adaptable energy–harvesting equipment: In order to absorb energy from ambient motions, conformable energy harvesting devices that may be incorporated into a variety of surfaces, including clothes or curved constructions, have been made possible by the flexibility of polymer nanocomposites.

The stress sensors: By adjusting the nanocomposite's composition, strain or deformation may be detected, which is helpful for creating smart materials or keeping an eye on structural integrity. A road safety sensor prototype has been created by researchers at the Centre for Nano and Soft Matter Sciences (CeNS), Bengaluru, using a polymer nanocomposite for pressure sensing and energy harvesting applications. The prototype can be installed in a movable ramp and fastened to the road just 100 meters before sharp and deadly turning points, alerting any vehicle approaching from the opposite side. It operates on the principle of the piezoelectric effect, which allows it to generate energy that can be stored and further used to power electronic devices. Vanadium disulfide (VS_2) with a very high surface charge was created by scientists Mr. Ankur Verma, Dr. Arjun Hari Madhu, and Dr. Subash Cherumannil Karumuthil [33]. This compound has the ability to enhance the piezoelectric properties of polymers. These nanoparticles were incorporated at different concentrations into Poly (Vinylidene Difluoride), a well-known piezoelectric polymer, to create polymer nanocomposite films (PVDF). They also looked at how the surface charge of the nanoparticles impacts the polymer nanocomposite's piezoelectric characteristics. Additionally, a smart door and road safety sensor demonstration was set up at the laboratory level using a pressure sensor prototype. This work shows that PVDF-VS_2 nanocomposites will be a valuable addition to flexible, long-term pressure sensing and energy generation applications. The Journal of Materials Chemistry A has recently published this paper. Thus, with their high flexibility, well-controlled compositions, large surface areas, low weight, and porous structures that allow for both liquid and air to pass through, polymer nanocomposite mesh-based electronic devices are gaining a lot of traction for use in flexible displays, implantable bioelectronics, energy harvesting and storage devices, electronic skin, and healthcare monitoring (Fig. **6**). The recent developments in mesh-based flexible electronic devices made of polymer nanocomposite and the associated processes behind the capabilities generated by the mesh structure have been well documented. In addition to encouraging the use and integration of polymer nanocomposite mesh scaffolds into multifunctional flexible electronic devices with exceptional performance and environmental friendliness for a variety of applications, the associated critical challenges and opportunities are presented to provide inspiration for the design of advanced polymer nanocomposite meshes.

Passive and active components within polymer nanocomposite meshes are illustrated in Fig. (**6**) above. Polymer nanocomposite mesh passive components (functional substrates, templates, carbonized precursors) as well as active components (friction layers for nanogenerators, electroactive materials, and separators in energy storage devices) are used in a number of high-performance flexible electronic devices, as shown in the literature review. They are also used as sensors for chemical detection. Nearly all existing flexible electronic devices are based on inorganic/organic composites. For example, there is the idea of using metals, metal oxide semiconductors, carbon materials, and polymers. Inorganic semiconductors and metals are widely chosen as desirable materials for manufacturing electronic devices because of their high conductivity and reproducible properties. However, these substances only possess relatively high rigidity and cannot withstand frequent stretches or bends to a large degree over time..Numerous advanced structural designs and optimizations are devoted to providing metal-based materials with enhanced flexibility. The ability of single bonds in polymers to undergo internal rotations plays a crucial role in the conformation changes of molecular chains, contributing to the intrinsic flexibility of polymers. This flexibility offers considerable potential for adjusting the properties and functions of polymers by fine-tuning their molecular structures. In broad terms, polymers can act as substrates for conductive components, protective coatings, adhesives, matrices, or active materials. Some of the most commonly used flexible substrates in electronic devices include Polydimethylsiloxane (PDMS), Poly(ethylene terephthalate) (PET), Polyimide (PI), and Ecoflex, thanks to their exceptional flexibility. Conductive polymers, especially those with unique π-conjugated structures like Polyaniline (PANI), Polypyrrole (PPy), and Polythiophene (PT), have garnered significant attention for their adjustable electronic properties, ease of processing from solution, and excellent mechanical compliance, which allows for large-area processing. Despite the superior conductivity and charge transport characteristics of inorganic materials, conjugated polymers remain ideal candidates for flexible electronics. The Young's modulus of semiconducting polymers is comparable to that of human skin (0.34 MPa), further highlighting their potential for use in wearable, on-skin electronics. In the field of flexible device engineering, one effective strategy for improving flexibility and stretchability is the careful design of materials with appropriate geometries. In this context, polymers can be easily processed into a variety of Two-Dimensional (2D) or Three-Dimensional (3D) micro/macrostructures [35], including porous, fibrous, woven, coiled, wavy, and other hierarchical or patterned configurations, using various fabrication methods. Unlike traditional polymer thin films, which often suffer from mechanical mismatches and restricted permeability, fibrous mesh scaffolds—created by interlacing fibers, filaments, or yarns—present exciting opportunities for flexible electronics. These scaffolds offer benefits such as controlled compositions, softness, larger surface areas, and porous structures that allow air and liquids to pass through, along with a higher resistance to damage.

Furthermore, the mesh architecture is crucial in the development of ultrathin, lightweight, wearable, and conformal electronic devices, which are not only promising for improving the functionality of existing technologies but also for enabling a broader range of new applications. Natural polymer fibers, such as those found in spider webs made of silk or in paper and textiles derived from cellulose, are often used to create these intricate mesh structures. On the other hand, advances in polymer science and engineering have made it possible to produce a wide variety of synthetic polymer meshes, further benefiting from improved fiber-forming and weaving technologies.

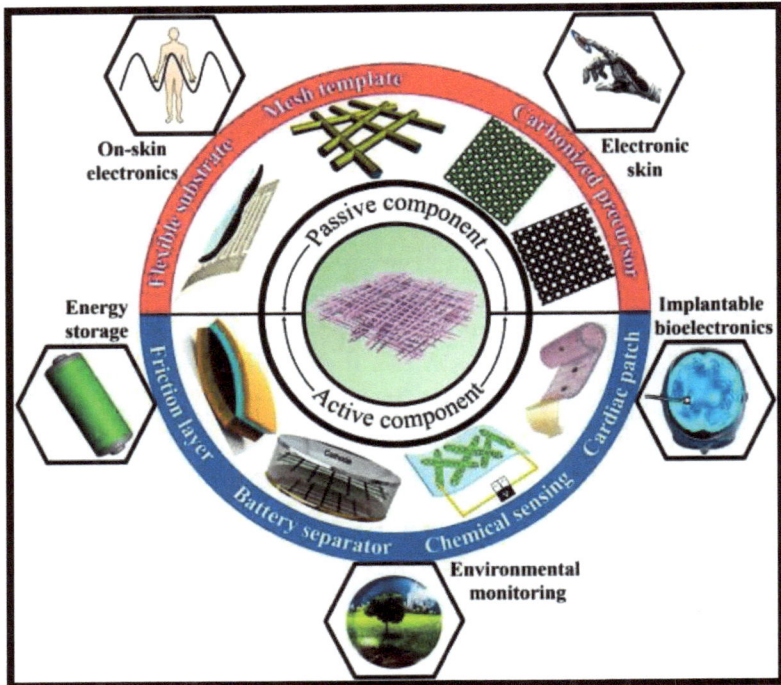

Fig. (6). Mesh-based nanocomposites that use polymers as their organic component and inorganic materials often show good performance in flexible electronic devices [35].

The incorporation of nano-scale conductive materials has also been shown to enhance both flexibility and electrical performance. As a result, there has been growing interest over the past two decades in combining soft polymers with active nanomaterials to create polymer nanocomposite meshes for high-performance electronic devices. This development has laid the groundwork for new approaches to mesh-based polymer nanocomposite electronic materials, as depicted in Fig. (**1**). Several reviews have previously examined the use of fibers and textiles, including those based on metals, carbon materials, and conductive polymers, in applications like energy harvesting, storage systems, and wearable electronics. However, a comprehensive review that focuses specifically on the use of polymer mesh composites in diverse flexible electronics—such as strain/pressure sensors, electronic skin, transparent conductors, energy systems, flexible displays, and bioelectronics—has been lacking. These strategies include the use of natural-resource-based meshes (such as cellulose paper, cotton fabric, and silk fabric), electrospinning, *in situ* polymerization with simultaneous deposition, and photolithography [36].

CONCLUSION

Despite its potential, a number of obstacles need to be overcome before polymer nanocomposites can be widely used in energy harvesting devices. Despite recent developments, polymer nanocomposites' energy output might still be somewhat low compared to other energy harvesting methods like lithium-ion batteries, fuel cells, and nanocomposites incorporating graphene or carbon nanotubes, which have different energy storage capabilities but do not always outperform conventional methods. It can be difficult to keep polymer nanocomposites performing well over time, particularly in severe environments.

Affordable, expandable production techniques must be developed to promote industrial uptake. There is also an issue of performance optimization, which is the process of improving the stability and energy conversion efficiency of devices based on nanocomposite technology. Furthermore, it must be ensured that the materials and fabrication techniques are environmentally friendly. By addressing these issues, polymer nanocomposites will find wider use and make a substantial contribution to worldwide efforts toward sustainable energy.

REFERENCES

[1] G.A. Olah, "Beyond oil and gas: The methanol economy", *Angew. Chem. Int. Ed.,* vol. 44, no. 18, pp. 2636-2639, 2005.
[http://dx.doi.org/10.1002/anie.200462121]

[2] S. Rajput, M. Averbukh, A. Yahalom, and T. Minav, "An approval of MPPT based on PV cell's simplified equivalent circuit during fast-shading conditions", *Electronics (Basel),* vol. 8, no. 9, 2019.
[http://dx.doi.org/10.3390/electronics8091060]

[3] A. Singh, A. Sharma, S. Rajput, A.K. Mondal, A. Bose, and M. Ram, "Parameter extraction of solar module using the sooty tern optimization algorithm", *Electronics (Basel),* vol. 11, no. 4, p. 564, 2022.

[4] B. Li, L. Wang, B. Kang, P. Wang, and Y. Qiu, "Review of recent progress in solid-state dye-sensitized solar cells", *Sol. Energy Mater. Sol. Cells,* vol. 90, no. 5, pp. 549-573, 2006.
[http://dx.doi.org/10.1016/j.solmat.2005.04.039]

[5] M.K. Siddiki, J. Li, D. Galipeau, and Q. Qiao, "A review of polymer multijunction solar cells", *Energy Environ. Sci.,* vol. 3, no. 7, pp. 867-883, 2010.
[http://dx.doi.org/10.1039/b926255p]

[6] P.V. Kamat, "Meeting the clean energy demand: nanostructure architectures for solar energy conversion", *J. Phys. Chem. C,* vol. 111, no. 7, pp. 2834-2860, 2007.

[7] J. Zhu, "Progress in TENG technology—A journey from energy harvesting to nanoenergy and nanosystem", *EcoMat,* vol. 2, no. 4, p. e12058, 2020.

[8] P. Kar, A. Choudhury, and S.K. Verma, "Conjugated polymer nanocomposites-based chemical sensors," in Fundamentals of Conjugated Polymer Blends, Copolymers, and Composites: Synthesis, Properties and Applications. Wiley: NJ, USA, 2015, pp. 619-686.

[9] R.S. Potember, "Conducting organics and polymers for electronic and optical devices", *Polymer (Guildf.),* vol. 28, no. 4, pp. 574-580, 1987.
[http://dx.doi.org/10.1016/0032-3861(87)90472-1]

[10] T.A. Skotheim, *Electroresponsive Molecular and Polymeric Systems.* vol. Vol. 2. CRC Press, 2021.

[11] F. Croce, "Properties and applications of lithium ion-conducting polymers", *Solid State Ion.,* vol. 40-41, pp. 375-379, 1990.
[http://dx.doi.org/10.1016/0167-2738(90)90362-u]

[12] J.M. Czajkowski, T. Baszczyk, and D. Kaźmierczak, "Automatic apparatus for precise measuring and recording of PZC value of liquid electrodes and its application", *Electrochim. Acta,* vol. 29, no. 4, pp. 439-443, 1984.
[http://dx.doi.org/10.1016/0013-4686(84)87091-7]

[13] R. Dupon, B.L. Papke, M.A. Ratner, and D.F. Shriver, "Ion transport in the polymer electrolytes formed between poly (ethylene succinate) and lithium tetrafluoroborate", *J. Electrochem. Soc.,* vol. 131, no. 3, pp. 586-589, 1984.
[http://dx.doi.org/10.1149/1.2115630]

[14] T.A. Skotheim, *Handbook of Conducting Polymers.* CRC Press, 1997.

[15] H.C. Nayak, S.S. Parmar, R.P. Kumhar, and S. Rajput, "Modulation in electric conduction of PVK and ferrocene-doped PVK thin films", *Electron. Mater.,* vol. 3, pp. 53-62, 2022.

[16] P.G. Pickup, "Alternating current impedance study of a polypyrrole-based anion-exchange polymer", *J. Chem. Soc., Faraday Trans.,* vol. 86, no. 21, pp. 3631-3636, 1990.
[http://dx.doi.org/10.1039/ft9908603631]

[17] Z. Qiu, B.A. Hammer, and K. Müllen, "Conjugated polymers–problems and promises", *Prog. Polym. Sci.,* vol. 100, p. 101179, 2020.

[18] S. Sonika, "Conducting polymer nanocomposite for energy storage and energy harvesting systems", *Adv. Mater. Sci. Eng.,* vol. 2022, no. 1, p. 2266899, 2022.

[19] R. Erlandsson, W.R. Salaneck, and I. Lundström, "Electrically conducting organic polymer materials: defects make them better", *Mater. Des.,* vol. 7, no. 5, pp. 246-251, 1986.

[20] R. Greene, G.B. Street, and L.J. Suter, "Superconductivity in polysulfur nitride (SN)X", *Phys. Rev. Lett.,* vol. 34, no. 10, p. 577, 1975.

[21] J. Murphy, *Additives for Plastics Handbook.* Elsevier, 2001.

[22] X. Hu, "Electrostrain enhancement at tricritical point for BaTi1−XHfXO3 ceramics", *J. Mater. Eng. Perform.,* vol. 29, pp. 5388-5394, 2020.

[23] G. Bidan, "Electroconducting conjugated polymers: new sensitive matrices to build up chemical or electrochemical sensors. A review", *Sens. Actuators B Chem.,* vol. 6, no. 1-3, pp. 45-56, 1992.

[24] S. Sundarram, Y.H. Kim, and W. Li, *"Preparation and characterization of poly (ether imide) nanocomposites and nanocomposite foams,"* in *Manufacturing of Nanocomposites with Engineering Plastics.* Woodhead Publishing, 2015, pp. 61-85.

[25] W. Gacitua, A. Ballerini, and J. Zhang, "Polymer nanocomposites: synthetic and natural fillers—a review", *Maderas Cienc. Tecnol.,* vol. 7, no. 3, pp. 159-178, 2005.

[26] J.N. Coleman, U. Khan, and Y.K. Gun'ko, "Mechanical reinforcement of polymers using carbon nanotubes", *Adv. Mater.,* vol. 18, no. 6, pp. 689-706, 2006.

[27] T.W. Chamberlain, A.M. Popov, A.A. Knizhnik, G.E. Samoilov, and A.N. Khlobystov, "The role of molecular clusters in the filling of carbon nanotubes", *ACS Nano,* vol. 4, no. 9, pp. 5203-5210, 2010.

[28] H. Kolya, T. Kuila, N.H. Kim, and J.H. Lee, *"Polymer nanocomposites for energy-related applications,"* in *Polymer-Based Advanced Functional Composites for Optoelectronic and Energy Applications.* Elsevier, 2021, pp. 215-248.

[29] Z. Peng, E. Wang, and S. Dong, "Incorporation of surface-derivatized gold nanoparticles into electrochemically generated polymer films", *Electrochem. Commun.,* vol. 4, no. 3, pp. 210-213, 2002.

[30] B.C. Sih, A. Teichert, and M.O. Wolf, "Electrodeposition of oligothiophene-linked gold nanoparticle films", *Chem. Mater.,* vol. 16, no. 14, pp. 2712-2718, 2004.

[31] A.A. Athawale, S.V. Bhagwat, P.P. Katre, A.J. Chandwadkar, and P. Karandikar, "Aniline as a stabilizer for metal nanoparticles", *Mater. Lett.,* vol. 57, no. 24–25, pp. 3889-3894, 2003.

[32] A. Mathur, ""An enzymatic multiplexed impedimetric sensor based on α-MnO2/GQD nano-composite for the detection of diabetes and diabetic foot ulcer using micro-fluidic platform," Chemosensors, vol. 9, no. 12, p. 339, 2021[33] A. Verma and S. C. Karumuthil, "Vanadium disulfide-incorporated polymer nanocomposites for flexible piezoelectric energy generators and road safety sensors,"", *J. Mater. Chem. A Mater. Energy Sustain.,* 2024.

[34] G. Mukherjee, A. Verma, A.H. Madhu, B.L. Prasad, and S.C. Cherumannil Karumuthil, "Polymer nanocomposites with UiO-derived zirconia fillers for energy generation and pressure-sensing devices: The role of crystal structure and surface characteristics", *ACS Appl. Nano Mater.,* vol. 7, no. 6, pp. 5809-5818, 2024.

[35] W. Ni, and L. Shi, "Layer-structured carbonaceous materials for advanced Li-ion and Na-ion batteries: Beyond graphene", *J. Vac. Sci. Technol. A,* vol. 37, no. 4, 2019.

[36] M. Gong, L. Zhang, and P. Wan, "Polymer nanocomposite meshes for flexible electronic devices", *Prog. Polym. Sci.,* vol. 107, p. 101279, 2020.

[37] A. D. De Oliveira, and C. A. G. Beatrice, *Polymer nanocomposites with different types of nanofiller,* 2018.

[38] P. Kumar, K.P. Sandeep, S. Alavi, V.D. Truong, and R.E. Gorga, "Preparation and characterization of bio-nanocomposite films based on soy protein isolate and montmorillonite using melt extrusion", *J. Food Eng.,* vol. 100, no. 3, pp. 480-489, 2010.

CHAPTER 13

A Brief Overview of Energy Harvesting in Advanced Sustainable Polymers

Akash Ranjan[1,*], Sabira Sultana Khadim[2] and Sonika[3]

[1] Faculty of Education, Banaras Hindu University, Kamachha, Varanasi, Uttar Pradesh, India

[2] Department of Education, Lima Aier Higher Secondary School, Dimapur, Nagaland, India

[3] Department of Physics, Rajiv Gandhi University, Rono Hills, Doimukh, Papumpare 791112, Arunachal Pradesh, India

Abstract: Energy harvesting represents a transformative strategy for capturing ambient energy and converting it into usable electrical power, thereby enabling the operation of electronic systems without reliance on conventional energy supplies such as batteries or wired connections. Polymers have emerged as a pivotal class of materials in this domain owing to their structural versatility, tunable properties, and potential alignment with sustainability goals. With the growing global demand for renewable and decentralized energy solutions—particularly for autonomous, wireless, and portable electronics—polymer-based energy harvesters are gaining increasing prominence. Sustainable polymers, derived from renewable or recycled precursors, offer distinct advantages including mechanical robustness, corrosion resistance, and biodegradability, thus meeting both environmental and economic imperatives. Nevertheless, conventional synthetic polymer production and disposal practices remain environmentally unsustainable, exacerbating resource depletion and pollution. To address these limitations, current research is advancing the development of polymers engineered within closed-loop life cycles to minimize ecological burden. Emerging innovations, such as photovoltaic-integrated sound barriers, polymer-based nanomaterials, thermoelectric generators, and induction-driven energy systems, highlight the expanding scope of polymer applications in this field. This chapter concludes by critically evaluating key challenges—most notably the enhancement of energy conversion efficiency, scalability, and techno-economic viability—while underscoring the pivotal role of advanced sustainable polymers in shaping the next generation of energy harvesting technologies.

Keywords: Biodegradability, Energy harvesting, Polymer-based devices, Renewable energy, Sustainable polymers.

---

[*] **Corresponding author Akash Ranjan:** Faculty of Education, Banaras Hindu University, Kamachha, Varanasi, Uttar Pradesh, India; E-mail: 1aranjanmedmphil@gmail.com

INTRODUCTION

Energy harvesting, broadly defined, is the process of capturing and converting ambient energy—derived from sources such as solar radiation, wind, thermal gradients, and mechanical vibrations—into usable electricity for powering electronic systems. Against the backdrop of rising global energy consumption and its adverse implications for energy security and environmental sustainability, energy harvesting technologies have emerged as indispensable strategies for sustainable power generation. Often referred to as "energy scavenging," these systems capitalize on otherwise wasted or underutilized energy—including temperature fluctuations, ambient vibrations, or human motion—while ensuring efficient storage and regulated delivery, particularly in low-power or off-grid applications. Unlike narrow definitions restricted to transducer-level conversion, energy harvesting can be more comprehensively viewed as an integrated approach encompassing energy capture, storage, and power management to provide electricity in scenarios where conventional energy infrastructure is impractical or inefficient [1].

A wide spectrum of energy sources has been explored for this purpose, ranging from environmental vibrations and water vapor motion to atmospheric sunlight and even biological systems such as the human body. Recent advances in materials science have introduced hyperbranched polymers capable of photon absorption and inward energy transfer, mimicking capacitor-like behavior for efficient energy storage. Similarly, kinetic energy from vapor-driven motion and piezoelectric responses from structural vibrations have opened new avenues for harvesting dispersed ambient energy. Collectively, these innovations highlight the potential of polymeric and nanostructured materials to enhance efficiency and sustainability in energy conversion processes [2]. The impetus for energy harvesting research has been further strengthened by the rapid proliferation of wireless technologies and autonomous devices. Conventional batteries, with their limited lifespans, environmental hazards, and frequent replacement requirements, are poorly suited for long-term or remote applications. In contrast, energy harvesting enables wireless sensors and distributed devices to operate autonomously over extended durations, even in hostile or inaccessible environments. The integration of compact harvesters with advanced power management units not only reduces dependence on bulky conductors but also enhances the feasibility of remote operation, thereby disrupting conventional paradigms of energy sourcing [2].

As depicted schematically (Fig. **1**), a typical energy harvesting system comprises three fundamental components: (i) the harvester, responsible for converting ambient energy into electrical signals; (ii) a power management circuit, which regulates and conditions the output; and (iii) an energy storage module or direct load connection. While the architecture remains broadly similar across platforms, the specific technologies employed—such as photovoltaic, thermoelectric, piezoelectric, or triboelectric transducers—determine overall efficiency and applicability. Ongoing research is thus increasingly directed toward improving conversion efficiency, miniaturization, and cost-effectiveness, with particular emphasis on polymer- and nanomaterial-based solutions.

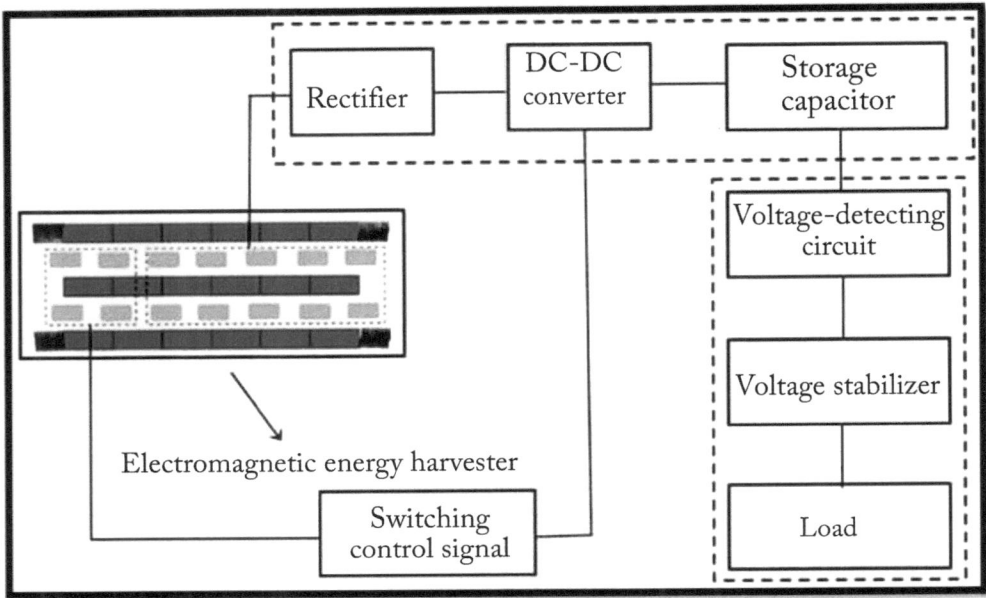

Fig. (1). Block diagram of an energy harvesting system [3].

SIGNIFICANCE OF SUSTAINABLE POLYMERS IN MODERN ENERGY APPLICATIONS

As shown in Fig. (2), energy harvesting enables the use of ambient energy sources to power small electronic devices. These sources include industrial heat, heat from vehicles (such as buses and motors), and even energy from the human body. Energy transducers convert this ambient energy into usable electrical power. The primary goals of energy harvesting are to extend the lifespan of energy storage devices, minimize the frequency of battery recharges, and ultimately eliminate the need for battery replacement or dependence on traditional power lines. One of the earliest large-scale applications is in building automation, particularly in self-powered light switches, commonly known as piezo switches. Future applications include structural health monitoring of large infrastructures and condition monitoring in industrial systems. Ambient energy sources are also increasingly used to power wireless sensors and sensor networks. In consumer products, embedded energy transducers—such as solar cells integrated into clothing, bags, or other items—can recharge portable devices like mobile phones, representing another exciting sector of development [4].

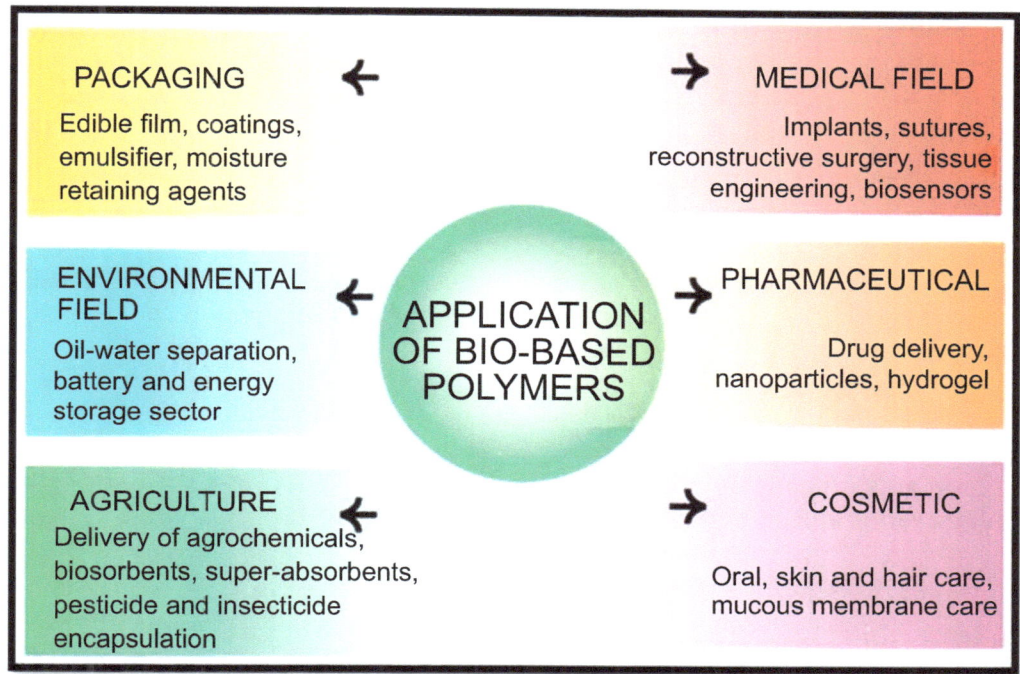

Fig. (2). Importance of sustainable polymers for modern society and development [4].

In today's modern world, polymers are among the most significant and widely used materials. To enhance their sustainability, scientists and researchers continue to improve polymer properties through ongoing studies. This chapter highlights the importance of sustainable polymers for contemporary development and society. It also discusses the current applications of synthetic and biodegradable polymers, along with their environmental advantages and limitations. Furthermore, strategies for polymer waste management and the sustainability of biodegradable polymers are examined [5].

There is an increasing shift toward using renewable resources in polymer production. Raw materials such as carbon dioxide, terpenes, vegetable oils, and carbohydrates can serve as feedstocks for producing a wide variety of eco-friendly products, including elastomers, plastics, hydrogels, flexible electronics, resins, engineering polymers, and composites. Effective catalysis plays a crucial role in creating monomers, promoting selective polymerization, and enabling the recycling or transformation of waste materials. Sustainable polymers, therefore, hold promise for applications in both high-value sectors (such as electronics and engineering) and simpler uses like packaging. Life-cycle assessment methods can be applied to evaluate the environmental benefits of these polymers [4, 5].

Sustainable polymers are materials made from biodegradable or renewable resources that assist technological growth while addressing environmental issues. The following succinctly describes their significance in contemporary energy applications [6]:

Renewable Energy Technologies

- Solar Panels: Photovoltaic cells employ polymers as coatings, substrates, or encapsulants. Solar energy systems' carbon footprint is decreased by sustainable alternatives.
- Wind Turbines: To increase energy efficiency, turbine blades are made of lightweight, robust polymers that are frequently strengthened with bio-based composites.
- Hydrogen Storage: Polymers with specific qualities, such as high strength-t--weight ratios and energy efficiency, are essential for hydrogen storage systems.

Energy Storage Systems

- Batteries: Lithium-ion and solid-state batteries operate better with sustainable polymer electrolytes, which are both safer and better for the environment.
- Supercapacitors: High conductivity and energy density in energy storage devices are facilitated by polymers such as polypyrrole and polyaniline.
- Thermal Energy Storage: Phase change materials are encapsulated in biodegradable polymers to provide effective thermal regulation.

Energy Efficiency

- Insulation: To reduce energy loss in industrial systems and buildings, sustainable polymers are used in innovative insulation materials.
- Lightweight Structures: Bio-composites minimize pollutants and improve fuel efficiency by consuming less energy during transportation.

Waste-to-Energy Conversion

- Biodegradable Plastics: By using pyrolysis or composting, these polymers can be transformed into energy, lowering the need for fossil fuels.

- Polymer Recycling: Cutting-edge recycling technologies make it possible to turn waste polymers into new materials or fuel, promoting a circular economy.

Environmental Benefits

- Less Carbon Footprint: Greenhouse gas emissions are decreased when bio-based polymers are used in energy applications.
- Degradability: Biodegradable polymers combine energy developments with ecological sustainability by preventing long-term environmental pollution.

ROLE OF POLYMERS IN ENERGY HARVESTING

Polymer material is lightweight, low-cost, and flexible, making it ideal for a variety of uses. With its intricately structured macromolecular structure, this material is especially suitable for energy harvesting. This chapter discusses solar, thermoelectric, piezoelectric, and triboelectric devices, as well as the polymer materials used in these applications. It begins with the fundamentals of energy harvesting devices and then goes on to detail the requirements for the materials utilized in terms of structure and morphology. The use of polymers in these devices, as well as recent advances in structural design, is highlighted. The remaining difficulties and future orientation are also addressed in the end (Peng, Sun, Weng, & Fang, 2017) [6, 7].

Because they are inexpensive, flexible, lightweight, and have adjustable qualities, polymers are essential to energy harvesting [8]. Their uses cover a wide range of energy-harvesting technologies, increasing productivity and opening the door for advancements in renewable energy systems.

Photovoltaics (Solar Energy Harvesting)

- Organic Photovoltaics (OPVs): Because polymers like Poly(3-hexylthiophene) (P3HT) absorb light and conduct energy, they are utilized in organic solar cells.
- Perovskite solar cells: Polymers enhance stability and energy conversion efficiency by serving as encapsulants or layers for electron transport.
- Flexible solar panels: They are made by polymer-based substrates, which make them lightweight and pliable enough for wearable and portable applications.

Piezoelectric Energy Harvesting

- Polymer Piezoelectrics: Polymers with piezoelectric qualities, such as PVDF (Polyvinylidene Fluoride), can transform mechanical energy into electrical energy.
- Wearable Technology: Energy harvesters and sensors built into clothes or shoes use flexible piezoelectric polymers.

Thermoelectric Energy Harvesting

- Conductive Polymers: Substances that transform temperature gradients into electrical energy include Polyaniline (PANI) and derivatives of polythiophene.
- Flexible Thermoelectric Devices: Polymer-based thermoelectrics are perfect for wearable electronics since they are lightweight and can adjust to uneven surfaces.

Bioenergy Harvesting

- Microbial Fuel Cells: To improve the efficiency of microbial fuel cells, which turn organic matter into electricity, polymers are employed as proton exchange membranes.
- Biocompatible Polymers: These substances make it easier to capture energy from biological systems, like body heat or motion.

Triboelectric Energy Harvesting

- Triboelectric Nanogenerators (TENGs): These devices use polymers with high triboelectric activity, such as Polydimethylsiloxane (PDMS), to transform mechanical motion—such as waves, vibrations, and wind—into electrical energy.
- Self-Powered Sensors: Polymer-based TENGs are frequently seen in wearable electronics and self-sufficient sensors for Internet of Things devices.

Wind and Wave Energy

- Flexible Blades and Membranes: Lightweight, flexible parts of wind or wave energy harvesting systems are made of polymers.
- Durability in Adversities: Energy-harvesting systems are shielded from environmental deterioration by polymer coatings.

Energy Storage Integration

- In energy harvesting and storage systems, polymers frequently play a dual role, allowing for hybrid devices that effectively capture and store energy.

SUSTAINABLE POLYMERS

Materials made with sustainable polymers are intended to have as little of an impact on the environment as possible throughout the course of their lifetime. These polymers are environmentally beneficial substitutes for conventional petroleum-based plastics since they are either made from renewable resources, biodegradable material, or both. They are designed to lessen pollution and reliance on non-renewable resources [5].

Materials made from renewable, recycled, or other lower-carbon feedstocks that are managed sustainably through recycling and biodegradation at the end of their useful lives (EoL) are known as sustainable polymers. These consist of synthetic, natural, and modified polymers that, in accordance with the circular economy paradigm, are sustainable throughout their life cycle. A schematic of the life cycle of sustainable polymers is shown in Fig. (**3**), from feedstock to regeneration. Although sustainable polymers have been used by humans for a long time, recent research on them and their commercialization has been spurred by the serious environmental issues brought on by the persistence and buildup of plastic pollution on land and in the oceans [8, 9].

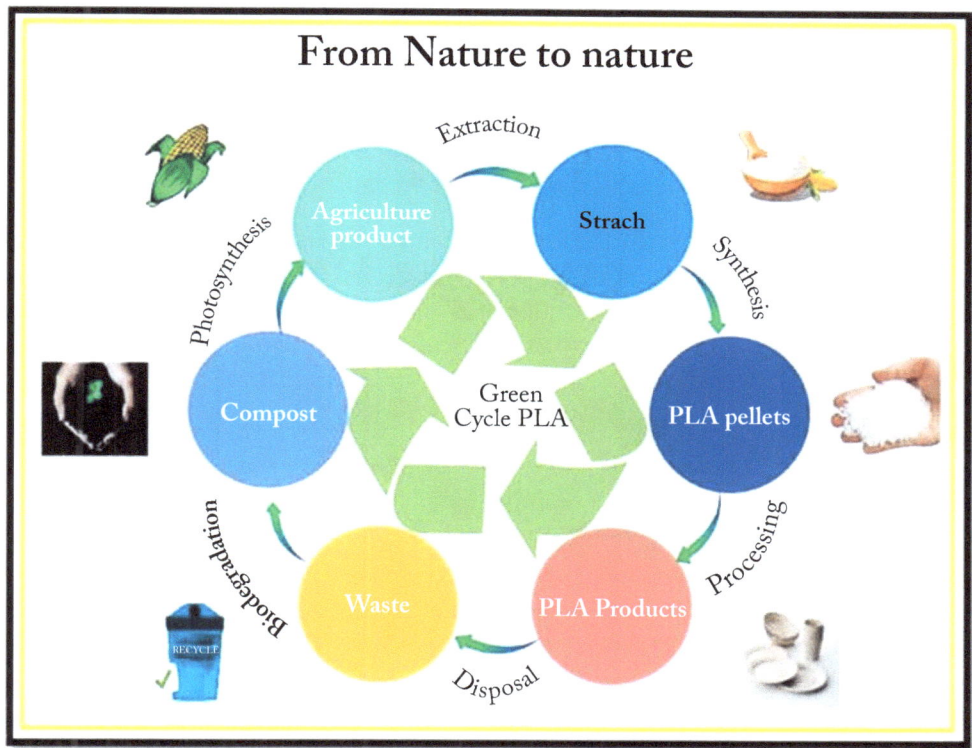

Fig. (3). Life cycle of sustainable polymers [10].

TYPES OF SUSTAINABLE POLYMERS

Bio-based Polymers

Global awareness of material sustainability has raised demand for bio-based polymers, given that the massive expansion in the use of materials derived from fossil fuels (due to their low density, low cost, and simplicity of processing) is causing environmental issues [9]. Biodegradable polymers derived from microbial, plant, and animal sources are known as bio-based polymers [12]. Biodegradability and biocompatibility are characteristics of bio-based polymers. Bio-based polymers are a significant advancement in industrial applications due to their qualities, which include low density, recyclability, minimal health hazard, biodegradability, specific strength, good thermal characteristics, reduced tool wear, no skin irradiation, and high toughness. Because of the serious global depletion of fossil fuel supplies and growing worries about environmental degradation, chemists from a wide range of fields have shown a great deal of interest in bio-based polymeric materials [8, 9]. Bio-based polymers are derived from renewable biological sources such as plants, animals, and microorganisms. These polymers are valued for their low density, recyclability, and environmental compatibility. Common properties include biocompatibility, reduced toxicity, good thermal characteristics, high toughness, and ease of processing. Due to the depletion of fossil fuels and environmental concerns, bio-based polymers have gained considerable attention in both research and industry [8, 9].

Fig. (4) shows a characteristic depiction of sustainability as three intersecting circles. In order to replace traditional petroleum-based polymers, researchers have worked very hard to create and employ bio-based polymers. Polymeric biomolecules created by living things are known as bio-based polymers. One significant class of biopolymers that has garnered a lot of study interest is polysaccharides [6 - 10].

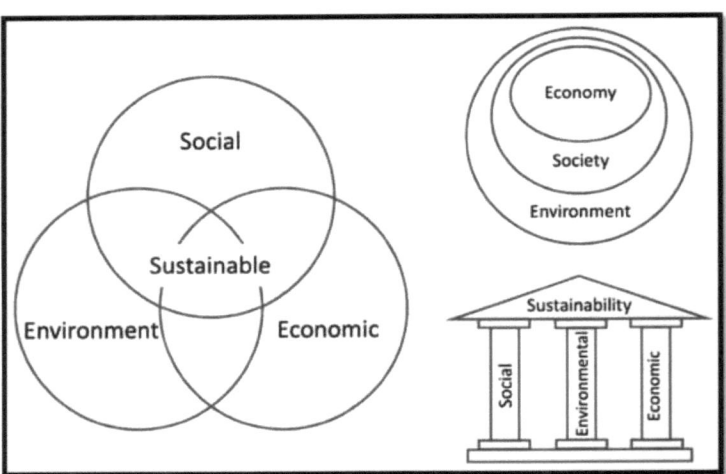

Fig. (4). A characteristic depiction of sustainability as three intersecting circles, (right): alternative depictions: below as literal "pillars" and above as concentric circles. Diagram of Sustainability [11].

Biodegradable Polymers

The field of biodegradable polymers is a recent development. Recently, a huge variety of biodegradable polymers have been created, and some microbes and enzymes that can break them down have been identified. Synthetic polymer pollution of the environment has reached alarming levels in developing nations. Consequently, efforts have been undertaken to address these issues by incorporating biodegradability into commonly used polymers by making minor structural changes. These polymers are defined by their ability to break down naturally through the action of microorganisms and enzymes. Although many bio-based polymers are biodegradable, not all biodegradable polymers are bio-based—some are synthetically derived but designed with structural features that enable biodegradation. This field has expanded rapidly, with ongoing research focused on modifying the structures of conventional polymers to enhance biodegradability and reduce synthetic polymer pollution, which is particularly concerning in developing countries [7, 12, 13].

Recycled Polymers

Plastic products that have been reprocessed from trash or discarded plastics into usable raw materials are known as recycled polymers. These materials are obtained through recycling procedures that entail gathering, washing, and mechanically or chemically reprocessing waste plastic. By using the recovered polymers to create new goods, the need for virgin polymers is decreased, and plastic waste and pollution are reduced. Examples include recycled HDPE (High-Density Polyethylene), which is used in pipes and containers, and recycled PET (rPET), which is used to make bottles and textiles [13].

Hybrid Polymers

A material that has two or more distinct types of molecules is called a hybrid polymer. A polymer will be present in at least one of these. Two distinct polymers can be combined to create hybrid polymers. Usually, they entail mixing a polymer with either other organic materials or inorganic ones. One famous example is the rubber found in automobile tires. Isoprene polymers are mixed with other organic materials, like natural rubber, to create this.

Manufacturers can produce new materials by combining the performance attributes of various components thanks to hybrid polymers. Applications in a variety of industries, including paints, medications, and semiconductors, are made possible by this. Because of their many significant uses, hybrid polymers are a top research topic [13].

BENEFITS OVER TRADITIONAL POLYMERS

Advantages of sustainable polymers over conventional polymers are given below [14]:

- **Environmental impact**: Because fossil fuels are used less frequently, there is less carbon imprint. They naturally break down, reducing waste pollution over time.
- **Resource efficiency**: Ensures sustainability by using renewable feed stocks. Encourages the circular economy by being recyclable and reusable.
- **Safety and health**: Degradation and production-related harmful emissions were decreased; safer substitutes for commodities derived from petroleum.
- **Energy efficiency**: Products are frequently produced using less energy. It can be integrated with green energy technology to improve sustainability in general.
- **Financial sustainability**: Provides long-term cost benefits by reducing waste and using resources efficiently. Encourages recycling and bio-economy-focused industries.

MECHANISMS OF ENERGY HARVESTING IN POLYMERS

Energy harvesting in polymers refers to processes that use the inherent characteristics of polymeric materials to transform external energy sources such as mechanical, thermal, or light energy into useful electrical energy. When mechanical stress is applied, polymers with piezoelectric characteristics like Polyvinylidene Fluoride (PVDF) and its copolymers produce electric charges. Sensors and energy-collecting devices make extensive use of this piezoelectric property. Furthermore, thermoelectric polymers of which conducting polymers like polyaniline and polypyrrole are well-known examples can use temperature gradients to produce energy through the Seebeck effect [11, 12, 15].

By producing excitons and then separating them into free charge carriers, photovoltaic polymers including conjugated polymers like P3HT (Poly(3-Hexylthiophene) can absorb solar radiation and transform it into electrical energy [16]. Polymers are potential materials for sustainable energy solutions in wearable technology, sensors, and portable devices because of these mechanisms, which take advantage of their flexibility, lightweight nature, and affordability, as represented in Fig. (**5**).

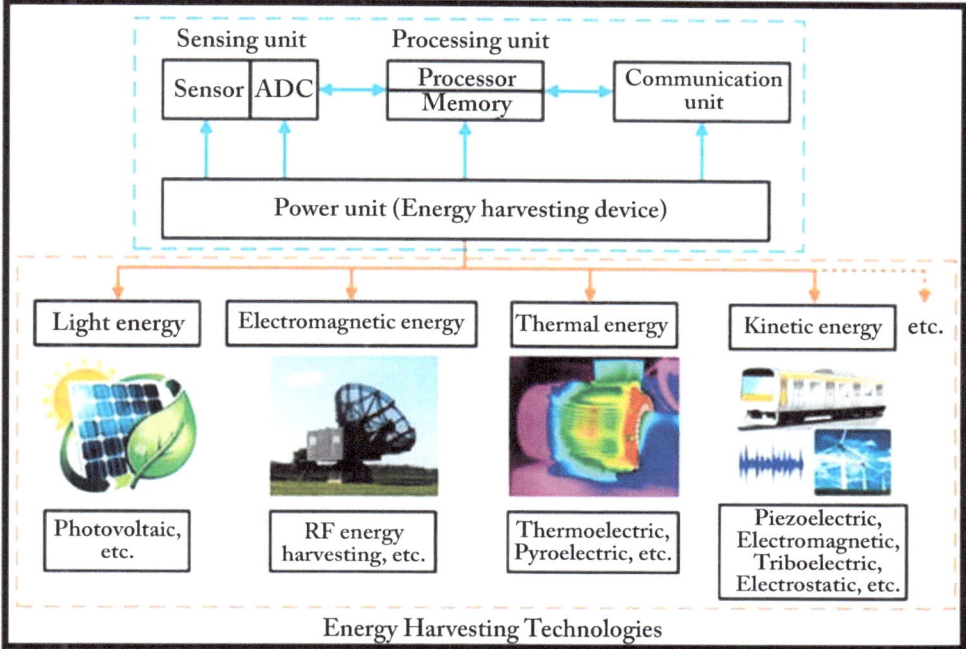

Fig. (5). Energy harvesting techniques [17].

Because of their adaptability, affordability, and adjustable qualities, polymers are useful materials for energy harvesting. The ability of polymers to transform mechanical, thermal, or light energy from external stimuli into electrical energy is the foundation of their energy harvesting processes. The main mechanisms and the kinds of polymers involved are explained in detail below [17]:

Piezoelectric Polymers

Mechanism: Piezoelectric polymers use their capacity to polarize under strain to produce electrical charges when mechanical stress is applied. The alignment of dipoles within the polymer matrix is what causes this phenomenon.

Procedure:

- Mechanical Stress: The polymer deforms as a result of an outside force.
- Dipole Realignment: A charge separation is produced when dipoles inside the polymer realign.
- Charge Generation: This produces a detectable voltage and an electric field.

Important Ingredients:

- Polyvinylidene Fluoride (PVDF): Renowned for its flexibility and excellent piezoelectric coefficients.
- Polyvinylidene Fluoride (Trifluoroethylene) (PVDF-TrFE): Improves thermal stability and piezoelectric response [[14]].
- Polyimides: Wearable and flexible piezoelectric devices use polyimides.

Applications include:

- Using vibration energy to power devices like shoes and equipment.
- Medical equipment and wearable sensors.
- Electronic devices that run on their own power.

Thermoelectric Polymers

Mechanism: By using the Seebeck effect, which creates a potential difference between two surfaces of the material with different temperatures, thermoelectric polymers produce energy when subjected to a temperature gradient [18].

Procedure:

- Heat Gradient: When heat is applied, a temperature differential is produced throughout the material.
- Diffusion of Charge Carrier: Electrons or holes migrate from the heated side to the cold side.
- Generation of Voltage: A voltage is produced when charge carriers travel.

Key Materials:

- Polyaniline (PANI): A conductive polymer with good thermoelectric properties.
- Poly(3,4-ethylenedioxythiophene) (PEDOT): High electrical conductivity and tunable thermoelectric performance.
- Polythiophene Derivatives: Known for flexibility and thermal stability.

Applications include:

- Recovering heat from waste in industrial operations.
- Wearable technology that collects body heat.
- Providing power to distant, low-energy devices.

Photovoltaic Polymers

Mechanism: Through the photovoltaic effect, which includes light absorption, exciton production, and charge separation, photovoltaic polymers transform light energy into electrical energy [18].

Process:

- Light Absorption: Sunlight's photons excite the polymer's electrons, forming electron-hole pairs (excitons).
- Exciton Diffusion: Excitons move toward the donor-acceptor material interface.
- Charge Separation: While holes stay in the donor, electrons are moved to the acceptor at the donor-acceptor contact.
- Charge Transport: A current is produced when distinct charges are moved to the appropriate electrodes.

Important Ingredients:

- Poly(3-hexylthiophene) (P3HT): A popular donor substance with superior light absorption properties.
- PCBM and other fullerene derivatives are utilized as acceptors to improve charge separation.
- Non-Fullerene Acceptors (NFAs): New substitutes that increase stability and efficiency.
- Polythiophene and its derivatives: Offer electronic characteristics that can be adjusted for various spectrum bands.

Applications include:

- Flexible and lightweight solar panels using Organic Photovoltaic (OPV) cells.
- Photovoltaics integrated into buildings, such as semi-transparent solar windows.
- Devices that run on solar power are portable.

RECENT ADVANCES IN POLYMER-BASED ENERGY HARVESTING

In order to satisfy the increasing demand for sustainable energy solutions, recent developments in polymer-based energy harvesting have concentrated on enhancing efficiency, stability, and multifunctionality. Innovations such as the addition of nanostructures (such as graphene and carbon nanotubes) and sophisticated processing methods have greatly increased piezoelectric coefficients in piezoelectric polymers, allowing for more efficient mechanical-to-electrical energy conversion, as depicted in Fig. (**6**). The creation of hybrid materials that combine polymers with inorganic nanoparticles has enhanced thermoelectric polymers' thermal conductivity and power conversion efficiency; conducting polymers such as PEDOT:PSS have demonstrated notable advancements in this area. The power conversion efficiency of polymer-based Organic Photovoltaics (OPVs) has surpassed 19% due to advancements in tandem solar cell topologies and non-fullerene acceptors in photovoltaic polymers [19, 20].

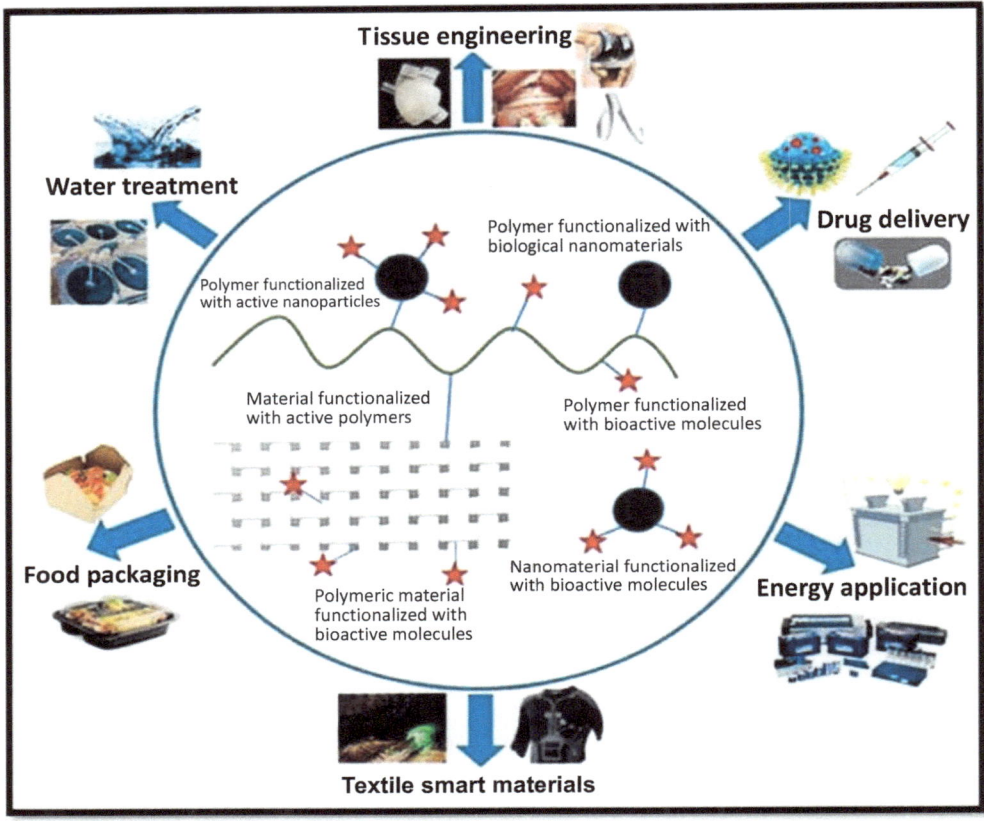

Fig. (6). Recent Advances in Polymer-Based Energy Harvesting [21].

Furthermore, integration into flexible electronics is made possible by developments in wearable and stretchable polymer energy harvesters, which are made possible by breakthroughs in device architecture and molecular design. Additionally, biodegradable and self-healing polymers are being developed, guaranteeing the longevity of devices and the sustainability of the environment. The promise of polymer-based energy harvesting technologies for use in biomedical applications, renewable energy systems, and Internet of Things devices is being fuelled by these advancements as well as scalable production techniques [21].

Efficiency, adaptability, and integration across a range of applications have significantly improved as a result of recent developments in polymer-based energy harvesting [13, 22]. Important advancements include:

- **Conducting Polymer-Based Nanogenerators:** Piezoelectric Nanogenerators (PENGs) and Triboelectric Nanogenerators (TENGs) have been used for biomechanical energy harvesting. Conducting polymers like Polyaniline (PANI), Polypyrrole (PPy), and Poly(3,4-Ethylenedioxythiophene) (PEDOT) have been used in these devices. These materials can be used in wearable energy harvesting systems because of their electrical characteristics and flexibility.
- **Flexible PVDF-Based Piezoelectric Devices:** The piezoelectric qualities of Polyvinylidene Fluoride (PVDF) and its copolymers have been thoroughly studied. The goal of recent research has been to improve the energy conversion efficiency and flexibility of PVDF-based devices so that they can be used in wearable sensors and flexible electronics.
- **Polymer Thermoelectric Composites:** By adding conductive fillers like graphene and carbon nanotubes to polymer matrices, advances in organic polymer thermoelectric composites have been made. These composites are potential options for turning waste heat into electrical energy because of their enhanced thermoelectric qualities.
- **High-Performance Thermoelectric Materials:** Jaime C. Grunlan's research has produced organic thermoelectric coatings for textiles that have the ability to use body heat as a source of energy. These materials, which are appropriate for wearable applications, exhibit high electrical conductivity and notable thermoelectric power factors by employing polyaniline, graphene, and double-walled carbon nanotubes.
- **Nanocomposite Films for Solar Energy Harvesting:** Research has shown promise in solar energy harvesting by the incorporation of materials such as manganese ferrite nanoparticles into polymer matrices like Poly(Vinylidene Fluoride-co-Trifluoroethylene) (P(VDF-TrFE)). These films' pyroelectric and magnetoelectric characteristics suggest that they are appropriate for use in spintronic devices and smart materials.
- **Polymer Nanocomposites for Energy Storage and Harvesting** Studies on fillers based on dopamine that are added to PVDF have demonstrated promise for a variety of adaptable energy storage and harvesting uses. These nanocomposites are appropriate for multipurpose energy devices because of their enhanced dielectric qualities, breakdown strength, and energy storage density.

APPLICATIONS OF ENERGY HARVESTING POLYMERS

Energy harvesting polymers are lightweight, cost-effective, and versatile, making them ideal for diverse applications [1]. In wearable electronics and biomedical devices, piezoelectric and thermoelectric polymers convert body motion and heat into power for sensors, smart textiles, and implants like pacemakers. Photovoltaic polymers enable flexible, lightweight solar panels used in portable devices, solar vehicles, and building-integrated photovoltaics.

In wireless sensor networks, particularly for agriculture, environmental monitoring, and industrial automation, these polymers provide off-grid power to remote sensors. They also enhance consumer electronics by reducing battery dependency in smart devices. Recent advances in biodegradable and self-healing polymers support sustainable, long-lasting energy systems. The main application areas are listed below [23]:

Electronics that are worn

- Wearable technology, such as fitness trackers, smartwatches, and medical sensors, is powered by piezoelectric polymers, which capture energy from physical activities like walking and running.
- Thermoelectric Polymers: Produce energy from body heat to sustain low-power gadgets' continuous operation.
- Photovoltaic Polymers: Make it possible for lightweight, flexible solar cells to be incorporated into apparel.

Medical Equipment

- Implantable Energy Harvesters: Pacemakers and other medical implants are powered by polymers like PVDF, which are employed in systems that transform body movements or heartbeats into electrical energy.
- Self-Powered Sensors: These sensors do not require external batteries and are utilized in real-time health monitoring systems.

Systems of Renewable Energy

- Adaptable Solar Panels: Organic Photovoltaic (OPV) polymers are used in portable, lightweight solar cells that can be used as part of building-integrated photovoltaics or to power small gadgets.

- Wind and Wave Energy: Flexible structures that gather energy from wind or water flow use durable polymer composites.

Electronics for consumers

- Self-Charging Devices: Polymer-based Triboelectric Nanogenerators (TENGs) power portable gadgets, such as cell phones, minimizing the need for external charging.
- Sustainable Charging Stations: In isolated locations, eco-friendly ways to charge devices can be found with devices that use energy-harvesting polymers.

Uses in Automobiles

- Energy-Scavenging Sensors: Tire pressure and engine health sensors are powered by piezoelectric polymers implanted in the tires or chassis of vehicles, which capture energy from mechanical vibrations.
- Waste Heat Recovery: By capturing heat from exhaust systems, thermoelectric polymers increase vehicle efficiency by producing energy.

Internet of Things (IoT)

- Self-Powered IoT Sensors: Solar and triboelectric polymers allow for self-sustaining sensors for industrial automation, smart homes, and agriculture.
- Energy Harvesting Networks: By reducing dependency on external power sources, polymers incorporated into IoT devices allow for widespread adoption.

Monitoring of the Environment

- Remote Sensors: In remote or inaccessible locations, sensors that measure environmental parameters like temperature, humidity, and pollution levels are powered by energy-harvesting polymers.
- Ocean and Marine Monitoring: Autonomous sensors for oceanographic research are powered by polymers in TENGs that absorb energy from water waves.

Intelligent Textiles

- Energy-Generating Clothing: Textiles made of polymers like PVDF are woven to capture motion or heat energy, allowing for uses such as portable device charging.
- Embedded Sensors: Supply energy to sensors incorporated into textiles for use in sports, the military, and healthcare.

Use in Industry

- Machine Vibration Energy Harvesting: Diagnostic sensors are powered by electrical energy generated from mechanical vibrations in industrial machinery using piezoelectric polymers.

- Waste Heat Recovery Systems: To increase overall energy efficiency, thermoelectric polymers harvest energy from industrial processes.

Applications in Aerospace

- Lightweight Power Sources: Photovoltaic and piezoelectric polymers offer energy solutions for spacecraft and airplanes' lightweight sensors and systems.
- Structural Health Monitoring: Embedded sensors that use energy harvesting polymers to power them keep an eye on the structural integrity of spacecraft and airplanes.

Agriculture

- Smart Farming Sensors: In distant agricultural fields, sensors that measure temperature, crop health, and soil moisture are powered by photovoltaic and piezoelectric polymers.
- Wind and Water Energy Harvesting: Small-scale energy harvesting for irrigation system powering is made possible by polymers in flexible structures.

CHALLENGES AND FUTURE DIRECTIONS IN POLYMER-BASED ENERGY HARVESTING

Despite the enormous potential of polymer-based energy harvesting, a number of issues must be resolved to increase its effectiveness, scalability, and commercial viability. The main issues in this area are listed below [24 - 27]:

Low Efficiency of Energy Conversion

- Challenge: Compared to conventional inorganic materials (such as silicon for solar cells and metal-based thermoelectric), many polymer-based energy harvesting devices, particularly Organic Photovoltaics (OPVs) and thermoelectric, continue to have low efficiency.
- Solution: Research aims to improve the intrinsic qualities of polymers by employing cutting-edge nanocomposites to boost energy conversion efficiency, improve charge transport, and optimize material interfaces.

Stability and Durability

- Challenge: Polymers, particularly those utilized in flexible applications (such as textiles and wearable technology), may deteriorate over time as a result of mechanical stress or exposure to environmental factors, including UV light, temperature, and humidity, impacting their long-term performance.
- Solution: Durability and resistance to environmental deterioration can be

increased with the development of more stable and robust polymer materials, such as better encapsulation methods and the use of hybrid polymers.

Manufacturing and Scalability

- Challenge: Large-scale production and integration present difficulties for a number of energy-harvesting polymer technologies, especially in organic photovoltaics and piezoelectric. Inconsistent quality control and high manufacturing costs continue to be obstacles.
- Solution: Developments in inkjet printing, roll-to-roll processing, and other scalable manufacturing methods can lower production costs and enhance material homogeneity. The secret to commercialization is the development of economical and effective processing methods.

Integration of Mechanical and Electrical

- Challenge: Efficient coupling of mechanical, electrical, and thermal properties is necessary to incorporate energy-harvesting polymers into useful devices. Many polymers that exhibit high performance in one area (such as thermoelectric or piezoelectric) might not mix well with other elements.
- Solution: New developments in hybrid systems that integrate several energy-harvesting technologies (such as thermoelectric and piezoelectric) may improve overall efficiency and result in multipurpose gadgets.

Density of Power

- Challenge: Polymer-based energy harvesters are less suited for high-power applications since they frequently have lower power densities than conventional materials.
- Solution: By improving the energy density and output power of polymer-based devices, research into nanomaterial fillers, composite materials, and innovative hybrid structures can broaden the variety of applications for these devices.

Availability and Cost of Materials

- Challenge: Some polymer-based energy harvesters, such as high-performance conductive polymers, can have expensive raw material costs. Furthermore, widespread adoption may be constrained by the availability of some of these materials.
- Solution: Increasing the availability of essential ingredients and creating alternative, affordable feedstocks (such as employing bio-based precursors or agricultural waste) would help lower costs and increase the accessibility of polymer-based devices.

FUTURE DIRECTIONS

Sustainable polymers in energy harvesting have been a recent trend that enables sustainable development, thus enhancing the future of the environment and the well-being of all living beings. Fig. (7) depicts forthcoming polymer-based sustainability and global forums.

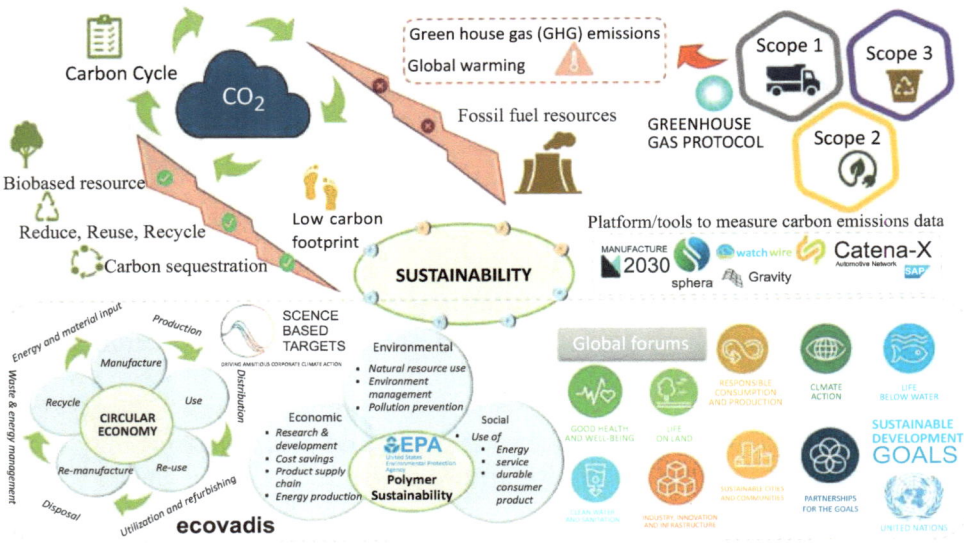

Fig. (7). Forthcoming polymer-based sustainability and global forums [26].

Following are some of the points highlighting the potential direction of sustainable polymers in energy harvesting practices [26, 27]:

High-Performance Hybrid Materials:-

These materials combine organic polymers with inorganic components (such as metal nanoparticles, graphene, and carbon nanotubes) to improve electrical conductivity, thermal stability, and piezoelectric response. Higher energy conversion efficiency and more durable gadgets may result from this.

Flexible and Transparent Solar Cells

Organic Photovoltaics (OPVs) with longer lifespans and better power conversion efficiencies are still being developed for use in flexible, lightweight, and transparent solar cells. Wearable solar cells, building-integrated photovoltaics, and portable electronics are just a few of the novel uses for these.

Multipurpose Equipment

The creation of multipurpose polymer devices that can simultaneously capture energy from several sources, such as heat, light, and mechanical motion. By eliminating the requirement for external power sources, such devices might be included in wearable electronics, self-powered sensors, and even autonomous systems.

Smart Textiles

Energy-harvesting polymers are incorporated into smart textiles for use in sports, healthcare, and the military. By using human movements, heat, or ambient light, these fabrics can power sensors and wearable technology.

Better Thermoelectric Materials

Studies on adding conductive fillers, such as graphene or carbon nanotubes, to polymers to improve their thermoelectric qualities may result in waste heat recovery systems that are more effective. These materials have the potential to be employed in automotive or industrial settings to capture energy from exhaust heat.

Sustainable Energy Harvesting Polymers

Recyclable, biobased, and biodegradable polymers have been created for energy harvesting applications. These sustainable polymers have the potential to promote a circular economy and lessen their negative effects on the environment, especially in long-term use.

Self-Healing Polymers

Increasing the lifespan and functionality of energy-harvesting devices, especially in severe conditions, will need the development of self-healing polymers that can fix themselves after mechanical injury.

Implantable and Wearable Medical Equipment

The use of energy-harvesting polymers in medical devices is promoted to develop wearable health monitoring systems or self-powered implants (like pacemakers), which lessen the need for batteries and improve patient safety and comfort.

Economic Feasibility and Market Readiness

While scalability remains a technical challenge, the economic viability of polymer-based energy-harvesting technologies also requires critical evaluation. High material costs, limited manufacturing throughput, and inconsistent product quality hinder the cost-effectiveness of large-scale production. Moreover, the lack of established supply chains and market infrastructure slows commercialization. Addressing these economic barriers involves investing in scalable fabrication techniques—such as roll-to-roll printing and inkjet deposition—that can reduce per-unit costs and improve process consistency. Strategic partnerships between academia, industry, and government can also foster a favorable ecosystem for market readiness and adoption [28].

CONCLUSION

Utilizing sophisticated, sustainable polymers for energy harvesting holds promise as a game-changing strategy for clean energy. Because of their light weight, flexibility, and adaptability, these materials provide a sustainable substitute for traditional energy technologies by allowing the conversion of ambient energy sources, including mechanical stress, temperature gradients, and sunlight, into electrical energy. Developments in thermoelectric, photovoltaic, and piezoelectric polymers, as well as the incorporation of nanotechnology and hybrid materials, have greatly increased energy conversion efficiency and broadened their range of applications. These polymers' compatibility with flexible and wearable electronics is one of their main advantages; this makes them perfect for powering gadgets like implanted medical sensors, smart textiles, and health monitors. Photovoltaic polymers have transformed organic solar cell technology in the field of renewable energy, resulting in solar power systems that are portable, lightweight, and scalable. In the meantime, the creation of self-healing and biodegradable polymers highlights the dedication to environmental sustainability by lowering e-waste and prolonging gadget lifespans. Energy harvesting polymers' future depends on resolving issues with durability, efficiency, and large-scale manufacturing. The goal of ongoing research is to improve these materials' performance using sophisticated production techniques, molecular design, and nanocomposites. To spur advances and hasten the implementation of these technologies, cooperation between academia, business, and policymakers will be essential. To sum up, energy harvesting in cutting-edge sustainable polymers has enormous potential to power the upcoming generation of gadgets while resolving environmental issues. These materials have the potential to become essential components of sustainable energy systems as they develop further, opening the door to a more environmentally friendly and clean future. They are essential to reaching global energy sustainability targets because of their adaptability and environmental friendliness.

REFERENCES

[1]   J. Palosaari, E. Virta, M. Miinala, and Y. Bai, "Method to monitor cough by employing piezoelectric energy harvesting configurations", In: *Digital Health and Wireless Solutions (NCDHWS 2024).*, M. Särestöniemi, Ed., vol. Vol. 2084. Springer: Cham, 2024, pp. 303-314.

[http://dx.doi.org/10.1007/978-3-031-59091-7_26]

[2] D. Zhu, *Advanced Energy Harvesting Technologies.* MDPI, 2022.
[http://dx.doi.org/10.3390/en15072366]

[3] K. Li, X. He, X. Wang, and S. Jiang, "A nonlinear electromagnetic energy harvesting system for self-powered wireless sensor nodes", *Journal of Sensor and Actuator Networks,* vol. 8, no. 1, p. 18, 2019.
[http://dx.doi.org/10.3390/jsan8010018]

[4] M.R. Rahman, N-A.A. Taib, M.K. Bakri, and S.N. Taib, "Importance of sustainable polymers for modern society and development", In: *Advanced Sustainable Polymer Composites.* Springer, 2021, pp. 1-35.
[http://dx.doi.org/10.1016/B978-0-12-820338-5.00001-1]

[5] Y. Zhu, C. Romain, and C.K. Williams, "Sustainable polymers from renewable resources", *Nature,* vol. 540, no. 7633, pp. 354-362, 2016.
[http://dx.doi.org/10.1038/nature21001] [PMID: 27974763]

[6] H. Peng, X. Sun, W. Weng, and X. Fang, "Polymer materials for energy and electronic applications",
[http://dx.doi.org/10.1016/B978-0-12-811091-1.00005-7]

[7] A. Kumar, V.K. Thakur, H.Y. Nezhad, and K.S. Lee, "Prospects of sustainable polymers", *Sci. Rep.,* vol. 14, no. 1, p. 9430, 2024.
[http://dx.doi.org/10.1038/s41598-024-59439-z] [PMID: 38658595]

[8] S.D. Sadhu, S. Chakraborty, M. Garg, and S.G. Varmani, "Polymers in energy harvesting", *Int. J. Eng. Sci. Invent.,* vol. 3, no. 4, pp. 21-28, 2014.

[9] G.I.C. Righetti, F. Faedi, and A. Famulari, "Embracing sustainability: The world of bio-based polymers in a mini review", *Polymers (Basel),* vol. 16, no. 7, p. 950, 2024.
[http://dx.doi.org/10.3390/polym16070950] [PMID: 38611207]

[10] A. Visco, C. Scolaro, M. Facchin, S. Brahimi, H. Belhamdi, V. Gatto, and V. Beghetto, "Agri-food wastes for bioplastics: European perspective on possible applications in their second life for a circular economy", *Polymers (Basel),* vol. 14, no. 13, p. 2752, 2022.
[http://dx.doi.org/10.3390/polym14132752] [PMID: 35808796]

[11] J. Luukkanen, J. Vehmas, J. Kaivo-oja, and T. O'Mahony, "Towards a general theory of sustainable development: Using a sustainability window approach", *Sustainability (Basel),* vol. 16, no. 13, p. 5326, 2024.
[http://dx.doi.org/10.3390/su16135326]

[12] R. Rustgi, and R.C. Gupta, *Biodegradable Polymers.* Pergamon, 1998.

[13] Y. EL-Ghoul, F.M. Alminderej, F.M. Alsubaie, R. Alrasheed, and N.H. Almousa, "Recent advances in functional polymer materials for energy, water, and biomedical applications: A review", *Polymers (Basel),* vol. 13, no. 24, p. 4327, 2021.
[http://dx.doi.org/10.3390/polym13244327] [PMID: 34960878]

[14] D.M. Taylor, S.G. Lu, and Q.M. Zhang, "Dielectric and thermal properties of ferroelectric poly(vinylidene fluoride-trifluoroethylene) copolymers", *IEEE Trans. Dielectr. Electr. Insul.,* vol. 11, no. 2, pp. 230-235, 2004.
[http://dx.doi.org/10.1109/TDEI.2004.1282454]

[15] X. Tang, X. Wang, R. Cattley, F. Gu, and A.D. Ball, "Energy harvesting technologies for achieving self-powered wireless sensor networks in machine condition monitoring: A review", *Sensors (Basel),* vol. 18, no. 12, p. 4113, 2018.
[http://dx.doi.org/10.3390/s18124113] [PMID: 30477176]

[16] K.M. Zia, and M.U. Akbar, *Processing Technology for Bio-Based Polymers.* Elsevier, 2021.

[17] S. Sojan, and R.K. Kulkarni, "A comprehensive review of energy harvesting techniques and its potential applications", *Int. J. Comput. Appl.,* vol. 139, no. 3, pp. 14-19, 2016.

[http://dx.doi.org/10.5120/ijca2016909120]

[18] T. Rodrigues-Marinho, N. Perinka, P. Costa, and S. Lanceros-Mendez, "Printable lightweight polymer-based energy harvesting systems: materials, processes, and applications", *Materials Today Sustainability,* vol. 21, p. 100292, 2023.
[http://dx.doi.org/10.1016/j.mtsust.2022.100292]

[19] P.B.V. Scholten, J. Cai, and R.T. Mathers, "Polymers for a sustainable future", *Macromol. Rapid Commun.,* vol. 42, no. 3, p. 2000745, 2021.
[http://dx.doi.org/10.1002/marc.202000745] [PMID: 33543832]

[20] C. Cazan, "Advances in sustainable polymeric materials", *Polymers (Basel),* vol. 14, no. 22, p. 4972, 2022.
[http://dx.doi.org/10.3390/polym14224972] [PMID: 36433099]

[21] Y. Meng, "Sustainable polymer & energy: A new open-access journal to share your research", *Sustainable Polymer & Energy,* vol. 1, no. 1, p. 10001, 2023.
[http://dx.doi.org/10.35534/spe.2023.10001]

[22] A. Samir, F. H. Ashour, A. A. Hakim, and M. Bassyouni, "Recent advances in biodegradable polymers for sustainability," Mater. Degrad., 2022
[http://dx.doi.org/10.1038/s41529-022-00277-7]

[23] A.K. Mohanty, F. Wu, R. Mincheva, M. Hakkarainen, J-M. Raquez, D.F. Mielewski, R. Narayan, A.N. Netravali, and M. Misra, "Sustainable polymers", *Nature Reviews Methods Primers,* vol. 2, no. 1, p. 46, 2022.
[http://dx.doi.org/10.1038/s43586-022-00124-8]

[24] N.A. Tarazona, R. Machatschek, J. Balcucho, J.L. Castro-Mayorga, J.F. Saldarriaga, and A. Lendlein, "Opportunities and challenges for integrating the development of sustainable polymer materials within an international circular (bio)economy concept", *MRS Energy & Sustainability,* vol. 9, no. 1, pp. 28-34, 2022.
[http://dx.doi.org/10.1557/s43581-021-00015-7] [PMID: 37521367]

[25] S.A. Miller, "Sustainable polymers: Opportunities for the next decade", *ACS Macro Lett.,* vol. 2, no. 6, pp. 550-554, 2013.
[http://dx.doi.org/10.1021/mz400207g] [PMID: 35581816]

[26] M. Hong, and E.Y-X. Chen, *Future Directions for Sustainable Polymers.* Elsevier, 2019.
[http://dx.doi.org/10.1016/j.trechm.2019.03.004]

[27] J. Thomas, R.S. Patil, M. Patil, and J. John, "Addressing the sustainability conundrums and challenges within the polymer value chain", *Sustainability (Basel),* vol. 15, no. 22, p. 15758, 2023.
[http://dx.doi.org/10.3390/su152215758]

[28] H. Zhang, L. Wang, and Y. Yang, "Recent progress in flexible and wearable piezoelectric devices for energy harvesting, sensing and actuation", *Nano Energy,* vol. 74, p. 104878, 2020.

CHAPTER 14

Biobased Resorbable Polymeric Nanocomposites for Sustainable Healthcare Applications

Sumit Bhowmik[1], Debasish Banerjee[1], Arbind Prasad[2,*], Souvik Debnath[3] and Sudipto Datta[3]

[1] *Mechanical Engineering Department, Omdayal Group of Institutions, Howrah 711316, West Bengal, India*

[2] *Mechanical Engineering Department, Katihar Engineering College (Under Department of Science, Technology and Technical Education, Government of Bihar), Katihar 854109, Bihar, India*

[3] *Department of Materials Engineering, Indian Institute of Science, Bangalore 560012, Karnataka, India*

Abstract: Biobased resorbable polymeric nanocomposites are the biggest breakthroughs in the search for green and eco-friendly healthcare compounds. These biobased materials incorporate biodegradability characteristics with good mechanical, thermal, and biological properties, which are attributable to the nanoscale reinforcements. These nanocomposites help meet the fast-emerging need for green and biocompatible materials in applications such as drug delivery systems, tissue engineering scaffolds, and medical implants. Their ability to not cause any harm upon being metabolized in the human body and maintain their structure during their utilitarian lifespan keeps their environmental footprint and healthcare visitors' waste level to a bare minimum. This paper aims to review the state of existing knowledge, limitations, and opportunities associated with the use of resorbable polymeric biobased nanocomposites in healthcare to establish them as a part of the sustainable future in medical technology.

Keywords: Biobased polymeric nanocomposites, Biodegradable polymers, Nanotechnology, Resorbable materials, Sustainable healthcare.

INTRODUCTION

Anticipating the growing importance of ecological materials and resources in contemporary medical practice, the healthcare industry has spurred a significant advancement in the biosynthesis of resorbable polymeric nanocomposites based

---

* **Corresponding author Arbind Prasad:** Mechanical Engineering Department, Katihar Engineering College (Under Department of Science, Technology and Technical Education, Government of Bihar), Katihar 854109, Bihar, India; E-mail: arbind.iitg@gmail.com

Sushil Kumar Verma, Sonika & Arbind Prasad (Eds.)
All rights reserved-© 2026 Bentham Science Publishers

on biobased materials [1 - 3]. These are new materials derived from renewable biological resources, which are unique and environmentally friendly as opposed to petroleum-based polymers [4]. Polymers, mainly obtained from renewable sources, are classified into biodegradable and non-biodegradable categories [5].

Among other polymers, resorbable polymers are equally preferable for uses in medicine since they can disintegrate in the body and do not need to be surgically removed [5]. As a result, they can be left in the human body for long periods as they end up being medical waste [6]. Polymeric nanocomposites are developed by identifying a nanoscale filler, which, when added to the biobased polymers, results in improved mechanical strength, thermal stability, and biological performance [7]. The combination of biobased polymers and nanotechnology enables the design and development of materials that conform to the strict test requirements of a medical device while also addressing sustainable environmental and application needs [8]. The application of these nanocomposites include drug delivery to control the release of therapeutic substances and tissue engineering scaffolds that enable cell growth and regeneration [9].

Additionally, the use of polymeric nanocomposites in medical products helps reduce the product's environmental impact; these polymeric nanocomposites break down into harmless substances after use [10, 11]. This biodegradability is of paramount importance for minimizing the stockpiling of medical waste and lessening the effects on the environment caused by disposable medical products [11]. However, there are some issues that require further investigation, including the maximization of the degradation rates, controlling the mechanical properties, and fabrication at a commercial scale [12, 13]. These issues must be addressed in order to properly integrate biobased resorbable polymeric nanocomposites into clinical practice. This chapter presents general features of the importance of biobased resorbable polymeric nanocomposites to enhance sustainable healthcare solutions [14]. With the help of renewable resources and nanotechnology applications, these materials provide a unique opportunity for the development of environmentally friendly yet high-performance medical materials.

BIOBASED RESORBABLE POLYMERIC NANOCOMPOSITES

The everyday emerging environmental and biomedical demands require healthcare to embrace sustainable materials. In these, biobased resorbable polymeric nanocomposites have proven to be revolutionary materials because of their biocompatibility, degradability, and increased functionality [15]. This chapter delves into the basic concepts and their development, ushering the reader through the possibilities of these materials, together with their weaknesses and the contemplated options for the future [16]. Biobased resorbable polymeric nanocomposites are nanocomposite materials that can be synthesized using renewable resources and are biodegradable and non-toxic in physiological environments [17]. Such materials are special in healthcare applications because biobased polymers are biodegradable, while nanoscale reinforcements improve structure and function [18]. This synergy has made them usefulin the development of drugs, engineering of tissues, and the production of medical implants [19]. This provides another stimulus for developing biopolymers that are dispatched worldwide, such as PLA and PHA, which can be functional with the help of nanotechnology. The study findings are relevant to the UN's Sustainable Development Goals (SDGs) regarding resource consumption and production in healthcare [20].

PLA, PHA, and Polycaprolactone (PCL) are examples of biopolymers that provide the basis of these composite materials [21]. These polymers are produced from renewable resources, for instance, corn starch and sugarcane, and through microbial feeding processes [22]. This reduces environmental problems associated with disposal, and in medical applications, the surgical expulsion of materials used is not required due to their non-biodegradability [23]. Nanofillers, such as nanoclays, carbon nanotubes, and hydroxyapatite, increase the mechanical properties, thermal properties, and biological activity of polymers. These nanofillers are also used to control the degradation rate of the composites to suit a given biomedical application [23, 24]. Resorbable polymeric nanocomposites based on biopolymers are a common element in controlled-release drug delivery systems. For instance, it is possible to develop nano clay-PLA composites that present effective drug delivery of anti-inflammatory medicines, with a reduced percentage of side effects and increased patient compliance [25]. There is increasing utilization of biobased nanocomposites in tissue engineering scaffolds because of their biocompatibility and promotion of cell attachment, growth, and differentiation. Nanocomposite scaffolds of PCL-hydroxyapatite have the potential to be used in bone tissue engineering [26]. Biodegradable implants, such as screws and plates prepared from poly lactic acid-carbon nanotube composites, have the benefit of degrading in the body if their function of holding the bone together is served without having to remove them through another operation [27].

APPLICATIONS IN HEALTHCARE

The uses of biobased resorbable polymeric nanocomposites in the healthcare field are broad and impactful, meeting modern medical needs while aiming for a greener world. Below are detailed explorations of their applications across various domains:

Advanced Drug Delivery Systems

Biobased polymeric nanocomposites have been found to transform drug delivery systems as they are capable of encapsulating and releasing drugs. Depending on the choice of polymer matrix and nanofillers, these systems can create the desired kinetics of drug release and adjust them to therapeutic indications.

For example, chemotherapeutic agents such as PLA-nano clays have been engineered to target cancer cells, enhancing efficiency and reducing side effects [28]. Composites containing Polyhydroxyalkanoates (PHA) and carbon nanotubes can be used to enhance insulin release, improve stability, and control the release rate over longer durations in comparison with earlier work from the same group. The designed composite has a significantly larger size, 200 nm in diameter, according to a study by Kim *et al.*, 2022 [3].

Tissue Engineering Scaffolds

In tissue engineering, the EF focuses on scaffolds that resemble the ECM found in regenerative medicine [29]. Biobased resorbable polymeric nanocomposites are rheologically suitable for scaffold fabrication, and the mechanical and biological characteristics of these systems can be manipulated.

The blended nanocomposite of PLA and hydroxyapatite has been proven to work effectively in bone tissue engineering due to increased osteoconductivity and mechanical strength [30]. Likewise, PCL/graphene oxide nanofiber scaffolds have shown biocompatibility for nerve tissue regeneration, along with encouraging neurite extension and electrical conduction [5].

Resorbable Medical Implants

Bio-based nanocomposites are gaining popularity in designing resorbable implants, including screws, plates, stents, *etc.*, that are designed to dissolve after fulfilling their function. This eliminates the need for secondary surgery removal in patients, thereby cutting the risks and expenses associated with the patient's medical bills.

Poly(lactic acid)-carbon nanotube composites have been used in orthopedic implants as they improve the mechanical properties and biocompatibility of the material [30]. Moreover, magnesium-based polymeric nanocomposite stents are under development for cardiovascular use, where degradation of the stent is planned to support vessel healing [15].

Wound Healing and Skin Regeneration

Additional nanocomposites used in wound healing applications include biobased resorbable nanocomposites. Heterogeneous hydrogels with antimicrobial activity in combination with chitosan-based polymers, in which reinforcement involves silver nanoparticles, stimulate tissue regeneration [30]. Previous research on burn treatment using nanocomposite dressings made of polylactic acid and zinc oxide nanoparticles has proved that they facilitate faster healing by creating a moist environment and minimizing bacterial infection [31].

Biodegradable Biosensors

Currently, there has been development of biodegradable biosensors derived from biobased resorbable polymeric nanocomposites for use in tracking physiological status. The benefits include the sensor being implantable and resorbable, hence no need for the surgical removal of the device.

For example, biosensors using PHA-carbon nanotube composites have been designed to measure glucose levels in diabetic patients, which do not require additional invasive procedures but give continuous results [32].

Dental Applications

Other applications of resorbable nanocomposites include the manufacture of temporary dental filling materials, guided tissue regeneration membranes, and other resorbable orthodontic elements. Some PLA-hydroxyapatite nanocomposites, for instance, are used in dental implants to improve insertion torque and stimulate bone formation [16].

Antimicrobial Coatings for Medical Devices

Nanocomposites with antimicrobial activity derived from biological sources are used as coatings for medical equipment and instruments, such as catheters and surgical tools. These coatings, which contain silver or copper nanoparticles, help prevent biofilm formation and lower the incidence of HAIs [33].

Controlled Release of Growth Factors

In regenerative therapies, the ability to deliver the controlled release of growth factors is critical to tissues. Polymeric nanocomposites prepared from bioresources containing growth factors such as VEGF have been developed to stimulate new blood vessel formation and enhance the rate of wound healing [26].

Injectable Nanocomposites

Hydrogels for injectable therapies characterized by biobased nanocomposites have been studied for minimally invasive procedures. These hydrogels with nanofillers, such as graphene oxide or carbon nanotubes, are utilized in soft tissue transplantation and site-specific chemotherapeutic drug delivery applications [12].

Anticancer Applications

Biodegradable polymeric nanocomposites based on biopolymers and containing anticancer agents, such as curcumin or gold nanoparticles, have proven to be effective in targeted cancer treatment. These systems enable controlled drug delivery to cancer cells without harm to the normal cells surrounding the tumors [16].

CHALLENGES AND LIMITATIONS

Limitations of biobased resorbable polymeric nanocomposites are as follows:

Degradation Control

It is always a challenge to get a highly predictable degradation rate. The properties of the polymer and the conditions within its environment can influence the rate of degradation [22].

Cost and Scalability

Biobased polymers and nanocomposites are often costly to manufacture, and this has predefined their consumption. Moreover, the challenges of manufacturing nanocomposites at a commercial level are technical [31 - 33].

Toxicological Concerns

Although these materials are intended to be biocompatible, the long-term behavior of nanofillers, especially carbon-based ones, has not been fully investigated for safety [27]

FUTURE DIRECTIONS OF BIOBASED RESORBABLE POLYMERIC NANOCOMPOSITES IN HEALTHCARE

The prospect for future development of biobased resorbable polymeric nanocomposites lies in the use of high technologies, including AI and ML, for the prediction of the material properties during their creation [24]. Furthermore, the synthesis of bio-inspired nanocomposites that resemble human tissues such as bone and cartilage will be a key area of application in regenerative medicine [34 - 36].

This is a relatively young discipline of biobased resorbable polymeric nanocomposites, where new opportunities are being sought due to the transition to the use of environmentally friendly materials. Such specific materials have already proven to be the materials of the future in numerous healthcare-related applications, but the future has even more to offer. This essay identifies key areas to brainstorm in the quest to identify trends that are likely to define the future of these materials in advancing healthcare challenges globally.

Advanced Material Design with Artificial Intelligence

AI and ML, particularly, have become the key drivers for promoting material design at a higher rate [23]. These technologies allow for the prediction of the best balances between polymers and nanofillers, resulting in biobased nanocomposites with specific characteristics for application. For instance, AI stimulation means can be used to enhance the degradation rate, tensile strength, and cell compatibility of nanocomposites [37]. Furthermore, the use of the technological processes of machine learning makes it possible to implement efficient, less costly, and scalable manufacturing processes. This inclusion of AI into material design is a major advancement in increasing the speed and accuracy of creating future healthcare materials [38].

Development of Smart Nanocomposites

Self-assembled nanocomposites with the ability to change their properties or behavior by adjusting to changes in their environment, for example, variation in pH, temperature, or the presence of a magnetic field, are still in development. These materials offer potential for uses where there is a need to program the material's behavior. For instance, pH-sensitive nanocomposites can release drugs selectively in acidic tumor sites, resulting in improved therapeutic outcomes but reduced toxicity [37 - 40]. Likewise, thermoresponsive nanocomposites could also be applied in self-healing implants where the material is capable of healing itself in a given physiological environment. Thus, smart nanocomposites are foundational to future advancements in healthcare because of their flexibility, which allows the body to respond and adapt to its environment [17].

Bioinspired Nanocomposites

Whenever dynamic advancements are observed in the field of material science, nature has often been seen as a reference point, hence the formation of bioinspired nanocomposites. The intended nanocomposites replicate the forms and functions of natural composites, such as nacre and spider silk, to possess unique mechanical and functional characteristics [41]. These designs may pave the way to develop harder and more robust scaffolds for tissue engineering and other uses in the medical field. For instance, nacre's composite structure may be used to make composites with excellent damage tolerance, and the elasticity of spider silk may be used for the creation of stretchable materials for soft tissue engineering applications, as noted by Patel & Kumar (2023) [30].

Green Manufacturing Techniques

Biodegradability is the key selling point of resorbable polymeric nanocomposites; therefore, future work efforts must be directed at increasing ecological compatibility. The treatment of these materials and the production of new ones with reduced toxicity are being achieved through green manufacturing processes, solvent-free production, and bio-degradable additives. Moreover, an effort to scrutinize low-energy fabrication techniques is being made to minimize energy consumption in mass production. These developments correlate with the United Nations' sustainable development goals in managing healthcare solutions with enhanced features rather than having benefits limited to their usage [15].

Multifunctional Healthcare Devices

The healthcare devices that will be developed in the next generation are expected to have multifunctional biobased nanocomposite functions that will enable multiple tasks at once. For instance, a conjoint could work as a template on which tissue can reproduce, as a probe to track the healing progress, or as a system through which medicine can be administered singularly to the damaged site. Besides improving patient care and engagement, this multifunctionality serves the purpose of eliminating the need for many individual devices that streamline the overall medical process, thus redefining the development and utilization of medical devices and hence, future arrangements for remedial interventions [16].

Exploration of Novel Biopolymers and Nanofillers

To further enhance the application of nanocomposites derived from biobased materials, researchers sought other ranges of biopolymers and nanofillers. New materials include alginate, cellulose derivatives, or fibroin derived from silkworm cocoons due to their extraordinary characteristics. Similarly, nanofillers of quantum dots, Metal-Organic Frameworks (MOFs), and biomimicking nanoparticles are used as nanofillers in polymer matrices. Such development may result in enhanced mechanical properties, degradability, and functionality of nanocomposites, which will expand their utilization in the healthcare sector, as reported in a study [42].

PERSONALIZATION THROUGH 3D PRINTING

Long-Term Biocompatibility and Toxicity Studies

Nevertheless, it is important to note that the long-term applications of biobased nanocomposites, as well as their safety and biocompatibility, remain underexplored in most cases. More investigative research is required to investigate the effects of this class of material on biosystems over long-term intervals. More studies should be directed to the possible impact of degradation by-products and the actual behavior of nanoscale fillers in elaborated tissues, chronic swelling, or other unfavorable responses. These concerns must be addressed to pave the way for the use of these materials in clinical environments [43].

Applications in Emerging Fields

The application area of biobased nanocomposites is progressively moving towards the development of new areas like biosensing, bioelectronics, and regenerative medicine [44 - 47]. Thus, biosensors in the form of clothing made from these materials can continuouslyprovide data on the patient's health status, including their vitals and the status of the disease [48]. For bioelectronics, nanocomposites can serve as materials that help create devices that integrate with neural tissue for applications in prostheses and other forms of brain-computer interfaces. It is important to note that all of these applications point to the androgynous uses of biobased nanocomposites for a broad spectrum of healthcare needs [49, 50].

Global Collaboration and Standardization

The focus, in turn, suggests that increased global cooperation of researchers, industries, and regulatory bodies will facilitate the increased use of biobased nanocomposites [51, 52]. This will help establish the required protocols for the synthesis, testing, and approval of these materials to ensure they remain safe and effective over the continuum of large-scale manufacture and use. Furthermore, collaborations between academia and industrial sectors may help in fast-tracking the uptake of some of the laboratory research findings into practical use. To eliminate existing barriers and reach its full potential, the field must support cooperation in training and improve comprehensiveness [27].

CONCLUSION

Bio-derived, resorbable polymeric nanocomposites are at the cross-section of sustainability and medical advancement. Due to their utilization of renewable resources and nanotechnology features, these materials provide solutions to significant environmental and medical problems. Challenges such as costs and scalability remain even today, but further research and development in this area will guarantee the enhancement of potential medical applications within materials science for the future of sustainability.

The various uses of biobased resorbable polymeric nanocomposites further demonstrate how they can revolutionize healthcare. Whether used as drug delivery systems or tissue engineering scaffolds, the importance of the combination of sustainability and enhanced product functionality cannot be overemphasized in contemporary medicine. Future work is expected to enhance their applications in both environmental and patient concerns, which will define the new paradigm for environmentally sustainable and patient-centered medical technologies.

REFERENCES

[1]   L. Garcia, and R. Martinez, "Enhancing mechanical properties of biobased polymers through nanocomposite formation", *J. Sustainable Mater.*, vol. 15, no. 3, pp. 245-260, 2023.

[2]   Y. Huang, and X. Chen, "Challenges in scaling up production of resorbable polymeric nanocomposites for medical applications", *Adv. Manuf. Process.*, vol. 10, no. 2, pp. 112-128, 2022.

[3]   S. Kim, and J. Lee, "Applications of polymeric nanocomposites in tissue engineering and drug delivery systems", *Biomed. Nanotechnol.*, vol. 8, no. 4, pp. 301-315, 2022.

[4]   H. Lee, S. Park, and D. Lee, "Biodegradable polymers for resorbable medical devices: A comprehensive review", *Mater. Sci. Med.*, vol. 32, no. 1, pp. 89-104, 2021.

[5]   V. Oliver-Cuenca, "Bio-based and biodegradable polymeric materials for a circular economy", *Polymers (Basel)*, vol. 16, no. 21, p. 3015, 2024.

[6]   T. Nguyen, P. Tran, and Q. Le, "Environmental impact of biodegradable medical polymers: Degradation pathways and byproducts", *Environ. Polym. J.*, vol. 19, no. 1, pp. 58-75, 2023.

[7]   M. Patel, and A. Singh, "Reducing medical waste through biodegradable polymers: Current trends and future prospects", *J. Green Chem.*, vol. 27, no. 5, pp. 403-420, 2021.

[8]   R. Barua, and S. Datta, "Revolutionizing nerve repair: The transformative role of nanoparticles in peripheral nerve regeneration", In: *Nanomater. Nerv. Syst.* IGI Global, 2025, pp. 275-300.

[9]   E. Azadi, M. Dinari, M. Derakhshani, K.R. Reid, and B. Karimi, "Sources and extraction of biopolymers and manufacturing of bio-based nanocomposites for different applications", *Molecules*, vol. 29, no. 18, p. 4406, 2024.

[10] N. Jain, "Bioinspired nanocomposites: Multifunctional materials towards sustainable alternatives", In: *Nanomanufacturing Techniques Sustainable Healthcare Applications.* CRC Press, 2024, pp. 78-93.

[11] Y.O. Waidi, N. Jain, R. Barua, and S. Datta, "Clinical translational studies and challenges of bioabsorbable polymeric composites", In: *Apple Academic Press eBooks*, 2025, pp. 267-282.

[12] R. Barua, S. Datta, and N. Jain, "Novel power generation technology based on renewable energy", In: *Optim. Tech. Hybrid Power Syst.: Renew. Energy, Electric Vehicles, and Smart Grid.* IGI Global, 2024, pp. 407-425.
[http://dx.doi.org/110.1201/9781003569817-13]

[13] J. Smith, and K. Johnson, "Renewable resources in polymer science: Innovations and applications", *Polym. Chem. Today*, vol. 14, no. 2, pp. 134-150, 2022.

[14] Y. Zhang, H. Liu, and Q. Wang, "Nanotechnology in sustainable healthcare: The role of polymeric nanocomposites", *Nano Rev.*, vol. 5, no. 3, pp. 210-225, 2020.

[15] S. Das, S. Datta, A. Barman, and R. Barua, "Smart Biodegradable and Bio-Based polymeric biomaterials for biomedical applications", In: *Advances in chemical and materials engineering book series*, 2023, pp. 56-82.
[http://dx.doi.org/10.4018/978-1-6684-9224-6.ch003]

[16] T. Ahmed, and R. Gupta, "Challenges in scaling biobased nanocomposite production for medical applications", *J. Adv. Mater.*, vol. 10, no. 2, pp. 112-120, 2023.

[17] X. Chen, and J. Lin, "Hydroxyapatite-based nanocomposites for bone tissue regeneration", *Biomed. Eng. Res.*, vol. 8, no. 4, pp. 201-215, 2023.

[18] R. Barua, S. Das, A.R. Chowdhury, and P. Datta, "Experimental and simulation investigation of surgical needle insertion into soft tissue mimic biomaterial for minimally invasive surgery (MIS)", *Proc. Inst. Mech. Eng. H*, vol. 237, no. 2, pp. 254-264, 2022.
[http://dx.doi.org/10.1177/09544119221143860]

[19] R. Barua, "Advanced biomimetic compound continuum robot for minimally invasive surgical applications", In: *Modeling, Simulation, and Control of AI Robotics and Autonomous Systems*. IGI Global, 2024, pp. 213-231.

[20] S. Chowdhury, "Biological smart biomaterials: Materials for biomedical applications", In: *Applications Biotribology Biomedical Systems*. Springer Nature Switzerland, 2024, pp. 313-325.

[21] R. Gonzalez, and H. Martinez, "The role of biobased polymers in achieving sustainable development goals", *J. Green Chem.*, vol. 12, no. 5, pp. 345-362, 2023.

[22] A. Banerjee, and R. Barua, "A comprehensive overview of carbon-based nanofluids and related progress for heat transfer uses", In: *Environ. Appl. Carbon-Based Mater.* IGI Global, 2024, pp. 196-217.

[23] R. Barua, "The emerging role of artificial intelligence in organ-on-a-chip (OOAC) biomedical devices", In: *Reshaping Healthcare Cutting-Edge Biomedical Advancements*. IGI Global, 2024, pp. 369-381.

[24] J. Kurowiak, T. Klekiel, and R. Będziński, "Biodegradable polymers in biomedical applications: A review—Developments, perspectives and future challenges", *Int. J. Mol. Sci.*, vol. 24, no. 23, p. 16952, 2023.

[25] R. Barua, "An investigation of AI techniques for detecting kidney stones in CT scan images through advanced image processing", In: *Enhancing Med. Imaging Emerging Technol.* IGI Global, 2024, pp. 133-150.

[26] Y. Huang, and Z. Wang, "Artificial intelligence in the design of polymeric nanocomposites", *Mater. Today*, vol. 36, no. 3, pp. 58-74, 2023.

[27] R. Barua, "Exploring artificial intelligence in evolving healthcare environments: A comprehensive analysis", In: *Advances Comput. Intell. Healthcare Industry 4.0*. IGI Global, 2024, pp. 123-138.

[28] J. Lee, and H. Park, "Advances in nanoclay-polymer drug delivery systems", *J. Nanomed.*, vol. 17, no. 3, pp. 102-118, 2023.

[29] P. Nguyen, and Q. Tran, "Biodegradable nanocomposites for resorbable implants: A review", *Polym. Sci. Med.*, vol. 25, no. 1, pp. 90-105, 2023.

[30] S. Patel, and A. Kumar, "Biobased polymers: A review of sources, properties, and applications", *Int. J. Polym. Sci.*, vol. 15, no. 6, pp. 291-312, 2023.

[31] T. Rahman, and X. Zhao, "Nanofiller-enhanced biopolymer composites for healthcare", *Comp. Med.*, vol. 7, no. 2, pp. 165-180, 2022.

[32] N. Jain, Y. O. Waidi, R. Barua, V. Vannaladsaysy, A. Prasad, S. Datta, "Plastic recycling for energy

production. Biodegradable Waste Processing for Sustainable Developments", 113. 2024

[33] S. K. Verma, A. Prasad, & V. Katiyar, "State of art review on sustainable biodegradable polymers with a market overview for sustainability packaging". Materials Today Sustainability, 100776, 2024.

[34] S. Datta, R. Barua, and A. Prasad, "Additive Manufacturing for the Development of Artificial Organs. Advanced Materials and Manufacturing Techniques for Biomedical Applications", 411-427, 2023.

[35] A. Prasad, S. Datta, A. Kumar and M. Gupta, "Introduction to Next-Generation Materials for Biomedical Applications". "Advanced Materials and Manufacturing Techniques for Biomedical Applications", 1-24, 2023.

[36] G. Chakraborty, V. Pandey, A. Prasad, and A. Kumar, "Introduction to Sustainable Manufacturing for Industries 4.0", In: *Sustainable Smart Manufacturing Processes in Industry 4.0*. CRC Press, 2023, pp. 1-17.

[37] A. Prasad, S.M. Bhasney, V. Prasannavenkadesan, M.R. Sankar, and V. Katiyar, "Polylactic acid reinforced with nano-hydroxyapatite bioabsorbable cortical screws for bone fracture treatment", *J. Polym. Res.*, vol. 30, no. 5, p. 177, 2023.

[38] A. Prasad, S.M. Bhasney, V. Prasannavenkadesan, M.R. Sankar, and V. Katiyar, "Nano-hydroxyapatite reinforced polylactic acid bioabsorbable cancellous screws for bone fracture fixations", *J. Appl. Polym. Sci.*, vol. 140, no. 43, p. e54577, 2023.

[39] G. Chakraborty, A. Prasad, and A. Kumar, *Processing of biodegradable composites. In Biodegradable Composites for Packaging Applications*. CRC Press, 2022, pp. 33-48.

[40] A. Prasad, A. Kumar, K.K. Gajrani, Ed., *Biodegradable Composites for Packaging Applications.* CRC Press, 2022.

[41] A. Prasad, S. Datta, S. De, P. Singh, and B. Mahto, "Bioresorbable Composite for Orthopedics and Drug Delivery Applications", In: *Applications of Biotribology in Biomedical Systems*. Springer Nature Switzerland: Cham, 2024, pp. 327-344.

[42] P. Singh, and R. Kumar, "Renewable biopolymers in sustainable healthcare", *J. Biomater.*, vol. 28, no. 4, pp. 234-250, 2023.

[43] Y. Wang, and M. Luo, "Controlling degradation rates in polymeric nanocomposites", *Sci. Degrad. Mater.*, vol. 18, no. 2, pp. 78-91, 2023.

[44] R. Barua, "Unleashing human potential: Exploring the advantage of biomechanics in sport performance", In: *Global Innov. Phys. Educ. Health*. IGI Global, 2025, pp. 409-436.

[45] H. Zhang, and Y. Li, "Toxicological evaluation of nanofillers in medical composites", *J. Toxicol. Mater.*, vol. 15, no. 3, pp. 85-97, 2023.

[46] R. Barua, "Revolutionizing Medicine: Exploring Cutting-Edge Technologies in Computer-Aided Medicinal Analysis", In: *Computer-Assisted Analysis for Digital Medicinal Imagery*. IGI Global, 2025, pp. 365-392.

[47] N. Jain, R. Barua, and S. Datta, "Functional Nanomaterials and their Applications in Biomedicine", In: *Apple Academic Press eBooks*, 2025, pp. 61-79.
[http://dx.doi.org/10.1201/9781003569817-3]

[48] S. Datta, Y.O. Waidi, and R. Barua, "3D printing of bioabsorbable polymeric composites in biomedical applications", In: *Apple Academic Press eBooks*, 2025, pp. 243-266.
[http://dx.doi.org/10.1201/9781003569817-12]

[49] R. Barua, "Neuroengineering Horizons," in Advances in medical education, research, and ethics (AMERE) book series, 2025, pp. 347–374.
[http://dx.doi.org/10.4018/979-8-3693-5464-3.ch013]

[50] D. Banerjee, R. Barua, S. Datta, and D. Pathote, "The use of artificial neural networks in biomedical engineering," in Advances in chemical and materials engineering book series, 2025, pp.

33–50
[http://dx.doi.org/10.4018/979-8-3693-7250-0.ch002]

[51] R. Barua and S. Datta, "Tailored healthcare," in Advances in medical technologies and clinical practice book series, 2024, pp. 17–36.
[http://dx.doi.org/10.4018/979-8-3693-4422-4.ch002]

[52] N. Jain, R. Barua, Y. O. Waidi, and S. Datta, "Nanowires for Bio-Sensing applications," in Advances in chemical and materials engineering book series, 2024, pp. 205–219.
[http://dx.doi.org/10.4018/979-8-3693-1306-0.ch010]

CHAPTER 15

Carbon-Based Nanocomposites (CBNs) for Environmental Energy Harvesting Applications

Sagar Vikal[1], Durvesh Gautam[1], Ajay Kumar[2], Rizwan Khan[1], Yogendra K. Gautam[1,*], Ashwani Kumar[3] and Neetu Singh[2]

[1] *Smart Materials and Sensor Laboratory, Department of Physics, Ch. Charan Singh University, Meerut 250004, Uttar Pradesh, India*

[2] *Department. of Biotechnology, Mewar University, Chittorgarh 312901, Rajasthan, India*

[3] *Department of Physics, Regional Institute of Education (NCERT), Bhubaneswar 751022, Odisha, India*

Abstract: Rapid urbanization has a significant impact on various components of the environment, putting stress on energy reservoirs. These effects targeting human health risks, in particular, include air pollutants, wastewater load, poor sanitation, *etc*. Unsustainable urbanization has also led to a surge in energy requirements for domestic, industrial, road, and transport use. To address such issues, different methodologies have been employed to treat different wastewater types, remove gaseous pollutants, and develop technological interventions that lead to energy harvesting. In recent years, among several approaches, Carbon-Based Nanocomposites (CBNs) have been recognized by the scientific community in several fields, including energy and the environment, because of their increased "surface area to volume ratio", good chemical stability, higher conductivity, reduced toxicity, high dispersibility, excellent thermal, mechanical, and electrical properties, cost-effectivity, facile synthesis, the scope of surface functionalization, biocompatibility, *etc*.

This chapter draws attention to the application of various Carbon-Based Nanomaterials (CBNs) (such as polymer nanocomposites, graphene, Graphene Oxide (GO), reduced GO (rGO), graphite, activated carbon, *etc*.) (i) for the detection of heavy metals, pesticides, bacteria, *etc*., in contaminated water and their efficient removal, (ii) for the degradation and removal of organic species from wastewater, (iii) as environmental sensors for detecting toxic gases, and (iv) as mechanical sensors and energy-harvesting devices. Further, the latest developments, current challenges, and future scope of research in CBNs for enhanced environmental and energy-harvesting applications have also been discussed.

---

* Corresponding author Yogendra K. Gautam: Smart Materials and Sensor Laboratory, Department of Physics, Ch. Charan Singh University, Meerut 250004, Uttar Pradesh, India; E-mail: ykg.iitr@gmail.com

Sushil Kumar Verma, Sonika & Arbind Prasad (Eds.)
All rights reserved-© 2026 Bentham Science Publishers

Keywords: Carbon-based nanocomposites, Environment, Energy conversion, Energy storage, Wastewater.

INTRODUCTION

The major driving forces aiding water pollution include unsustainable urbanization, population growth, rapid industrialization, poor sanitation, climate change, *etc* [1]. Every day about two million tons of industrial, sewage, and agricultural waste are globally discharged into the water [2]. The major types of contaminants associated with water pollution include organic, inorganic, radiological, and biological, which are categorized into natural (higher than acceptable limits of nitrate, fluoride, arsenate, chloride, *etc.*), anthropogenic (heavy metals, dyes, pesticides, antibiotics, paints, oils, detergents, *etc.*), and pathogenic microbes (like viruses, bacteria, parasitic protozoa, and worms) [2]. As per the WHO, about 80% of human diseases are linked to water pollution [1]. Some of the reported diseases include cancer (liver, lungs, kidney, prostate, bladder, nasal passages, skin, *etc.*) and non-cancer ailments (diarrhea, vomiting, nausea, stomach pain, typhoid, severe headaches, cholera, skin discoloration, numbness in feet and hands, blindness, partial paralysis, *etc.*) [2]. The different techniques mentioned in the literature for the removal of physicochemical contaminants and disinfection of water from microbial pathogens are physical (sedimentation, filtration, boiling, distillation), chemical (chlorination, adsorption, coagulation, ozonation, flocculation), and biological methods (biologically-activated carbon, bioremediation, *etc.*). The above-mentioned techniques of wastewater treatment have both advantages and limitations. Apart from this, the magnitude of the problem is so vast that the above-mentioned selective treatments are not enough. Under such circumstances, exploring cost-effective and novel techniques for wastewater treatment has become a priority.

Nanotechnology-inspired routes have recently attracted attention among researchers, and in this direction, different types of CBNs (Fig. **1**) are prominent candidates due to their excellent structural and surface properties [3]. The diverse applications of CBNs are attributed to their high thermal conductivity, physical strength, current density, electronic conductivity, electron mobility, affinity binding, surface area, strong adsorption, ion absorption, catalysis support, light weight, strong and slim structure, metallic/semi-conductive properties, *etc* [4].

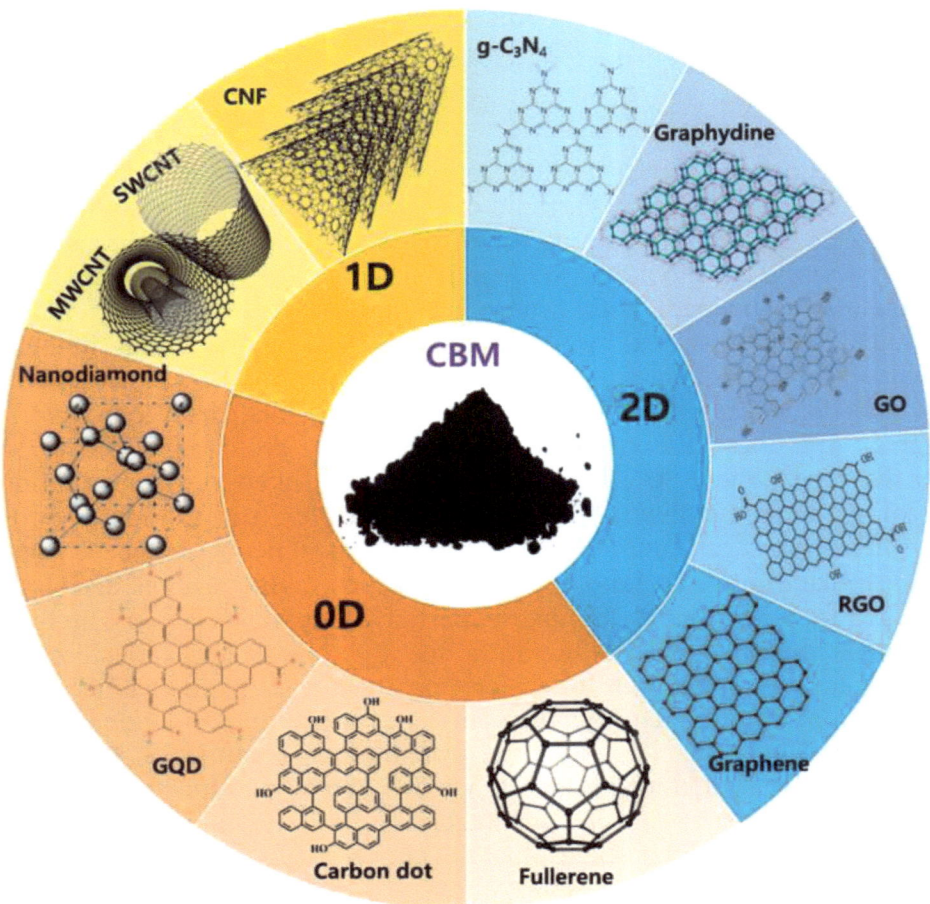

Fig. (1). Types of carbon-based materials [7].

Composites made of Graphene-based NMs (GNMs) can be used to absorb various water contaminants, including colors, heavy metals, antibiotics, and pesticides [5, 6]. The biocompatibility, pH sensitivity, and improved surface properties of CBNs qualify them as potential antimicrobial agents for water decontamination. Several studies have been reported on CBNs decorated with noble metals (Au, Ag, Pt, Pd, *etc.*) for enhanced heterogeneous photocatalysis, leading to the efficient treatment of dye-contaminated textile effluents. Carbon-based hybrid materials offer superior antimicrobial and photocatalytic properties compared to pure metal/metal oxide Nanomaterials (NMs). The capacity to manipulate the composition, structure, and morphology of CBNs to modify their attributes further enhances their performance and expands their application possibilities. Hence, developing

CBNs with superior antimicrobial and photocatalytic properties can be a novel platform for an eco-friendly and inexpensive approach to environmental clean-up.

CBNs also play a significant role is energy-harvesting devices. The need for sustainable and clean energy sources has led to intense research on energy harvesting, where the goal is to capture and convert ambient energy from the environment into usable electrical energy. CBNs have become a potential class of materials for energy harvesting due to their exceptional properties and versatility [8].

Integrating different materials in CBNs enhances electrical, mechanical, and thermal properties, making them highly suitable for energy harvesting devices. CBNs possess unique characteristics that make them well-suited for energy conversion processes. For instance, CNTs exhibit high mechanical strength, remarkable thermal stability, and excellent electrical conductivity. These properties enable CNT-based nanocomposites to efficiently harvest energy from various sources, including thermal gradients, vibrations, solar radiation, and electromagnetic fields. Additionally, with its high carrier mobility and exceptional electrical conductivity, graphene holds great potential for energy storage applications, such as supercapacitors. The compatibility of CBNs with various fabrication techniques allows for the creation of flexible, lightweight, and scalable energy harvesting devices. Significant advancements have been achieved recently in designing and synthesizing CBNs for energy harvesting applications. Researchers have demonstrated their effectiveness in diverse energy harvesting mechanisms, including piezoelectric, thermoelectric, and photovoltaic systems. Moreover, exploring novel nanocomposite architectures and their hybridization with other functional materials has opened up new avenues for optimizing energy conversion efficiency and stability [9].

This book chapter thoroughly summarizes the recent advancements in CBNs for environmental and energy-harvesting applications. By delving into these materials' fundamental principles, fabrication techniques, and performance characteristics, we seek to offer valuable insights into their potential and lay the foundation for future environmental clean-up and energy harvesting breakthroughs. We analyze the existing literature, incorporating insights from prominent studies and research papers.

FABRICATION AND CHARACTERIZATION OF CBNS

General approaches to NM fabrication include top-down and bottom-up. The former process includes converting bulk materials into monomeric units followed by NMs synthesis using techniques such as laser ablation, ball milling, sputtering, chemical etching, explosion process, *etc*. [10, 11]. In the bottom-up approach, atoms or molecules are converted into nuclei followed by NMs using techniques like sol-gel synthesis, hydrothermal synthesis, biological reduction, chemical reduction, vapor deposition, chemical/electrochemical deposition, atomic/molecular condensation, aerosol process, *etc* [10, 11]. Carbon nanocomposites are synthesized by covalent functionalization (example: single-walled and multi-walled CNTs) and non-covalent functionalization (example: reduced graphene oxide) [12]. The wet synthesis of CBNs involved laser ablation, chemical, and biological reduction. Another versatile technique is dry synthesis, which requires no additional reducing agents, solvents, or applied electric current; instead, simple mixing of a precursor salt with CNTs and then inert atmospheric heating is sufficient [12, 13]. CNTs are ideal for fabricating polymer nanocomposites and possess remarkable mechanical, electrical, and electromagnetic properties [12]. CNTs are categorized as fullerene, single-walled CNTs, single-walled carbon nanocorn, multi-walled CNTs, nongraphite, carbon black nanoparticle, *etc.*, depending on their structural characteristics [13]. Chemical Vapor Deposition (CVD), spray pyrolysis, hydrothermal treatment, laser ablation, electric-arc discharge, *etc.*, are all methods used to synthesize CNTs. Among them, CVD is preferred due to its higher production yields, reduced impurities, and lower temperature requirements (<800 °C) [14].

Graphene is another carbon-based material preferred in fabricating "thermal conductive nanocomposites" owing to its excellent thermal efficiency and electrical conductivity [12]. The commonly applied GNMs are single-layered graphene, GO, rGO, *etc* [15], and among them, rGO is preferred due to its ease of fabrication, high electrical conductivity, reduced cost, and enhanced availability [15]. Pure graphene sheets are synthesized using the exfoliation/electrolytic exfoliation method [14]. The rGO can be produced by chemical reduction, photocatalytic reduction, electrochemical reduction, *etc.*, of which the latter is preferred [14]. Another carbon nanomaterial is carbon dots, possessing novel optical properties, good stability, high electron mobility, eco-friendliness, low cost, *etc.*, and are recognized in energy storage and catalysis applications [16]. Carbon nanodots are synthesized *via* exfoliation, electrochemical method, laser ablation, pyrolysis, *etc* [14]. The CBNs, such as CNTs, graphene, GO, rGO, and carbon nanodots, are also synthesized by green methods using bio-based precursors like walnut, *T. orientalis* biomass, bagasse, peanut shells, *H. pennisetum* biomass, tamarind pods, wool fibres, prawn shells, *etc* [14]. To improve the functionality of CBNs (such as GNMs), surface functionalization can beachieved*via* covalent (like amide and carbamate ester bonds) and non-covalent approaches (like electrostatic interactions, π-electrons stacking, weak van der Waals forces) [15]. Hubbard *et al.* [17] developed Eu-doped Y_2O_3-CNTs nanocomposites for enhanced electrical conductivity. The developed nanocomposite allowed improved contrast and better detection of the underlying

CNT network. Zhang et al. [18] also developed metal oxide (TiO_2) nanorods-rGO hybrid nanocomposites for enhanced microwave absorption. Amorphous and crystalline nanocomposites were obtained using Tetrabutyl Titanate (TBT) and glycerol at varied temperatures.

CARBON-BASED NANOCOMPOSITES FOR ENVIRONMENTAL APPLICATIONS

This section discusses the application of various carbon-based nanocomposites in environmental cleanup (such as removing pesticides, heavy metals, toxic dyes, volatile organic/inorganic compounds, and air pollutants), along with the latest developments in the area and future scope.

Detection and Removal of Heavy Metals

Graphene-Based Nanocomposites (GNCs) have been investigated extensively for detecting heavy metals. He et al. [19] developed GO/aptamer hybrids with excellent Hg^{2+} ions sensitivity (quantification limit of 30 nM) via self-assembly between the graphene base plane and the DNA. The developed hybrids also showed significant selectivity to other metal ions. Wen et al. [20] constructed graphene nanohybrids using "silver-specific cytosine-rich oligonucleotide" to detect silver ions (Ag^+) with higher sensitivity (quantification limit= 20 nM; detection limit= 5 nm. The graphene nanohybrid sensors detected Ag^+ in real environmental samples (like river water). Chandra et al. [21] achieved a 99.9 percent reduction of arsenic (As) (within one ppb concentration) using magnetite/rGO nanocomposites, with greater elimination capacity for As (III) compared to As (V). Fan et al. [22] fabricated a "magnetic cyclodextrin/GO nanocomposite" for speedy chromium (Cr) removal from wastewater (<120 mg/g within 60 mins). The adsorption of Cr on the surface of nanocomposites may be due to the surface envelope behavior. On the bare graphene portion, there is a single-layer adsorption process. The recovered fabricated nanocomposites are reused with a Cr removal efficiency of ~82% in the fifth treatment cycle. In order to remove Hg^{2+} ions in a highly effective and selective manner, Chandra and Kim [23] synthesized graphene/polypyrrole nanocomposites, which had an adsorption capacity of 980 mg/g in water. Patil et al. [24] developed Ag-reduced rGO nanocomposites through green synthesis and noted their substantial levels of selectivity towards Hg^{2+}, Pb^{2+}, and Cr (VI) with recognition limits of 15 nM, 50 nM, and 500 nM, respectively.

Environmental Gas Sensor for Inorganic and Organic Vapor

Inorganic and organic vapors in the environment are another persistent issue in present-day society. Graphene can detect NO, NO_2, CO_2, CO, SO_2, H_2, NH_3, Cl_2, benzene, acetone, toluene, *etc* [25 - 32]. Pristine graphene has been previously reported for higher sensitivity towards NO_2 molecules [25]. Later, rGO was developed for detecting dinitrotoluene, hydrogen cyanide, dimethyl methyl phosphonate, chloroethyl ethyl sulfide, NH_3, NO_2, Cl_2, *etc* [29 - 31]. GNCs for gas sensing applications have recently attracted the attention of researchers [25 - 28]. Ji *et al.* [28] reported a graphene/ionic liquid hybrid nanocomposite for the detection of toxic hydrocarbon vapors like benzene, toluene, pyridine, cyclohexane, hexane, acetone, ethanol, *etc*. The developed nanocomposites showed a higher affinity for aromatic compounds than aliphatic compounds. Li *et al.* [27] fabricated an rGO-decorated palladium composite to detect NO gas with a 2 to 420 ppb sensitivity. The Pd decoration on rGO significantly increased the sensitivity of the developed nanocomposite for NO detection. Singh *et al.* [25] reported Metal Oxide (MO) nanocrystal/graphene composites for sensing CO, NO, and NH_3 with a sensitivity of ~1 ppm at RT. Yadav and Kim [26] prepared anatase-TiO_2 loaded GO (0.25 wt%) and observed significant improvements in photocatalytic activity for the degradation of benzene gas. Likewise, Zang *et al.* [33] cited the role of TiO_2/GNCs in gas-phase benzene degradation. Hence, CBNs have a broad spectrum of actions against toxic gases regarding detection and remediation.

Bacterial Detection and Eradication

Bacterial detection has been established using graphene and GNCs. Wan *et al.* [34] mentioned GO sheet-inspired Ag enhancement functionalized with an "anti-sulfate-reducing bacteria antibody" as a potential biosensor for rapidly detecting pathogens. Jung *et al.* [35] developed a GO array functionalized with rotavirus-specific antibodies using carbodiimide chemistry for rapid, highly sensitive, and selective rotavirus detection. Zelada-Guillén *et al.* [36] developed SWCNTs functionalized with an aptamer for a highly selective biosensor to detect pathogenic microbes in real-time. In another study, Huang *et al.* [37] constructed a graphene-based, highly sensitive nanoelectronics biosensor using a CVD technique and functionalized it with *E. coli*-specific antibodies for facile and rapid detection of *E. coli*. Graphene/GNCs also showed significant antimicrobial activities *via* physical (direct contact) and chemical stress (generation of reactive oxygen species) induced by them [38] (Fig. **2**). GBNs are biocompatible and possess higher surface area, enhanced diffraction strength, fast ionic migration, and high Young's modulus, thereby having good solubility, dispensability, and stability, leading to enhanced antimicrobial activity [39]. Hu *et al.* [38] observed significant antibacterial activity of rGO and GO against *E. coli* with minimal cytotoxicity. Nanda *et al.* [40] demonstrated degradation of the inner and outer cell membranes of *E. coli* upon treatment with different doses of GO, resulting in the release of protein and adenine from the bacteria. Jaworski *et al.* [41]

synthesized a GO-AgNP nanocomposite for a broad spectrum of antimicrobial activity towards pathogens like *E. coli, C. albicans, S. aureus,* and *S. epidermidis*. The deformed morphology of microbial cells after 24 h of incubation at 37 °C is clearly demonstrated in the SEM micrographs. Li *et al.* [42] developed graphene films on Ge (G-Ge), Cu (G-Cu), and SiO_2 (G-S) and found substantial growth inhibition of G-Ge and G-Cu against *E. coli* and *S. aureus*. They reported that the antibacterial activity could be due to the physical stress induced by direct contact of developed graphene films with bacteria, leading to their membrane damage. Hseuh *et al.* [43] recommended rGO decorated with Ag/ZnO (3.47/34.91%; 7.08%/15.28%) nanoparticles as a potential antibacterial agent and catalyst for various applications. Likewise, the rGO-AgNPs nanocomposites were reported to reduce the growth of *E. coli* and *S. aureus* [44, 45].

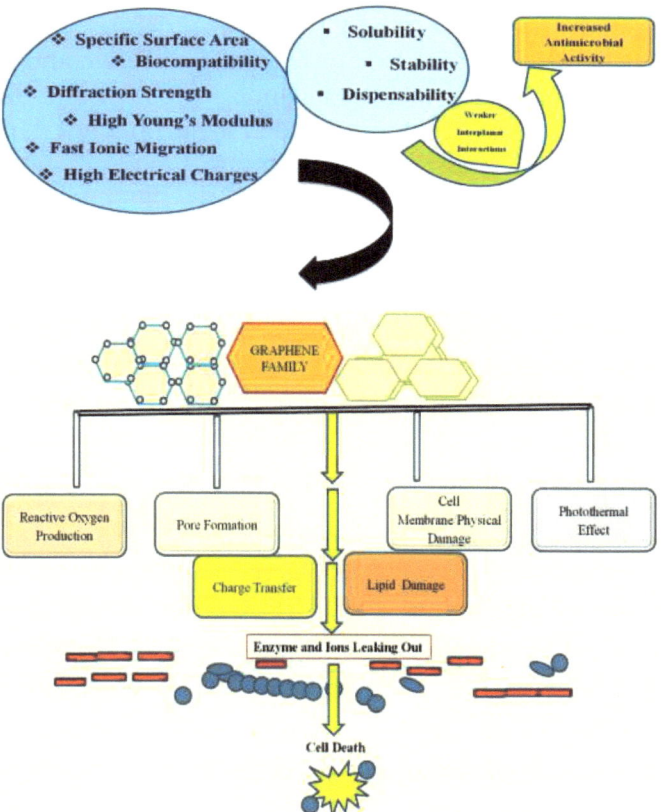

Fig. (2). Antimicrobial properties of GBNs and modes of microbial cell damage by these nanomaterials (compiled from [39]).

Degradation and Removal of Organic Species

CBNs have potential applications in photodegradation or direct removal of toxic dyes present in wastewater. Wei *et al.* [46] reported the application of multilayered GO in water treatment as a separation membrane using techniques like nanofiltration, forward osmosis, and pervaporation. Yee *et al.* [47] developed marine antifouling graphene-Ag (4.9 wt%) nanocomposites. They observed 99.6% biofilm (*Halo monas pacifica*) inhibition and also reported 80% growth inhibition of *Dunaliella tertiolecta* and *Isochrysis sp.* after 96 hrs. Using a "magnetic TiO_2-graphene" hybrid nanocomposite from wastewater, Tang *et al.* [48] eliminated 2,4-dichlorophenoxyacetic acid (2,4-D). The developed nanocomposite is a highly efficient and stable photocatalyst with 2,4-D removal of over 97% after nine successive treatment cycles.

Liu *et al.* [49] developed $Ag/AgSb_2O_{5.8}$-graphene nanocomposites (sintered at 500°C) and recorded an 80% reduction of Rhodamine B dye and 85% degradation of tetracycline hydrochloride in 120 min under visible light. Joshi *et al.* [50], using $rGO-Ag/Fe_3O_4$ nanocomposite, degraded 4-nitrophenol dye, observed their growth-inhibiting activity against *E. coli*, and encouraged its potential utilization in environmental remediation. Zhang *et al.* [51] developed TiO_2 (P25)-graphene nanocomposite for photocatalytic reduction of Methylene Blue (MB) dye. In this process, carbon was important for enhancing the photocatalysis of MB dye by effectively utilizing light for the catalyst, increasing the MB dye's absorptivity, and suppressing charge recombination. Zhang *et al.* [52] achieved excellent photocatalytic performance of "CNT-pillared rGO composites" synthesized *via* the Ni catalyst-based CVD method. The developed composites were porous, bearing a higher surface area (352 m^2/g) and showing significant degradation of Rhodamine-B dye. In another study, Fe_3O_4 NPs/rGO nanocomposites demonstrated rapid and efficient removal of rhodamine B (91%) and malachite green (94%) dyes [53]. Chen *et al.* [54] demonstrated the utilization of GO/Chitosan hydrogels for the removal of dyes (MB and Eosin Y; adsorption capacities >300 mg/g) and heavy metals (Pb(II) and Cu(II) ions at adsorption capacities of 90 and 70 mg/g) from water. The higher concentrations of GO in the composite promoted MB dye degradation, while an increase in chitosan concentration enabled more absorption of Eosin-Y dye.

Environmental and Health Implications of CBNs

Despite their promising role in sustainable technologies, Carbon-Based Nanomaterials (CBNs) such as graphene, Carbon Nanotubes (CNTs), and fullerenes raise significant concerns regarding their environmental and human health impacts, which must be critically addressed. The lifecycle of CBNs from synthesis to disposal can introduce risks related to ecotoxicity, bioaccumulation, and human exposure. Emerging evidence indicates that many CBNs exhibit persistent behavior in ecosystems and may induce oxidative stress, membrane damage, and DNA fragmentation in aquatic and terrestrial organisms, even at low concentrations [55]. During production, certain synthesis methods may release hazardous by-products, while at the disposal stage, nanoparticles can persist in the environment, potentially disrupting ecosystems. Moreover, inhalation or ingestion of CBNs has raised concerns in toxicological studies, indicating possible cytotoxic and genotoxic effects depending on the material's structure, surface chemistry, and concentration [55, 56]. To ensure the safe use of CBNs in real-world applications, Future work should prioritize green synthesis routes, comprehensive risk assessment protocols, and safe disposal or recycling strategies to minimize unintended consequences. Establishing regulatory frameworks and standardized ecotoxicological testing is essential to ensure that the deployment of CBNs in energy, sensing, and environmental remediation applications does not compromise ecological or public health. Addressing these concerns holistically is imperative to advancing CBNs as truly sustainable materials.

The literature review suggested that CBNs have great potential in the photocatalytic degradation of toxic dyes and other harmful organic species from wastewater. Further research is encouraged in this direction to develop carbon-composite-based novel catalysts for the rapid and efficient treatment of wastewater contaminants and gaseous air pollutants. Hence, it can become a prominent candidate for promoting environmental remediation *via* sustainable routes.

CARBON-BASED NANOCOMPOSITES FOR ENERGY HARVESTING APPLICATIONS

In addition to their superior mechanical, electrical, and thermal characteristics, carbon-based nanocomposites are showing immense potential as energy harvesting materials. These composites substantially enhance the performance of devices like triboelectric, piezoelectric, thermoelectric, and solar generators by combining nanostructures like graphene, Carbon Nanotubes (CNTs), and carbon dots with polymers or metal matrices (Fig. 3). They are perfect for wearable and self-powered devices due to their versatility, flexibility, and low weight. For broad use, however, concerns including consistent dispersion, stability, and scale production need to be resolved.

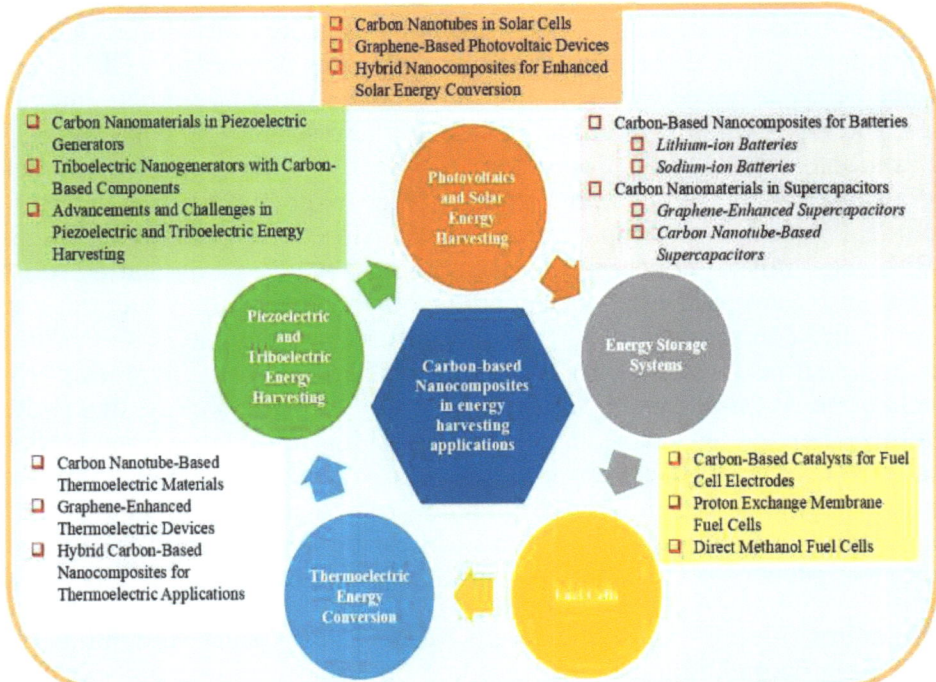

Fig. (3). Applications of CBNs in various sectors of energy conversion and storage [57].

Photovoltaics and Solar Energy Harvesting

Photovoltaics and solar energy harvesting have emerged as vital fields of research and development, aiming to harness the power of sunlight and convert it into usable electrical energy. Photovoltaic technologies are essential for reaching a more sustainable future owing to the higher demand for sustainable and clean energy sources. The continuous exploration and utilization of advanced materials and innovative techniques drive the advancements in this field. Researchers are actively investigating the potential of different materials, such as CNTs, graphene, and hybrid nanocomposites, to enhance solar energy conversion efficiency. CNTs, with their unique electrical and structural properties, offer promising opportunities for improving solar cells. These nanoscale structures possess high electrical conductivity and an extensive surface area, enabling efficient charge transport and enhanced light absorption. Integrating carbon nanotubes into solar cells has significantly improved device performance, leading to higher power conversion efficiencies.

Similarly, graphene, a two-dimensional material with outstanding transparency and electrical conductivity, has gained attention in photovoltaic research. Graphene-based photovoltaic devices hold the potential for achieving high carrier mobility, tunable bandgap, and improved charge transport, paving the way for next-generation solar cells. In addition to individual material advancements, integrating hybrid nanocomposites has emerged as a promising strategy for enhancing solar energy conversion. Hybrid nanocomposites combine semiconductors, nanoparticles, and organic compounds to synergistically improve light absorption, charge separation, and charge transport. Hybrid nanocomposites show enhanced solar cell performance by harnessing various materials' complementary properties. Researchers are exploring novel approaches, such as blending different semiconducting materials or coupling organic and inorganic components, to create efficient hybrid nanocomposite structures. These different material-based solar cells, including carbon nanotube-based, graphene-based, and hybrid nanocomposite-based solar cells, contribute to continuous advancements in photovoltaics and solar energy harvesting. They not only support the creation of solar cells that are more effective and affordable but also promote the wide use of renewable energy sources. With ongoing research and innovations in advanced materials and device architectures, the future holds great promise for further optimizing and commercializing photovoltaic technologies, enabling a sustainable and clean energy future.

CNTs in Solar Cells

CNTs' higher electrical conductivity and surface area make them ideal candidates for enhancing various aspects of solar cell components. Researchers have explored the integration of CNTs in different parts of solar cells, including the active layer, electrodes, and charge transport layers, to achieve higher energy conversion efficiencies. In a study by Khan *et al*. [58], CNTs were incorporated into the active layer of solar cells, significantly enhancing the device's performance. The efficient charge transport and reduced recombination losses provided by CNTs contributed to an improved power conversion efficiency of 10.4%, surpassing traditional silicon-based devices. This demonstrates the potential of CNTs in increasing the overall efficiency of solar cells. Additionally, transparent conductive electrodes for solar cells have been produced using CNTs. Wang *et al*. [59] successfully fabricated graphene-based transparent electrodes using CNTs. These electrodes exhibited high light transmittance, low sheet resistance, and excellent electrical conductivity. The integration of CNT-based transparent electrodes not only improved the efficiency of solar cells but also offered flexibility and durability, enabling their use in flexible and wearable solar devices. Furthermore, CNTs have been employed as charge transport layers in solar cells. Xu *et al*. [60] explored the use of CNT-based charge transport layers, which demonstrated enhanced electron mobility, efficient charge extraction, and reduced charge recombination losses. These properties led to improved overall device performance and higher energy conversion efficiencies. In conclusion, integrating carbon nanotubes in solar cells shows promising potential for enhancing efficiency. The unique characteristics of CNTs, such as their outstanding charge transport abilities, high electrical conductivity, and large surface area, contribute to improved light absorption, efficient charge separation, and enhanced charge collection. Continued research and development in utilizing carbon nanotubes in solar cells are promising for advancing photovoltaic technologies.

Graphene-Based Photovoltaic Devices

Graphene is a 2D material with outstanding transparency and electrical conductivity that has garnered significant recognition in photovoltaic devices. Researchers have explored the potential of Graphene-Based Materials (GBMs) and Architectures (GBAs) to enhance the performance of solar cells. Graphene offers several advantages for photovoltaics, including a tunable bandgap, high carrier mobility, and excellent charge transport properties. In a study by Safie *et al*. [61], graphene was integrated into the active layer of organic solar cells, leading to enhanced charge generation and transport. The high carrier mobility of graphene facilitated efficient charge collection, reducing recombination losses and improving the overall device performance. This study highlighted the potential of GBMs in achieving higher power conversion efficiencies in organic photovoltaic devices. Wang *et al*. [59] developed transparent electrodes based on graphene for dye-sensitized solar cells. These electrodes exhibited excellent electrical conductivity and optical transparency, allowing efficient light transmission and charge extraction. The integration of graphene electrodes improved the device's performance and offered mechanical flexibility and long-term stability. GBAs have also been explored in hybrid perovskite solar cells. Chen *et al*. [62] reported the incorporation of graphene as a charge transport layer and found improvements in charge extraction with reduced recombination losses. The unique properties of graphene enabled efficient charge transfer across the perovskite layer, leading to enhanced photovoltaic performance.

In summary, graphene-based photovoltaic devices show great promise in advancing solar cell technologies. Graphene's outstanding electrical conductivity, transparency, and charge transport capabilities are essential to improve light absorption, charge collection, and overall device performance. Continued research and development in GBAs and materials hold significant potential for achieving higher energy conversion efficiencies in photovoltaic devices.

Hybrid Nanocomposites for Enhanced Solar Energy Conversion

Hybrid nanocomposites have emerged as a promising strategy for enhancing solar energy conversion in photovoltaic devices. By combining different materials, such as semiconductors, nanoparticles, and organic compounds, researchers aim to leverage each component's synergistic effects and unique properties to improve charge separation, charge transfer, and light absorption in solar cells. In organic photovoltaics, the idea of hybrid nanocomposites was investigated by Liu *et al.* [63]. They demonstrated that incorporating nanoparticles, such as metal oxides or quantum dots, into organic semiconductor matrices can enhance light absorption by extending the spectral response range. Additionally, the presence of nanoparticles facilitates efficient charge separation due to the creation of nanoscale interfaces, leading to improved charge transport and reduced recombination losses. Such hybrid nanocomposites have shown promising advancements in organic photovoltaics, resulting in higher power conversion efficiencies. Furthermore, integrating hybrid nanocomposites in perovskite solar cells has shown significant potential. Chen *et al.* [60] investigated the use of hybrid perovskites comprising both organic and inorganic components. The combination of these materials resulted in improved light harvesting, charge carrier mobility, and stability. Hybrid nanocomposites in perovskite solar cells offer enhanced charge transport properties, reduced energy losses, and increased device efficiency, making them a promising avenue for future solar energy conversion technologies. Moreover, the development of nanostructured electrodes for solar cells has involved the exploration of hybrid nanocomposites. These electrodes typically combine conductive materials, such as Carbon Nanotubes (CNT) or graphene, with nanoparticles or polymers. Incorporating these materials enhances charge extraction and collection, leading to improved device performance. The synergistic effects of different components in the hybrid nanocomposite electrodes contribute to higher conductivity, increased surface area, and enhanced charge transport efficiency.

In conclusion, hybrid nanocomposites hold immense potential for enhancing solar energy conversion in photovoltaic devices. The combination of different materials in nanocomposite structures allows for improved light absorption, charge separation, and charge transport, leading to higher energy conversion efficiencies. Ongoing research and development in hybrid nanocomposites offer promising prospects for advancing the efficiency and performance of solar cells, contributing to the wider adoption of sustainable and renewable energy sources.

Energy Storage Systems

Energy Storage Systems (ESS) allow more efficient integration of renewable energy sources into the power grid. These systems are designed to store extra energy at times of high generation and release it during high energy demand or when renewable energy sources are unavailable. Various energy storage technologies have been developed to meet the diverse requirements of different applications. One widely used energy storage technology is battery storage systems [64]. Lithium-ion batteries have gained a lot of interest due to their high energy density, long cycle life, and fast response times. Battery storage systems provide a reliable and scalable solution for storing electricity from sustainable sources, such as solar and wind, and discharging it when needed. These systems find applications in various settings, from residential and commercial installations to utility-scale projects, contributing to grid stability and enabling the integration of intermittent renewables. Another prominent energy storage technology is pumped-hydroelectric storage. This technique uses surplus power to lift water from a region of the lower reservoir to the upper reservoir during low demand. When electricity demand increases, the stored water is released, flowing downhill and driving turbines to produce electricity. "Pumped hydro storage systems" have long cycle lives, high efficiency, and large-scale storage capacity, making them suitable for grid-level energy storage and balancing intermittent renewable energy generation. Other emerging energy storage technologies encompass a variety of systems, including hydrogen storage, thermal energy storage, flywheel energy storage, supercapacitors, Compressed Air Energy Storage (CAES), and advanced fuel cells [65]. Flywheel systems store kinetic energy in a spinning rotor and release it when needed, offering rapid response times and high-power output. CAES systems compress air using surplus electricity and store the compressed air in underground caverns, releasing it to run turbines for generating electricity during rising demands. Thermal energy storage systems store and release heat energy for various applications, while hydrogen storage, supercapacitors, and advanced fuel cells provide alternative methods for storing and releasing energy efficiently. The advancements in energy storage systems are driven by ongoing research and development efforts to enhance efficiency, cost-effectiveness, and sustainability. Integration of smart control systems, advanced materials, and new technologies is being explored to improve the performance and reliability of storage systems. For example, the progress of solid-state batteries, flow batteries, and other innovative approaches holds promise for achieving higher energy densities, longer cycle lives, and improved safety [66].

Energy storage systems encompass a wide range of technologies that address the unique needs of energy storage across different applications. Each technology offers specific advantages and characteristics, from battery and pumped hydroelectric storage to emerging systems like flywheel storage, CAES, thermal storage, hydrogen storage, supercapacitors, and advanced fuel cells. Continued advancements and innovations in energy storage systems are vital for maximizing the utilization of renewable energy sources, achieving a sustainable and efficient energy future, and reducing reliance on fossil fuels.

CBNs for Batteries

CBNs have gained significant devotion in the field of batteries due to their unique characteristics and potential for enhancing battery performance. These NCs, which combine carbon materials with other functional components, offer improved conductivity, enhanced structural stability, and increased capacity for energy storage. One promising example of CBNs is the incorporation of Carbon Nanotubes (CNTs) into battery electrodes. CNTs have remarkable electrical conductivity, strong mechanical properties, and higher surface area, which make them ideal for improving the performance of battery systems. Using CNTs in electrode architectures can increase the battery's internal ion and electron mobility, resulting in faster charge and discharge rates. Additionally, the high surface area of CNTs offers more active sites for electrochemical reactions, thus increasing the overall energy storage capacity of the battery [64].

Graphene, another carbon-based material, also has excellent potential for battery applications. By incorporating graphene into battery electrodes, researchers have improved various aspects of battery performance, including higher charge storage capacity, faster charging and discharging rates, and increased cycling stability [65]. Moreover, carbon-based nanocomposites can be tailored by combining carbon materials with other functional components such as polymers, metal oxides, or conductive additives. For example, carbon-based nanocomposites incorporating metal oxides, such as titanium dioxide (TiO_2), have enhanced battery performance. Combining carbon materials with metal oxides improves the conductivity and provides additional redox reactions, resulting in improved cycling stability and higher energy storage capacity [66]. Furthermore, the use of CBNs extends to different types of batteries, including Lithium-Ion (LIB), Sodium-Ion (SIB), and beyond lithium-ion systems. LIBs, owing to extended cycle life and high energy density, have been widely employed in portable electronic gadgets and electric vehicles. CBNs have been utilized to improve the effectiveness, capacity, charging rates, and overall efficiency of LIBs. SIBs, on the other hand, have emerged as potential alternatives to LIBs, driven by abundant and cheap sodium resources. CBNs have shown promise in SIBs, offering similar benefits in improved conductivity, capacity, and cycling performance. Research efforts are focused on optimizing the electrode materials and exploring novel CBNs further to enhance the performance and stability of sodium-ion batteries. Incorporating CNTs, graphene, and other carbon materials into battery electrodes has improved conductivity, charge storage capacity, and cycling stability. Researchers can further optimize their performance for specific battery applications by tailoring the composition and structure of carbon-based nanocomposites.

Lithium-Ion Batteries

CBNs have emerged as promising materials for increasing the performance of LIBs [67]. Incorporating CNTs into LIB electrodes improves conductivity, charge and discharge rates, and overall battery performance [68]. Graphene, a two-dimensional carbon material, has also shown significant potential for LIB applications, offering higher charge storage capacities, faster charge-discharge rates, and improved cycling stability [69]. Researchers have explored combining carbon materials with MOs, such as TiO_2, in CBNs for LIBs. These composites exhibit enhanced conductivity, energy storage capacity, and cycling stability [70]. Ongoing research focuses on developing novel fabrication methods, electrode architectures, and composite compositions to further optimize the performance of carbon-based nanocomposites for LIBs [71].

Sodium-Ion Batteries

CBNs have also shown promise for enhancing the performance of SIBs. Incorporating carbon materials with other functional components offers improved conductivity, stability, and energy storage capabilities. One example is the utilization of CNTs in SIB electrodes. CNTs provide a wide surface area and excellent electrical conductivity, facilitating efficient charge and ion transport within the battery. This enhances charge and discharge rates and improves overall battery performance [72]. With its exceptional electrical conductivity and mechanical strength, graphene has also been investigated for its potential in SIB applications. By incorporating graphene into SIB electrodes, researchers have achieved higher capacity, faster charge-discharge rates, and improved cycling stability [73]. The unique properties of graphene contribute to the overall performance enhancement of SIBs.

Lithium-sulfur Batteries

Beyond lithium-ion systems, CBNs are being investigated for next-generation battery technologies, such as Lithium-Sulfur Batteries (LSBs) and Solid-State Batteries (SSBs). LSBs have the potential for higher energy densities but face challenges related to the polysulfide dissolution and shuttle effects. CBNs are being explored as sulfur hosts and conductive additives to address these issues and improve the cycling stability and overall performance of LSBs. SSBs, which utilize solid electrolytes instead of liquid electrolytes, offer advantages in terms of safety, energy density, and lifespan. CBNs are being studied for their potential in SSBs to enhance the solid electrolytes' conductivity and improve the batteries' overall performance [74].

Moreover, CBNs can be tailored by combining carbon materials with metal oxides, polymers, or conductive additives to further enhance their performance in SIBs. For instance, incorporating carbon materials with metal oxides, such as TiO_2, has improved conductivity and redox reactions, as well as enhanced energy storage capacity and cycling stability in SIBs. The development and optimization of CBNs for SIBs is an active area of research. Researchers are exploring various synthesis methods, electrode designs, and composite compositions to further improve SIBs' performance. The goal is to achieve batteries with higher energy densities, longer cycle lives, and improved safety characteristics.

In conclusion, CBNs show promise for enhancing the performance of SIBs. Incorporating carbon nanotubes, graphene, and other carbon materials into SIB electrodes improves conductivity, charge storage capacity, and cycling stability. By tailoring the composition and structure of CBNs, researchers can further optimize their performance for specific SIB applications. Continued research and growth in this area will contribute to advancing high-performance and energy-efficient SIBs.

Carbon Nanomaterials in Supercapacitors

Carbon nanomaterials offer promising opportunities for enhancing supercapacitor devices' energy storage performance and efficiency. One widely studied carbon nanomaterial for supercapacitor applications is graphene. Graphene contains a single layer of carbon atoms, making up its 2D structure, and exhibits higher electrical conductivity, large surface area, and remarkable mechanical strength. These properties qualify graphene as an ideal material to fabricate electrodes in supercapacitors. Graphene-based electrodes have demonstrated enhanced energy storage capabilities, including higher capacitance, faster charge-discharge rates, and improved cycling stability [75]. CNTs are CBN that have shown promise in supercapacitor applications. The unique structure of CNTs allows for efficient ion transport and rapid charge storage, resulting in enhanced capacitance and power density in supercapacitor devices [76]. Activated Carbon (AC), derived from various carbon sources such as coconut shells or wood, is a commonly used carbon nanomaterial in commercial supercapacitors. AC exhibits a highly porous structure, providing a large surface area for electrochemical reactions and ion adsorption. This porous structure contributes to high capacitance and energy storage performance in supercapacitor devices [77]. Furthermore, other CBNs, such as carbon nanofibers, carbon aerogels, and carbon quantum dots, have also been investigated for their potential in supercapacitors. These materials offer unique properties and structural characteristics that can be tailored for specific supercapacitor applications, further expanding the range of available options for energy storage enhancement [78]. Continued research and development efforts in carbon nanomaterials hold great promise for the future development of high-performance supercapacitors.

Graphene-Enhanced Supercapacitors

Due to the special characteristics of graphene, Graphene-Enhanced Supercapacitors (GESs) have become a potential technique for effective energy storage. One significant advantage of GESs is their high energy density. The large surface area of graphene allows for a higher amount of charge to be stored, resulting in increased energy storage capacity. Graphene in supercapacitor electrodes enables efficient charge storage and rapid charge-discharge rates, enhancing power density and improving device performance [79]. GBMs, such as GO, rGO, and graphene nanoplatelets, can be included in supercapacitor electrodes in various ways. These graphene derivatives offer improved processability and compatibility with existing electrode fabrication techniques. The unique properties of graphene, such as its high electrical conductivity and large surface-to-volume ratio, contribute to the formation of thin and compact electrode structures with excellent electrochemical performance [80]. In addition to the improved energy storage capacity and power density, GESs exhibit enhanced cycling stability and longer cycle life. The enhanced durability of GBMs is crucial for the long-term reliability and practical application of supercapacitor devices [81]. Moreover, integrating graphene with other NMs, like MOs or conducting polymers, can further enhance the performance of supercapacitors. The synergistic effects between graphene and these materials lead to improved charge transfer kinetics, increased capacitance, and enhanced cycling stability. Graphene-based nanocomposites offer a versatile platform for tailoring the properties of supercapacitor electrodes to meet specific application requirements [82]. Developing graphene-enhanced supercapacitors is an active area of research, with ongoing efforts focused on optimizing electrode design, exploring scalable fabrication methods, and improving device integration. The potential applications of GESs span various fields like portable electronics, electric vehicles, and renewable energy storage. Continued advancements in GBMs and device engineering are expected to drive the commercialization and widespread adoption of high-performance supercapacitor technologies.

CNT-Based Supercapacitors

CNT-based supercapacitors have attracted significant attention in energy storage due to their exclusive properties and potential for high-performance devices. CNTs are cylindrical structures composed of rolled-up graphene sheets, offering strong mechanical properties, a wide surface area, and high electrical conductivity. One of the key advantages of CNT-based supercapacitors is their high-power density. The substantial electrical conductivity of CNTs enables efficient charge transfer and rapid charge-discharge rates, making them suitable for applications requiring quick energy release. The large surface area of CNTs allows for a higher amount of charge to be stored, resulting in increased energy storage capacity [83]. CNTs can be integrated into supercapacitor electrodes either as individual nanotubes or in the form of nanotube networks. Individual CNTs provide direct pathways for charge transport, enabling efficient electron transfer within the electrode material. On the other hand, nanotube networks offer interconnected conductive pathways, facilitating the movement of ions and reducing resistance within the electrode [84]. Additionally, CNT-based supercapacitors exhibit excellent cycling stability and long cycle life. The chemical stability and mechanical strength of CNTs contribute to the durability of the electrode materials, allowing them to endure repeated charge-discharge cycles without substantial degradation. This property is crucial for supercapacitor devices' reliable and long-term performance [85]. Moreover, integrating other materials, like conductive polymers or MOs, with CNTs can further enhance the electrochemical performance of supercapacitors. Combining CNTs with these materials creates hybrid nanocomposites, which offer synergistic effects, such as improved capacitance, enhanced charge storage capacity, and enhanced cycling stability. The exclusive properties of CNTs, coupled with the benefits of other materials, enable the development of high-performance supercapacitor devices [86]. CNTs-based supercapacitors find applications in electric vehicles, portable electronics, and renewable energy systems. The continuous research and development efforts focus on optimizing the synthesis and integration of carbon nanotubes, exploring scalable fabrication techniques, and improving the overall performance and efficiency of supercapacitor devices.

Fuel Cells

Fuel cells are cutting-edge devices that use an electrochemical process to transform a fuel's chemical energy into electrical energy. They offer clean and efficient power generation without combustion [87]. There are several varieties of fuel cells available for use in a variety of applications, including "Alkaline Fuel Cells" (AFCs), "Solid Oxide Fuel Cells" (SOFCs), "Proton Exchange Membrane Fuel Cells" (PEMFCs), "Molten Carbonate Fuel Cells" (MCFCs), and "Phosphoric Acid Fuel Cells" (PAFCs). Fuel cells have numerous advantages, including modularity, minimal emissions, and excellent efficiency [88 - 90]. Ongoing research focuses on improving durability, reducing costs, and exploring alternative fuel sources for wider adoption.

Carbon-Based Catalysts for Fuel Cell Electrodes

Carbon-Based Catalysts (CBCs) have emerged as promising alternatives to traditional platinum-based catalysts for fuel cell electrodes [91]. These catalysts, composed of graphene, CNTs, and carbon black, are examples of carbon-based compounds. They offer several advantages, including cost-effectiveness, abundance, and high catalytic activity. CBCs exhibit excellent electrical conductivity, large surface area, and tunable surface chemistry, making them ideal for promoting electrochemical reactions in fuel cells [92]. One of the key CBCs is graphene, a honeycomb-shaped, 2D sheet of carbon atoms. Functionalizing graphene with heteroatoms or doping it with other elements can further enhance its catalytic performance. Graphene-based catalysts have shown promise in various fuel cell reactions, including the Oxygen Reduction Reaction (ORR) at the cathode, where they exhibit comparable or even superior performance to platinum catalysts. CNTs are another carbon-based catalyst type that offers unique fuel cell electrode properties. CNT-based catalysts have demonstrated good catalytic activity in the oxygen reduction and hydrogen oxidation reactions, making them suitable for both anode and cathode applications in fuel cells. Carbon black, a form of amorphous carbon, is broadly used as a support material for catalysts in fuel cell electrodes. It provides a highly conductive framework and a large surface area for catalytic reactions. Carbon black can be functionalized or modified to improve its catalytic properties, such as enhancing the stability and selectivity of reactions. It is often combined with other carbon-based materials to create hybrid catalyst systems with enhanced performance.

The development of CBCs for fuel cell electrodes is an active area of research. Scientists are investigating various synthesis methods, surface modifications, and structural designs to optimize the catalytic activity, stability, and durability of these materials [93]. By tailoring the properties of carbon-based catalysts, such as their surface chemistry, morphology, and porosity, it is possible to enhance their performance and make them more suitable for practical fuel cell applications. In conclusion, CBCs offer promising prospects for fuel cell electrodes. Graphene, CNTs, and carbon black provide cost-effective alternatives to platinum-based catalysts while offering a larger surface area, excellent electrical conductivity, and enhanced catalytic activity [94]. Ongoing research aims to augment the performance and stability of CBCs further, contributing to fuel cell technology's advancement and its potential for clean and sustainable energy conversion.

Proton Exchange Membrane Fuel Cells (PEMFCs)

A polymer electrolyte membrane is used in PEMFCs (a form of fuel cell) to facilitate the electrochemical reactions involved in generating electricity.

PEMFCs offer several advantages, including high power density, rapid startup and shutdown times, and the capability to operate at comparatively low temperatures (typically below 100°C). These characteristics make PEMFCs appropriate for various applications, including automotive, portable, and stationary power systems [95]. The key component of a PEMFC is the Proton Exchange Membrane (PEM), also known as the electrolyte membrane. This membrane allows the transport of protons from the anode to the cathode while blocking the passage of electrons. Commonly used PEMs are Perfluoro Sulfonic Acid (PFSA) membranes, such as those with Na^+ ions, which exhibit good chemical stability and high proton conductivity [96]. However, the higher cost and limitations of PFSA membranes have led to ongoing research and development efforts to explore alternative materials with improved performance and lower costs. The anode of a PEMFC is typically made of a platinum-based catalyst supported on a carbon material, such as carbon black or CNTs. Hydrogen gas in the anode undergoes electrochemical oxidation and releases protons and electrons. The protons reach the cathode *via* an electrolyte membrane, and the electrons flow through an external circuit, generating an electrical current. The oxygen gas from the surrounding air combines with the electrons and protons to generate water as a byproduct at the cathode. To enhance the performance of PEMFCs, catalyst layers are typically coated on both the anode and cathode sides of the membrane [97]. These layers consist of the catalyst material, often Pt NPs dispersed on carbon support, possessing a higher surface area. The catalyst promotes the electrochemical reactions at a lower activation energy, improving the efficiency of the fuel cell. Gas diffusion layers are also used on both sides of the membrane to facilitate the distribution of reactant gases and ensure efficient mass transport. PEMFCs require a source of pure hydrogen fuel, which can be obtained through various methods such as reforming hydrocarbons, electrolysis of water, or utilizing hydrogen-rich fuels like methanol. Hydrogen as a fuel offers advantages in terms of high energy density and zero emissions, as the only byproduct is water. However, hydrogen production, storage, and distribution infrastructure remain challenging for the widespread adoption of PEMFCs.

Ongoing research and development efforts in PEMFC technology focus on improving key aspects such as reducing the amount of expensive platinum catalyst, enhancing the durability and lifetime of the membrane, increasing system efficiency, and developing efficient and cost-effective hydrogen production methods. Advances in materials science, membrane engineering, catalyst design, and system integration are driving the progress of PEMFCs toward commercial viability and broader application. In conclusion, PEMFCs, or proton exchange membrane fuel cells, are a potential technique for producing sustainable energy. With their high-power density, rapid response times, and suitability for various applications, PEMFCs offer significant potential for addressing energy needs while reducing greenhouse gas emissions. Continued research and development efforts are vital for enlightening the performance, durability, and cost-effectiveness of PEMFCs, ultimately paving the way for their widespread adoption as a sustainable energy solution [98].

Direct Methanol Fuel Cells (DMFCs)

Methanol is used as fuel in DMFCs, which do not require an external reformer to function. DMFCs offer several advantages, including ease of fuel storage and transportation, high energy density, and the potential for portable and mobile applications. These characteristics make DMFCs attractive for a variety of applications, such as portable electronic devices, automotive systems, and backup power sources. In DMFCs, methanol is directly supplied at the anode, where it undergoes electrochemical oxidation to release protons and electrons. The protons reach the cathode *via* a polymer electrolyte membrane, while the electrons flow through an external circuit, generating an electrical current. The oxygen gas from the surrounding air combines with the electrons and protons to generate water as a byproduct at the cathode. The overall reaction is:

At anode:

CH_3OH Methannol + H_2O water → CO_2 Carbon dioxide + $6H^+$ Protons + $6e^-$ electrons

At cathode:

$3 2O_2 + 6H^+$ Protons + $6e^-$ electrons → $3H_2O$ Water

The key component in a DMFC is the PEM, similar to that used in PEMFCs. The membrane allows the transport of protons while blocking the passage of methanol molecules. Commonly used proton exchange membranes for DMFCs include Na^+ ion and other Perfluorosulfonic Acid (PFSA) membranes, which offer good proton conductivity and chemical stability [99]. The anode catalyst in DMFCs is typically based on Pt NPs supported on a high-surface-area carbon material. The catalyst promotes the electrochemical oxidation of methanol, improving the efficiency of the fuel cell. The cathode catalyst, often platinum-based, facilitates the oxygen reduction reaction. To enhance the overall performance of DMFCs, gas diffusion layers and bipolar plates are used to distribute the fuel and oxidant, facilitate reactant transport, and provide electrical connections between individual cells.

DMFCs face several challenges that hinder their widespread commercialization. One major challenge is methanol crossover, where methanol molecules permeate through the membrane, reducing fuel efficiency and causing performance degradation. Efforts are underway to develop improved membrane materials and electrode structures to mitigate methanol crossover. Additionally, oxygen reduction reaction at the cathode has a slow reaction rate, and the high cost of catalyst materials is an area of active research to enhance the performance and reduce the cost of DMFCs [100].

Research and development efforts in DMFC technology aim to improve key aspects such as methanol crossover reduction, catalyst activity and durability, system efficiency, and cost-effectiveness. Advances in materials science, membrane engineering, catalyst design, and system integration are crucial for the successful commercialization of DMFCs and their widespread adoption as a sustainable and efficient energy solution. In conclusion, DMFCs hold promise as a compact and portable energy source that can operate on readily available methanol fuel. With the continuing research and development efforts to address challenges and optimize performance, DMFCs have the potential to provide clean and efficient power for various applications, contributing to a more sustainable energy future [101].

Thermoelectric Energy Conversion

Thermoelectric energy conversion is a fascinating field that focuses on directly converting temperature differences into electrical energy. This phenomenon, known as the Seebeck effect, occurs when a temperature gradient is applied across a thermoelectric material, generating an electric voltage. Conversely, the Peltier effect allows heat energy transfer when an electric current is passed through a thermoelectric device [102].

Thermoelectric materials possess unique properties that enable efficient energy conversion. These materials exhibit a phenomenon called the thermoelectric effect, where they have a high Seebeck coefficient and low electrical and thermal conductivity simultaneously. This combination of properties is essential for efficient thermoelectric energy conversion. In practical applications, thermoelectric materials are often used in modules or devices called Thermoelectric Generators (TEGs) and Thermoelectric Coolers (TECs) [102]. TEGs convert waste heat or temperature gradients into usable electrical power, making them valuable for waste heat recovery in industrial processes, automotive exhaust systems, and spacecraft. TECs, on the other hand, utilize the Peltier effect to provide localized cooling or heating in electronic devices, refrigeration systems, and thermal management applications. Thermoelectric energy conversion efficiency is characterized by a dimensionless figure of merit called the thermoelectric figure of merit, or ZT value. Higher ZT values indicate better performance, as they are directly related to the ability of a material to convert thermal energy into electrical energy efficiently. Advancements in thermoelectric materials research have focused on discovering new materials and optimizing their properties to achieve higher ZT values. Various strategies have been employed, including nano-structuring materials to enhance phonon scattering and reduce thermal conductivity, alloying to optimize electronic band structures, and utilizing advanced fabrication techniques to improve material performance [103]. While thermoelectric energy conversion holds significant potential for waste heat recovery and energy harvesting, there are challenges to overcome. One major challenge is the relatively low efficiency of current thermoelectric materials, limiting their widespread adoption. Additionally, the high cost and scarcity of certain elements used in thermoelectric materials pose obstacles to large-scale commercialization. Nonetheless, ongoing research and development efforts are dedicated to addressing these challenges and improving the performance and cost-effectiveness of thermoelectric energy conversion. Exploration of new materials, advances in nanotechnology, and optimization of device design and engineering are critical areas of focus in this field. Thermoelectric energy conversion offers a unique way to directly convert temperature differences into electrical energy. Ongoing research focuses on material-based approaches, such as carbon nanotube-based thermoelectric materials, graphene-enhanced thermoelectric devices, and hybrid CBNs. These advancements aim to improve the efficiency and performance of thermoelectric devices, enabling waste heat recovery, efficient thermal management, and enhanced energy harvesting capabilities. With continued research and development, thermoelectric technology holds great potential for a more sustainable and efficient energy landscape [104].

CNTs-Based Thermoelectric Materials

CNTs-based thermoelectric materials have gained significant attention in thermoelectric energy conversion due to their unique properties and potential for high performance. CNTs exhibit exceptional electrical and thermal conductivities, low thermal conductivity, and a high aspect ratio, making them promising candidates for enhancing the efficiency of thermoelectric devices [105]. The integration of CNTs into thermoelectric materials offers several advantages. CNTs provide enhanced electrical conductivity, thereby facilitating efficient charge transport and improving thermoelectric generators' overall power output [106]. Additionally, the low thermal conductivity of CNTs can help reduce heat loss, leading to higher thermoelectric conversion efficiency [107]. Research efforts have focused on synthesizing CNTs with controlled dimensions, doping them with different elements to tune their electronic properties, and optimizing their dispersion and alignment within the thermoelectric matrix. Various fabrication techniques, such as CVD, solution-based methods, and mechanical blending, have been employed to incorporate CNTs into thermoelectric materials and enhance their performance [8]. Studies have demonstrated the potential of CNT-based composites in achieving high thermoelectric figure of merit (ZT) values. The ZT value represents the material's efficiency in converting heat into electricity and is a key parameter for assessing thermoelectric performance. Researchers have enhanced ZT values in CNT-based thermoelectric materials by optimizing their composite architecture, content, and doping strategies [108]. Furthermore, advancements in CNT synthesis and manufacturing techniques have enabled the production of large-scale CNT-based thermoelectric materials with improved performance and stability. This progress paves the way for practical applications of carbon nanotube-based thermoelectric materials in waste heat recovery, power generation, and thermal management systems [109].

Graphene-Enhanced Thermoelectric Devices

Graphene-enhanced thermoelectric devices have emerged as a promising avenue for advancing thermoelectric energy conversion. Graphene, a 2D carbon material with exceptional thermal and electrical properties, offers unique advantages for improving the performance of thermoelectric devices [110].

Graphene in thermoelectric materials or device architectures can enhance thermoelectric properties. Graphene's high electrical conductivity allows efficient charge transport, facilitating improved power output in thermoelectric generators [111]. Additionally, graphene's high thermal conductivity can help reduce heat gradients and enhance heat dissipation in thermoelectric coolers [112]. Researchers have explored various strategies to introduce graphene into thermoelectric systems. One approach involves creating composites by incorporating graphene into the thermoelectric matrix, thereby improving electrical conductivity and charge carrier mobility [113]. Another approach focuses on utilizing graphene as a flexible and transparent electrode in thermoelectric devices, enabling efficient energy conversion while maintaining device functionality [114]. The unique properties of graphene, such as its large surface area, mechanical strength, and tunable electronic structure, offer opportunities for tailoring the thermoelectric performance of devices. Researchers have investigated methods to engineer graphene properties, such as chemical functionalization, doping, and patterning, to enhance its compatibility with thermoelectric materials and optimize energy conversion efficiency [115]. Graphene-enhanced thermoelectric devices have shown promise in various applications, including waste heat recovery, thermal management in electronic devices, and energy harvesting from temperature gradients. The ability of graphene to improve electrical and thermal transport properties, coupled with its scalability and manufacturability, makes it a compelling material for advancing thermoelectric technology [116].

Hybrid CBNs for Thermoelectric Applications

The hybridization of carbon-based materials, such as graphene or CNTs, with other materials like semiconductors, MOs, or polymers allows for tailoring the nanocomposites' electrical and thermal transport properties. This approach can enhance the power factor and reduce thermal conductivity, which are critical factors for improving thermoelectric performance [117]. The synergistic effects arising from the combination of different materials in hybrid CBNs enable enhanced electrical conductivity, improved charge carrier mobility, and efficient phonon scattering. These factors contribute to higher thermoelectric and conversion efficiency in thermoelectric generators and coolers [118]. Various fabrication techniques, including solution mixing, CVD, and electrodeposition, have been employed to synthesize hybrid CBNs. These techniques allow for precise control over the nanocomposites' composition, morphology, and interface characteristics, thereby further optimizing their thermoelectric performance [119]. Applying hybrid CBNs in thermoelectric devices can offer excellent waste heat recovery, energy harvesting from temperature gradients, and thermal management in electronic devices. Their tunable properties and compatibility with scalable manufacturing processes make them attractive for practical implementation [120]. Ongoing research and development efforts focus on exploring new combinations of carbon-based materials with other functional components, optimizing the nanocomposite structure and interface engineering, and improving the stability and durability of hybrid CBNs for long-term thermoelectric applications [121].

Piezoelectric and Triboelectric Energy Harvesting

Piezoelectric and triboelectric energy harvesting are promising technologies that convert mechanical energy into electrical energy. These approaches have gained significant attention in energy harvesting due to their ability to capture ambient vibrations and mechanical movements and convert them into usable electrical power [122]. Piezoelectric energy harvesting relies on the piezoelectric effect, which is the potential of certain materials to produce an electric charge in response to deformation or applied mechanical stress. When subjected to mechanical vibrations or impacts, piezoelectric materials, such as piezoelectric ceramics or polymers, produce an electric potential difference across their surfaces. This generated electrical energy can be stored and utilized for various applications, including powering small electronic devices or sensors.

Triboelectric energy harvesting, on the other hand, utilizes the triboelectric effect, which occurs when two materials with varied affinities for electrons come in contact and then separate. During the contact and separation process, electrons are transferred between the materials, generating an electric charge. This charge separation can be harnessed to produce electrical energy. Triboelectric energy harvesting can be achieved by various means, such as utilizing frictional motion, contact and separation of materials, or even capturing the energy from human motion or environmental vibrations. Both piezoelectric and triboelectric energy harvesting offer advantages such as scalability, versatility, and the ability to operate in various environments. These technologies have the potential to power small-scale wireless devices and wearable electronics or even contribute to energy harvesting in larger infrastructure systems [123]. Scientists are exploring new materials, such as advanced piezoelectric ceramics, nanomaterials, or flexible polymers, to enhance energy conversion efficiency and broaden the range of applications.

Additionally, innovative designs and integration strategies are being investigated to optimize the harvesting capabilities and increase the overall power output of these devices [124]. In conclusion, carbon nanomaterials play a vital role in enhancing the performance of piezoelectric generators and triboelectric nanogenerators in energy harvesting applications. Carbon Nanomaterials, including CNTs and graphene, have been incorporated into these devices to improve charge generation and transport, leading to higher energy conversion efficiency. However, challenges remain to be overcome, such as improving power density, durability, scalability, and system compatibility. Continued research and development efforts are necessary to advance these technologies further and make them more practical and efficient for a wide range of applications [125].

Carbon Nanomaterials in Piezoelectric Generators

Carbon nanomaterials have shown tremendous potential in improving the performance of piezoelectric generators for energy harvesting applications. These nanomaterials, such as carbon nanotubes and graphene, possess unique properties that make them suitable for enhancing charge generation and transfer in piezoelectric devices [122]. Incorporating carbon nanomaterials into the generator's structure enhances its mechanical flexibility, conductivity, and surface area, improving energy conversion efficiency [124]. Additionally, the high aspect ratio and exceptional electrical properties of carbon nanomaterials facilitate the effective alignment and orientation of piezoelectric materials, further enhancing the device's piezoelectric response. Using carbon nanomaterials in piezoelectric generators holds promise for developing efficient and flexible energy harvesting systems [125, 126].

Triboelectric Nanogenerators with Carbon-Based Components

With the ability to transform mechanical energy into electrical energy, Triboelectric Nanogenerators (TENGs) made of carbon-based materials have gained popularity. Carbon-based materials, including CNTs, graphene, and carbon nanofibers, play a vital role in enhancing the performance of TENGs [127, 128]. These materials exhibit significant electrical conductivity, mechanical flexibility, and high surface area, making them ideal for improving charge generation and transfer in TENGs. By incorporating carbon-based components into the triboelectric layers or electrodes of TENGs, the contact electrification and electrostatic induction processes can be enhanced, resulting in improved power output and efficiency. Furthermore, carbon-based materials' lightweight and scalable nature makes them suitable for various applications, including wearable devices, self-powered sensors, and portable electronics. Using carbon-based components in TENGs holds great promise for developing sustainable and efficient energy harvesting technologies [9].

Advancements and Challenges in Piezoelectric and Triboelectric Energy Harvesting

Advancements in piezoelectric and triboelectric energy harvesting technologies have significantly contributed to developing efficient and sustainable methods for converting mechanical energy into electrical energy. These technologies offer several advantages, including their ability to generate power from ambient vibrations and movements, making them suitable for a broad range of applications [6].

In piezoelectric energy harvesting, significant advancements have been made in developing high-performance piezoelectric materials, like lead-free alternatives, Lead Zirconate Titanate (PZT), and flexible polymer-based piezoelectric materials. These advancements have improved energy conversion efficiency and broadened the application possibilities of piezoelectric generators. Additionally, the integration of advanced nanomaterials, including carbon nanomaterials, into piezoelectric systems has shown promise in enhancing the devices' mechanical flexibility, durability, and overall performance. Similarly, in triboelectric energy harvesting, researchers have made notable progress in designing and optimizing Triboelectric Nanogenerators (TENGs) [129]. Developing novel triboelectric materials with high surface charge density, low friction properties, and excellent mechanical durability has improved power generation efficiency. Furthermore, incorporating carbon-based components, such as graphene, CNTs, and conductive polymers, has enhanced the charge transfer and output performance of TENGs. However, several challenges must be addressed in piezoelectric and triboelectric energy harvesting. One major challenge is enhancing energy conversion efficiency, as current systems often need help in extracting the maximum amount of energy from mechanical sources. Additionally, integrating energy harvesting devices into practical applications while ensuring their stability, durability, and miniaturization remains a significant challenge. Developing reliable and cost-effective manufacturing techniques and establishing industry standards will be crucial for their commercialization and large-scale implementation [130].

In conclusion, piezoelectric and triboelectric energy harvesting developments have opened up new possibilities for harvesting mechanical energy and generating electrical power. The integration of carbon nanomaterials, along with other advancements in materials and device design, has significantly contributed to the performance improvement of these technologies. However, efficiency, durability, and scalability challenges still need to be overcome to realize their full potential and enable their widespread adoption in various practical applications.

CONCLUSION AND FUTURE PERSPECTIVES

CBNs have emerged as highly promising materials with immense potential in environmental and energy-harvesting applications. Their exceptional properties, including high electrical conductivity, mechanical strength, and surface reactivity, make them versatile for addressing various challenges. Carbon-based nanocomposites exhibit remarkable adsorption capabilities, catalytic activity, and selective separation in environmental applications, effectively removing pollutants and contaminants from water sources and facilitating efficient water purification processes. Moreover, they play a crucial role in air pollution control by acting as efficient gas sensors and catalysts for degrading harmful pollutants. Additionally, they contribute to environmental sensing and monitoring by providing highly sensitive and selective detection platforms for monitoring environmental parameters. In the realm of energy harvesting, CBNs have revolutionized photovoltaics, enhancing the efficiency and performance of solar cells. Their incorporation into photovoltaic devices, such as carbon nanotube-based solar cells and graphene-based perovskite solar cells, has significantly improved energy conversion efficiency. Furthermore, carbon-based nanocomposites have played a vital role in energy storage systems, particularly in batteries and supercapacitors. They enhance the charge-discharge rates, energy storage capacity, and cycling stability of batteries, leading to more efficient and durable energy storage solutions.

In the context of supercapacitors, carbon nanomaterials, such as graphene and CNTs, have been instrumental in boosting the capacitance and power density, thus enabling rapid energy storage and release. Fuel cells also benefit from the utilization of CBNs as catalysts for fuel cell electrodes, enhancing the overall efficiency and performance of PEMFCs and DMFCs. Moreover, CBNs demonstrate promising prospects in thermoelectric energy conversion by efficiently converting waste heat into electricity, thus upgrading energy efficiency. Additionally, in piezoelectric and triboelectric energy harvesting, carbon nanomaterials are crucial in generating electricity from mechanical vibrations and frictional forces, offering potential applications in wearable electronics, self-powered sensors, and energy harvesting devices. While carbon-based nanocomposites have shown exceptional promise, challenges such as scalability, manufacturing processes, toxicity, and environmental impact remain.

Future research endeavors should address these challenges and develop sustainable synthesis methods, efficient manufacturing processes, and strategies for the safe disposal and recycling of these nanocomposites. With continued research, innovation, and collaboration, carbon-based nanocomposites hold tremendous potential to drive advancements in environmental sustainability and energy harvesting, ultimately contributing significantly to developing a greener, more sustainable future.

REFERENCES

[1] L. Jianping, L. Minrong, W. Jinnan, L. Jianjian, S. Hongwen, and H. Maoxing, "Global Environmental Issues and Human Wellbeing",

[2] S. Sharma, and A. Bhattacharya, "Drinking water contamination and treatment techniques", *Appl. Water Sci.*, vol. 7, pp. 1043-1067, 2017.
[http://dx.doi.org/10.1007/s13201-016-0455-7]

[3] X. Fang, Y. Bando, U.K. Gautam, C. Ye, and D. Golberg, "Inorganic semiconductor nanostructures and their field-emission applications", *J. Mater. Chem.*, vol. 18, pp. 509-522, 2008.
[http://dx.doi.org/10.1039/B712874F]

[4] S. Nasir, M. Hussein, Z. Zainal, and N. Yusof, "Carbon-based nanomaterials/allotropes: a glimpse of their synthesis, properties and some applications", *Materials (Basel)*, vol. 11, p. 295, 2018.
[http://dx.doi.org/10.3390/ma11020295]

[5] S. Mangala Nagasundari, K. Muthu, K. Kaviyarasu, D.A. Al Farraj, and R.M. Alkufeidy, "Current trends of Silver doped Zinc oxide nanowires photocatalytic degradation for energy and environmental application", *Surf. Interfaces*, vol. 23, p. 100931, 2021.
[http://dx.doi.org/10.1016/j.surfin.2021.100931]

[6] S.P. Muduli, L. Lipsa, A. Choudhary, S. Rajput, and S. Parida, "Modulation of electrical characteristics of polymer–ceramic–graphene hybrid composite for piezoelectric energy harvesting", *ACS Appl. Electron. Mater.*, vol. 5, pp. 3023-3037, 2023.
[http://dx.doi.org/10.1021/acsaelm.3c00078]

[7] L. Mohapatra, D. Cheon, and S.H. Yoo, "Carbon-based nanomaterials for catalytic wastewater treatment: a review", *Molecules*, vol. 28, p. 1805, 2023.
[http://dx.doi.org/10.3390/molecules28041805]

[8] L. Dai, D.W. Chang, J. Baek, and W. Lu, "Carbon nanomaterials for advanced energy conversion and storage", *Small*, vol. 8, pp. 1130-1166, 2012.
[http://dx.doi.org/10.1002/smll.201101594]

[9] R. Yu, S. Feng, Q. Sun, H. Xu, Q. Jiang, J. Guo, B. Dai, D. Cui, and K. Wang, "Ambient energy harvesters in wearable electronics: fundamentals, methodologies, and applications", *J. Nanobiotechnology,* vol. 22, p. 497, 2024.
[http://dx.doi.org/10.1186/s12951-024-02774-0]

[10] M. Huston, M. DeBella, M. DiBella, and A. Gupta, "Green synthesis of nanomaterials", *Nanomaterials (Basel),* vol. 11, p. 2130, 2021.
[http://dx.doi.org/10.3390/nano11082130]

[11] L. Joshi, S. Singh Rajput, and S. Keshri, "Structural and magneto-transport properties of LCMO–STO composites", *Phase Transit.,* vol. 83, pp. 482-490, 2010.
[http://dx.doi.org/10.1080/01411594.2010.492466]

[12] D. Veeman, M.V. Shree, P. Sureshkumar, T. Jagadeesha, L. Natrayan, M. Ravichandran, and P. Paramasivam, "Sustainable development of carbon nanocomposites: synthesis and classification for environmental remediation", *J. Nanomater.,* vol. 2021, pp. 1-21, 2021.
[http://dx.doi.org/10.1155/2021/5840645]

[13] X. Yuan, X. Zhang, L. Sun, Y. Wei, and X. Wei, "Cellular toxicity and immunological effects of carbon-based nanomaterials", *Part. Fibre Toxicol.,* vol. 16, p. 18, 2019.
[http://dx.doi.org/10.1186/s12989-019-0299-z]

[14] N. Mohammadian, S. Ghoreishi, S. Hafeziyeh, S. Saeidi, and D. Dionysiou, "Optimization of synthesis conditions of carbon nanotubes via ultrasonic-assisted floating catalyst deposition using response surface methodology", *Nanomaterials (Basel),* vol. 8, p. 316, 2018.
[http://dx.doi.org/10.3390/nano8050316]

[15] H.P. Bei, Y. Yang, Q. Zhang, Y. Tian, X. Luo, M. Yang, and X. Zhao, "Graphene-based nanocomposites for neural tissue engineering", *Molecules,* vol. 24, p. 658, 2019.
[http://dx.doi.org/10.3390/molecules24040658]

[16] J. Liu, R. Li, and B. Yang, "Carbon dots: a new type of carbon-based nanomaterial with wide applications", *ACS Cent. Sci.,* vol. 6, pp. 2179-2195, 2020.
[http://dx.doi.org/10.1021/acscentsci.0c01306]

[17] J. Hubbard, T. Isik, T.Y. Ansell, V. Ortalan, C. Luhrs, Introduction of rare-earth oxide nanoparticles in CNT-based nanocomposites for improved detection of underlying CNT network, Nanomaterials 11 (2021) 2168.
[http://dx.doi.org/10.3390/nano11092168]

[18] H. Zhang, Y. Zhao, X. Yang, G. Zhao, D. Zhang, H. Huang, S. Yang, N. Wen, M. Javid, Z. Fan, and L. Pan, "A facile synthesis of novel amorphous TiO_2 nanorods decorated rGO hybrid composites with wide band microwave absorption", *Nanomaterials (Basel),* vol. 10, p. 2141, 2020.
[http://dx.doi.org/10.3390/nano10112141]

[19] S. He, B. Song, D. Li, C. Zhu, W. Qi, Y. Wen, L. Wang, S. Song, H. Fang, and C. Fan, "A graphene nanoprobe for rapid, sensitive, and multicolor fluorescent DNA analysis", *Adv. Funct. Mater.,* vol. 20, pp. 453-459, 2010.
[http://dx.doi.org/10.1002/adfm.200901639]

[20] Y. Wen, F. Xing, S. He, S. Song, L. Wang, Y. Long, D. Li, and C. Fan, "A graphene-based fluorescent nanoprobe for silver(i) ions detection by using graphene oxide and a silver-specific oligonucleotide", *Chem. Commun.,* vol. 46, p. 2596, 2010.
[http://dx.doi.org/10.1039/b924832c]

[21] V. Chandra, J. Park, Y. Chun, J.W. Lee, I-C. Hwang, and K.S. Kim, "Water-dispersible magnetite-reduced graphene oxide composites for arsenic removal", *ACS Nano,* vol. 4, pp. 3979-3986, 2010.
[http://dx.doi.org/10.1021/nn1008897]

[22] L. Fan, C. Luo, M. Sun, and H. Qiu, "Synthesis of graphene oxide decorated with magnetic cyclodextrin for fast chromium removal", *J. Mater. Chem.,* vol. 22, p. 24577, 2012.

[http://dx.doi.org/10.1039/c2jm35378d]

[23] V. Chandra, and K.S. Kim, "Highly selective adsorption of Hg2+ by a polypyrrole–reduced graphene oxide composite", *Chem. Commun.*, vol. 47, p. 3942, 2011.
[http://dx.doi.org/10.1039/c1cc00005e]

[24] P. Onkar Patil, J.H. Patil, M.P. More, M.R. Mahajan, and A.G. Patil, "Green synthesis of graphene based manocomposite for sensing of heavy metals", *Journal of Pharmaceutical and Biological Sciences*, vol. 7, pp. 56-62, 2020.
[http://dx.doi.org/10.18231/j.jpbs.2019.010]

[25] G. Singh, A. Choudhary, D. Haranath, A.G. Joshi, N. Singh, S. Singh, and R. Pasricha, "ZnO decorated luminescent graphene as a potential gas sensor at room temperature", *Carbon N Y*, vol. 50, pp. 385-394, 2012.
[http://dx.doi.org/10.1016/j.carbon.2011.08.050]

[26] H.M. Yadav, and J-S. Kim, "Solvothermal synthesis of anatase TiO2-graphene oxide nanocomposites and their photocatalytic performance", *J. Alloys Compd.*, vol. 688, pp. 123-129, 2016.
[http://dx.doi.org/10.1016/j.jallcom.2016.07.133]

[27] W. Li, X. Geng, Y. Guo, J. Rong, Y. Gong, L. Wu, X. Zhang, P. Li, J. Xu, G. Cheng, M. Sun, and L. Liu, "Reduced graphene oxide electrically contacted graphene sensor for highly sensitive nitric oxide detection", *ACS Nano*, vol. 5, pp. 6955-6961, 2011.
[http://dx.doi.org/10.1021/nn201433r]

[28] Q. Ji, I. Honma, S. Paek, M. Akada, J.P. Hill, A. Vinu, K. Ariga, Layer-by-layer films of graphene and ionic liquids for highly selective gas sensing, Angewandte Chemie International Edition 49 (2010) 9737–9739.
[http://dx.doi.org/10.1002/anie.201004929]

[29] T.H. Han, Y-K. Huang, A.T.L. Tan, V.P. Dravid, and J. Huang, "Steam etched porous graphene oxide network for chemical sensing", *J. Am. Chem. Soc.*, vol. 133, pp. 15264-15267, 2011.
[http://dx.doi.org/10.1021/ja205693t]

[30] S.S. Rajput, R. Katoch, K.K. Sahoo, G.N. Sharma, S.K. Singh, R. Gupta, and A. Garg, "Enhanced electrical insulation and ferroelectricity in La and Ni co-doped BiFeO3 thin films", *J. Alloys Compd.*, vol. 621, pp. 339-344, 2015.
[http://dx.doi.org/10.1016/j.jallcom.2014.09.161]

[31] J.T. Robinson, F.K. Perkins, E.S. Snow, Z. Wei, and P.E. Sheehan, "Reduced graphene oxide molecular sensors", *Nano Lett.*, vol. 8, pp. 3137-3140, 2008.
[http://dx.doi.org/10.1021/nl8013007]

[32] F. Schedin, A.K. Geim, S.V. Morozov, E.W. Hill, P. Blake, M.I. Katsnelson, and K.S. Novoselov, "Detection of individual gas molecules adsorbed on graphene", *Nat. Mater.*, vol. 6, pp. 652-655, 2007.
[http://dx.doi.org/10.1038/nmat1967]

[33] Y. Zhang, Z-R. Tang, X. Fu, and Y-J. Xu, "TiO_2–graphene nanocomposites for gas-phase photocatalytic degradation of volatile aromatic pollutant: is TiO_2–graphene truly different from other TiO_2–carbon composite materials?", *ACS Nano*, vol. 4, pp. 7303-7314, 2010.
[http://dx.doi.org/10.1021/nn1024219]

[34] Y. Wan, Y. Wang, J. Wu, and D. Zhang, "Graphene oxide sheet-mediated silver enhancement for application to electrochemical biosensors", *Anal. Chem.*, vol. 83, pp. 648-653, 2011.
[http://dx.doi.org/10.1021/ac103047c]

[35] J.H. Jung, D.S. Cheon, F. Liu, K.B. Lee, and T.S. Seo, "A graphene oxide based immuno-biosensor for pathogen detection", *Angew. Chem. Int. Ed.*, vol. 49, pp. 5708-5711, 2010.
[http://dx.doi.org/10.1002/anie.201001428]

[36] G.A. Zelada-Guillén, J. Riu, A. Düzgün, F.X. Rius, Immediate detection of living bacteria at ultralow concentrations using a carbon nanotube based potentiometric aptasensor, Angewandte Chemie

International Edition 48 (2009) 7334-7337.
[http://dx.doi.org/10.1002/anie.200902090]

[37] Y. Huang, X. Dong, Y. Liu, L-J. Li, and P. Chen, "Graphene-based biosensors for detection of bacteria and their metabolic activities", *J. Mater. Chem.*, vol. 21, p. 12358, 2011.
[http://dx.doi.org/10.1039/c1jm11436k]

[38] W. Hu, C. Peng, W. Luo, M. Lv, X. Li, D. Li, Q. Huang, and C. Fan, "Graphene-based antibacterial paper", *ACS Nano*, vol. 4, pp. 4317-4323, 2010.
[http://dx.doi.org/10.1021/nn101097v]

[39] V. Ahmad, and M.O. Ansari, "Antimicrobial activity of graphene-based nanocomposites: synthesis, characterization, and their applications for human welfare", *Nanomaterials (Basel)*, vol. 12, p. 4002, 2022.
[http://dx.doi.org/10.3390/nano12224002]

[40] S.S. Nanda, D.K. Yi, and K. Kim, "Study of antibacterial mechanism of graphene oxide using Raman spectroscopy", *Sci. Rep.*, vol. 6, p. 28443, 2016.
[http://dx.doi.org/10.1038/srep28443]

[41] S. Jaworski, M. Wierzbicki, E. Sawosz, A. Jung, G. Gielerak, J. Biernat, H. Jaremek, W. Łojkowski, B. Woźniak, J. Wojnarowicz, L. Stobiński, A. Małolepszy, M. Mazurkiewicz-Pawlicka, M. Łojkowski, N. Kurantowicz, and A. Chwalibog, "Graphene oxide-based nanocomposites decorated with silver nanoparticles as an antibacterial agent", *Nanoscale Res. Lett.*, vol. 13, p. 116, 2018.
[http://dx.doi.org/10.1186/s11671-018-2533-2]

[42] J. Li, G. Wang, H. Zhu, M. Zhang, X. Zheng, Z. Di, X. Liu, and X. Wang, "Antibacterial activity of large-area monolayer graphene film manipulated by charge transfer", *Sci. Rep.*, vol. 4, p. 4359, 2014.
[http://dx.doi.org/10.1038/srep04359]

[43] Y-H. Hsueh, C-T. Hsieh, S-T. Chiu, P-H. Tsai, C-Y. Liu, and W-J. Ke, "Antibacterial property of composites of reduced graphene oxide with nano-silver and zinc oxide nanoparticles synthesized using a microwave-assisted approach", *Int. J. Mol. Sci.*, vol. 20, p. 5394, 2019.
[http://dx.doi.org/10.3390/ijms20215394]

[44] K. Prasad, G.S. Lekshmi, K. Ostrikov, V. Lussini, J. Blinco, M. Mohandas, K. Vasilev, S. Bottle, K. Bazaka, and K. Ostrikov, "Synergic bactericidal effects of reduced graphene oxide and silver nanoparticles against Gram-positive and Gram-negative bacteria", *Sci. Rep.*, vol. 7, p. 1591, 2017.
[http://dx.doi.org/10.1038/s41598-017-01669-5]

[45] T. Vi, S. Kumar, J-H. Pang, Y-K. Liu, D. Chen, and S. Lue, "Synergistic antibacterial activity of silver-loaded graphene oxide towards Staphylococcus aureus and Escherichia coli", *Nanomaterials (Basel)*, vol. 10, p. 366, 2020.
[http://dx.doi.org/10.3390/nano10020366]

[46] Y. Wei, Y. Zhang, X. Gao, Z. Ma, X. Wang, and C. Gao, "Multilayered graphene oxide membranes for water treatment: A review", *Carbon N Y*, vol. 139, pp. 964-981, 2018.
[http://dx.doi.org/10.1016/j.carbon.2018.07.040]

[47] M.S-L. Yee, P-S. Khiew, W.S. Chiu, Y.F. Tan, Y-Y. Kok, and C-O. Leong, "Green synthesis of graphene-silver nanocomposites and its application as a potent marine antifouling agent", *Colloids Surf. B Biointerfaces*, vol. 148, pp. 392-401, 2016.
[http://dx.doi.org/10.1016/j.colsurfb.2016.09.011]

[48] Y. Tang, G. Zhang, C. Liu, S. Luo, X. Xu, L. Chen, and B. Wang, "Magnetic TiO2-graphene composite as a high-performance and recyclable platform for efficient photocatalytic removal of herbicides from water", *J. Hazard. Mater.*, vol. 252–253, pp. 115-122, 2013.
[http://dx.doi.org/10.1016/j.jhazmat.2013.02.053]

[49] H. Liu, X. Hao, Y. Liu, and A. Yan, "Hydrothermal synthesis and photocatalytic properties of Graphene@Ag/AgSb2O5.8 composites: reaction laws of the composites in sintering process", *Adv. Mater. Sci. Eng.*, vol. 2022, pp. 1-11, 2022.

[http://dx.doi.org/10.1155/2022/3817050]

[50] M.K. Joshi, H.R. Pant, H.J. Kim, J.H. Kim, and C.S. Kim, "One-pot synthesis of Ag-iron oxide/reduced graphene oxide nanocomposite via hydrothermal treatment", *Colloids Surf. A Physicochem. Eng. Asp.*, vol. 446, pp. 102-108, 2014.
[http://dx.doi.org/10.1016/j.colsurfa.2014.01.058]

[51] H. Zhang, X. Lv, Y. Li, Y. Wang, and J. Li, "P25-graphene composite as a high performance photocatalyst", *ACS Nano*, vol. 4, pp. 380-386, 2010.
[http://dx.doi.org/10.1021/nn901221k]

[52] L.L. Zhang, Z. Xiong, and X.S. Zhao, "Pillaring chemically exfoliated graphene oxide with carbon nanotubes for photocatalytic degradation of dyes under visible light irradiation", *ACS Nano*, vol. 4, pp. 7030-7036, 2010.
[http://dx.doi.org/10.1021/nn102308r]

[53] H. Sun, L. Cao, and L. Lu, "Magnetite/reduced graphene oxide nanocomposites: One step solvothermal synthesis and use as a novel platform for removal of dye pollutants", *Nano Res.*, vol. 4, pp. 550-562, 2011.
[http://dx.doi.org/10.1007/s12274-011-0111-3]

[54] Y. Chen, L. Chen, H. Bai, and L. Li, "Graphene oxide–chitosan composite hydrogels as broad-spectrum adsorbents for water purification", *J. Mater. Chem. A Mater. Energy Sustain.*, vol. 1, pp. 1992-2001, 2013.
[http://dx.doi.org/10.1039/C2TA00406B]

[55] A.B. Seabra, A.J. Paula, R. de Lima, O.L. Alves, and N. Durán, "Nanotoxicity of graphene and graphene oxide", *Chem. Res. Toxicol.*, vol. 27, pp. 159-168, 2014.
[http://dx.doi.org/10.1021/tx400385x]

[56] L. Xuan, Z. Ju, M. Skonieczna, P. Zhou, and R. Huang, "Nanoparticles-induced potential toxicity on human health: Applications, toxicity mechanisms, and evaluation models", *MedComm (Beijing)*, vol. 4, 2023.
[http://dx.doi.org/10.1002/mco2.327]

[57] N. Rodoshi Khan, A. Bin Rashid, Carbon-Based Nanomaterials: a Paradigm Shift in Biofuel Synthesis and Processing for a Sustainable Energy Future, Energy Conversion and Management: X 22 (2024) 100590.
[http://dx.doi.org/10.1016/j.ecmx.2024.100590]

[58] D. Khan, Z. Ali, D. Asif, M. Kumar Panjwani, and I. Khan, "Incorporation of carbon nanotubes in photoactive layer of organic solar cells", *Ain Shams Eng. J.*, vol. 12, pp. 897-900, 2021.
[http://dx.doi.org/10.1016/j.asej.2020.06.002]

[59] X. Wang, L. Zhi, and K. Müllen, "Transparent, conductive graphene electrodes for dye-sensitized solar cells", *Nano Lett.*, vol. 8, pp. 323-327, 2008.
[http://dx.doi.org/10.1021/nl072838r]

[60] W-L. Xu, M-S. Niu, X-Y. Yang, J. Xiao, H-C. Yuan, C. Xiong, and X-T. Hao, "Carbon nanotubes as the effective charge transport pathways for planar perovskite photodetector", *Org. Electron.*, vol. 59, pp. 156-163, 2018.
[http://dx.doi.org/10.1016/j.orgel.2018.05.004]

[61] N.E. Safie, M.A. Azam, M.F.A. Aziz, and M. Ismail, "Recent progress of graphene-based materials for efficient charge transfer and device performance stability in perovskite solar cells", *Int. J. Energy Res.*, vol. 45, pp. 1347-1374, 2021.
[http://dx.doi.org/10.1002/er.5876]

[62] Q. Chen, and N. De Marco, "Y. (Michael) Yang, T.-B. Song, C.-C. Chen, H. Zhao, Z. Hong, H. Zhou, Y. Yang, Under the spotlight: The organic–inorganic hybrid halide perovskite for optoelectronic applications", *Nano Today*, vol. 10, pp. 355-396, 2015.
[http://dx.doi.org/10.1016/j.nantod.2015.04.009]

[63] R. Liu, "Hybrid organic/inorganic nanocomposites for photovoltaic cells", *Materials (Basel)*, vol. 7, pp. 2747-2771, 2014.
[http://dx.doi.org/10.3390/ma7042747]

[64] S. Rajput, A. Sharma, V. Jately, and M. Ram, *Recent advances in energy harvesting technologies.* River Publishers: New York, 2023.

[65] S. Rajput, A. Kuperman, A. Yahalom, and M. Averbukh, "Studies on dynamic properties of ultracapacitors using infinite r–C chain equivalent circuit and reverse fourier transform", *Energies*, vol. 13, p. 4583, 2020.
[http://dx.doi.org/10.3390/en13184583]

[66] S.K. Sonika, "Verma, S. Samanta, A.K. Srivastava, S. Biswas, R.M. Alsharabi, S. Rajput, Conducting polymer nanocomposite for energy storage and energy harvesting systems", *Adv. Mater. Sci. Eng.*, vol. 2022, pp. 1-23, 2022.
[http://dx.doi.org/10.1155/2022/2266899]

[67] D. Bresser, E. Paillard, R. Kloepsch, S. Krueger, M. Fiedler, R. Schmitz, D. Baither, M. Winter, and S. Passerini, "Carbon coated ZnFe 2 O 4 nanoparticles for advanced lithium-ion anodes", *Adv. Energy Mater.*, vol. 3, pp. 513-523, 2013.
[http://dx.doi.org/10.1002/aenm.201200735]

[68] L.S. Roselin, R-S. Juang, C-T. Hsieh, S. Sagadevan, A. Umar, R. Selvin, and H.H. Hegazy, "Recent advances and perspectives of carbon-based nanostructures as anode materials for Li-ion batteries", *Materials (Basel)*, vol. 12, p. 1229, 2019.
[http://dx.doi.org/10.3390/ma12081229]

[69] Q. Cheng, J. Tang, J. Ma, H. Zhang, N. Shinya, and L-C. Qin, "Graphene and carbon nanotube composite electrodes for supercapacitors with ultra-high energy density", *Phys. Chem. Chem. Phys.*, vol. 13, p. 17615, 2011.
[http://dx.doi.org/10.1039/c1cp21910c]

[70] J. Cai, C. Liu, S. Tao, Z. Cao, Z. Song, X. Xiao, W. Deng, H. Hou, and X. Ji, "MOFs-derived advanced heterostructure electrodes for energy storage", *Coord. Chem. Rev.*, vol. 479, p. 214985, 2023.
[http://dx.doi.org/10.1016/j.ccr.2022.214985]

[71] R. Febrian, N.L.W. Septiani, M. Iqbal, and B. Yuliarto, "Review—recent advances of carbon-based nanocomposites as the anode materials for lithium-ion batteries: synthesis and performance", *J. Electrochem. Soc.*, vol. 168, p. 110520, 2021.
[http://dx.doi.org/10.1149/1945-7111/ac3161]

[72] H. Zhang, Y. Huang, H. Ming, G. Cao, W. Zhang, J. Ming, and R. Chen, "Recent advances in nanostructured carbon for sodium-ion batteries, J Mater", *Chem. Mater.*, vol. 8, pp. 1604-1630, 2020.
[http://dx.doi.org/10.1039/C9TA09984K]

[73] I. Amiel, S. Rajput, and M. Averbukh, "Capacitive reactive power compensation to prevent voltage instabilities in distribution lines", *Int. J. Electr. Power Energy Syst.*, vol. 131, p. 107043, 2021.
[http://dx.doi.org/10.1016/j.ijepes.2021.107043]

[74] M. Zhao, B-Q. Li, X-Q. Zhang, J-Q. Huang, and Q. Zhang, "A perspective toward practical lithium–sulfur batteries", *ACS Cent. Sci.*, vol. 6, pp. 1095-1104, 2020.
[http://dx.doi.org/10.1021/acscentsci.0c00449]

[75] Y. Zhou, P. Jin, Y. Zhou, and Y. Zhu, "High-performance symmetric supercapacitors based on carbon nanotube/graphite nanofiber nanocomposites", *Sci. Rep.*, vol. 8, p. 9005, 2018.
[http://dx.doi.org/10.1038/s41598-018-27460-8]

[76] L.L. Zhang, and X.S. Zhao, "Carbon-based materials as supercapacitor electrodes", *Chem. Soc. Rev.*, vol. 38, p. 2520, 2009.
[http://dx.doi.org/10.1039/b813846j]

[77] S. Rajput, A. Lugovskoy, M. Averbukh, and A. Yahalom, "Porous metal-oxide based electrostatic energy generator", *2019 International IEEE Conference and Workshop in Óbuda on Electrical and Power Engineering (CANDO-EPE)*, 2019pp. 133-136

[78] C-F. Liu, Y-C. Liu, T-Y. Yi, and C-C. Hu, "Carbon materials for high-voltage supercapacitors", *Carbon N Y*, vol. 145, pp. 529-548, 2019.
[http://dx.doi.org/10.1016/j.carbon.2018.12.009]

[79] H. Wang, L-F. Cui, Y. Yang, H. Sanchez Casalongue, J.T. Robinson, Y. Liang, Y. Cui, and H. Dai, "Mn 3 O 4 −graphene hybrid as a high-capacity anode material for lithium ion batteries", *J. Am. Chem. Soc.*, vol. 132, pp. 13978-13980, 2010.
[http://dx.doi.org/10.1021/ja105296a]

[80] L.L. Zhang, R. Zhou, and X.S. Zhao, "Graphene-based materials as supercapacitor electrodes", *J. Mater. Chem.*, vol. 20, p. 5983, 2010.
[http://dx.doi.org/10.1039/c000417k]

[81] M. Li, Z. Tang, M. Leng, and J. Xue, "Flexible solid-state supercapacitor based on graphene-based hybrid films", *Adv. Funct. Mater.*, vol. 24, pp. 7495-7502, 2014.
[http://dx.doi.org/10.1002/adfm.201402442]

[82] Y. Zhu, S. Murali, M.D. Stoller, K.J. Ganesh, W. Cai, P.J. Ferreira, A. Pirkle, R.M. Wallace, K.A. Cychosz, M. Thommes, D. Su, E.A. Stach, R.S. Ruoff, Carbon-based supercapacitors produced by activation of graphene, Science (1979) 332 (2011) 1537–1541.
[http://dx.doi.org/10.1126/science.1200770]

[83] S. Zhu, J. Ni, and Y. Li, "Carbon nanotube-based electrodes for flexible supercapacitors", *Nano Res.*, vol. 13, pp. 1825-1841, 2020.
[http://dx.doi.org/10.1007/s12274-020-2729-5]

[84] Y. Huang, J. Liang, and Y. Chen, "An overview of the applications of graphene-based materials in supercapacitors", *Small*, vol. 8, pp. 1805-1834, 2012.
[http://dx.doi.org/10.1002/smll.201102635]

[85] Y. Abetbool, S. Rajput, A. Yahalom, M. Averbukh, Comprehensive study on dynamic parameters of symmetric and asymmetric ultracapacitors, Electronics (Basel) 8 (2019) 891.
[http://dx.doi.org/10.3390/electronics8080891]

[86] Z. Zhai, L. Zhang, T. Du, B. Ren, Y. Xu, S. Wang, J. Miao, and Z. Liu, "A review of carbon materials for supercapacitors", *Mater. Des.*, vol. 221, p. 111017, 2022.
[http://dx.doi.org/10.1016/j.matdes.2022.111017]

[87] L. Carrette, K.A. Friedrich, and U. Stimming, "Fuel cells - fundamentals and applications", *Fuel Cells (Weinh.)*, vol. 1, pp. 5-39, 2001.
[http://dx.doi.org/10.1002/1615-6854(200105)1:1<5:AID-FUCE5>3.0.CO;2-G]

[88] D.P. Wilkinson, J. Zhang, R. Hui, J. Fergus, X. Li, Ed., *Proton exchange membrane fuel cells.* CRC Press, 2009.

[89] A. Sharma, R.A. Khan, A. Sharma, D. Kashyap, and S. Rajput, "A novel opposition-based arithmetic optimization algorithm for parameter extraction of PEM fuel cell", *Electronics (Basel)*, vol. 10, p. 2834, 2021.
[http://dx.doi.org/10.3390/electronics10222834]

[90] R. Pachauri, A. Sharma, and S. Rajput, "Parametric effects on proton exchange membrane fuel cell performance: an analytical perspective, international journal of mathematical", *Engineering and Management Sciences*, vol. 5, pp. 926-938, 2020.
[http://dx.doi.org/10.33889/IJMEMS.2020.5.5.071]

[91] S. Li, C. Cheng, and A. Thomas, "Carbon-based microbial-fuel-cell electrodes: from conductive supports to active catalysts", *Adv. Mater.*, vol. 29, 2017.
[http://dx.doi.org/10.1002/adma.201602547]

[92] C. Huang, C. Li, and G. Shi, "Graphene based catalysts", *Energy Environ. Sci.,* vol. 5, p. 8848, 2012.
[http://dx.doi.org/10.1039/c2ee22238h]

[93] H. Singh, S. Zhuang, B. Ingis, B.B. Nunna, and E.S. Lee, "Carbon-based catalysts for oxygen reduction reaction: A review on degradation mechanisms", *Carbon N Y,* vol. 151, pp. 160-174, 2019.
[http://dx.doi.org/10.1016/j.carbon.2019.05.075]

[94] J. Yan, F. Ye, Q. Dai, X. Ma, Z. Fang, L. Dai, and C. Hu, "Recent progress in carbon-based electrochemical catalysts: From structure design to potential applications", *Nano Research Energy,* vol. 2, p. e9120047, 2023.
[http://dx.doi.org/10.26599/NRE.2023.9120047]

[95] W. Vielstich, A. Lamm, and H. Gasteiger, *Handbook of fuel cells.,* 2003.https://www.osti.gov/etdeweb/biblio/20480552

[96] Y. Zhao, X. Li, S. Wang, W. Li, X. Wang, S. Chen, J. Chen, and X. Xie, "Proton exchange membranes prepared via atom transfer radical polymerization for proton exchange membrane fuel cell: Recent advances and perspectives", *Int. J. Hydrogen Energy,* vol. 42, pp. 30013-30028, 2017.
[http://dx.doi.org/10.1016/j.ijhydene.2017.08.167]

[97] S. Ahmad, T. Nawaz, A. Ali, M.F. Orhan, A. Samreen, and A.M. Kannan, "An overview of proton exchange membranes for fuel cells: Materials and manufacturing", *Int. J. Hydrogen Energy,* vol. 47, pp. 19086-19131, 2022.
[http://dx.doi.org/10.1016/j.ijhydene.2022.04.099]

[98] Y. He, and G. Wu, "PGM-free oxygen-reduction catalyst development for proton-exchange membrane fuel cells: challenges, solutions, and promises, Acc", *Mater. Res.,* vol. 3, pp. 224-236, 2022.
[http://dx.doi.org/10.1021/accountsmr.1c00226]

[99] S. Sundarrajan, S.I. Allakhverdiev, and S. Ramakrishna, "Progress and perspectives in micro direct methanol fuel cell", *Int. J. Hydrogen Energy,* vol. 37, pp. 8765-8786, 2012.
[http://dx.doi.org/10.1016/j.ijhydene.2011.12.017]

[100] K. Scott, and A.K. Shukla,

[101] "W. Taama, J. Cruickshank, "Performance of a direct methanol fuel cell", *J. Appl. Electrochem.,* vol. 28, pp. 289-297, 1998.
[http://dx.doi.org/10.1023/A:1003263632683]

[102] G.J. Snyder, and E.S. Toberer, "Complex thermoelectric materials", *Nat. Mater.,* vol. 7, pp. 105-114, 2008.
[http://dx.doi.org/10.1038/nmat2090]

[103] M. Zebarjadi, G. Joshi, G. Zhu, B. Yu, A. Minnich, Y. Lan, X. Wang, M. Dresselhaus, Z. Ren, and G. Chen, "Power factor enhancement by modulation doping in bulk nanocomposites", *Nano Lett.,* vol. 11, pp. 2225-2230, 2011.
[http://dx.doi.org/10.1021/nl201206d]

[104] D.M. Rowe, Ed., *CRC Handbook of Thermoelectrics.* CRC Press, 2018.

[105] M.S. Dresselhaus, G. Dresselhaus, P. Avouris, Ed.,

[106] E. Pop, D. Mann, Q. Wang, K. Goodson, and H. Dai, "Thermal conductance of an individual single-wall carbon nanotube above room temperature", *Nano Lett.,* vol. 6, pp. 96-100, 2006.
[http://dx.doi.org/10.1021/nl052145f]

[107] J.L. Blackburn, A.J. Ferguson, C. Cho, and J.C. Grunlan, "Carbon-nanotube-based thermoelectric materials and devices", *Adv. Mater.,* vol. 30, 2018.
[http://dx.doi.org/10.1002/adma.201704386]

[108] J. Yun, S. Choi, and S.H. Im, "Advances in carbon-based thermoelectric materials for high-performance, flexible thermoelectric devices", *Carbon Energy,* vol. 3, pp. 667-708, 2021.
[http://dx.doi.org/10.1002/cey2.121]

[109] L. Lipsa, S. Rajput, S. Parida, S.P. Ghosh, S.K. Verma, and S. Gupta, "Influence of hot-press temperature on β-phase formation and electrical properties of solvent-casted PVDF-HFP co-polymer films prepared from two different solvents: a comparison study", *Macromol. Chem. Phys.*, vol. 224, 2023.
[http://dx.doi.org/10.1002/macp.202300204]

[110] K.S. Novoselov, "V.I. Fal'ko, L. Colombo, P.R. Gellert, M.G. Schwab, K. Kim, A roadmap for graphene", *Nature,* vol. 490, pp. 192-200, 2012.
[http://dx.doi.org/10.1038/nature11458]

[111] S.P. Muduli, S. Parida, S.K. Behura, S. Rajput, S.K. Rout, and S. Sareen, "Synergistic effect of graphene on dielectric and piezoelectric characteristic of PVDF -(BZT-BCT) composite for energy harvesting applications", *Polym. Adv. Technol.*, vol. 33, pp. 3628-3642, 2022.
[http://dx.doi.org/10.1002/pat.5816]

[112] V.H. Guarochico-Moreira, C.R. Anderson, V. Fal'ko, I.V. Grigorieva, E. Tóvári, M. Hamer, R. Gorbachev, S. Liu, J.H. Edgar, A. Principi, A.V. Kretinin, and I.J. Vera-Marun, "Thermopower in hBN/graphene/hBN superlattices", *Phys. Rev. B*, vol. 108, p. 115418, 2023.
[http://dx.doi.org/10.1103/PhysRevB.108.115418]

[113] J. Wu, H.A. Becerril, Z. Bao, Z. Liu, Y. Chen, and P. Peumans, "Organic solar cells with solution-processed graphene transparent electrodes", *Appl. Phys. Lett.*, vol. 92, 2008.
[http://dx.doi.org/10.1063/1.2924771]

[114] J. Wu, W. Pisula, and K. Müllen, "Graphenes as potential material for electronics", *Chem. Rev.*, vol. 107, pp. 718-747, 2007.
[http://dx.doi.org/10.1021/cr068010r]

[115] S. Chen, L. Brown, M. Levendorf, W. Cai, S-Y. Ju, J. Edgeworth, X. Li, C.W. Magnuson, A. Velamakanni, R.D. Piner, J. Kang, J. Park, and R.S. Ruoff, "Oxidation resistance of graphene-coated Cu and Cu/Ni alloy", *ACS Nano*, vol. 5, pp. 1321-1327, 2011.
[http://dx.doi.org/10.1021/nn103028d]

[116] J.H. Seol, I. Jo, A.L. Moore, L. Lindsay, Z.H. Aitken, M.T. Pettes, X. Li, Z. Yao, R. Huang, D. Broido, N. Mingo, R.S. Ruoff, L. Shi, Two-dimensional phonon transport in supported graphene, Science (1979) 328 (2010) 213–216.
[http://dx.doi.org/10.1126/science.1184014]

[117] P. Chakraborty, T. Ma, A.H. Zahiri, L. Cao, and Y. Wang, "Carbon-based materials for thermoelectrics", *Adv. Condens. Matter Phys.*, vol. 2018, pp. 1-29, 2018.
[http://dx.doi.org/10.1155/2018/3898479]

[118] Y. Nonoguchi, "Recent progress in thermoelectric materials based on single-wall carbon nanotubes", *Carbon N Y,* vol. 176, p. 657, 2021.
[http://dx.doi.org/10.1016/j.carbon.2021.01.105]

[119] N. Bisht, P. More, P.K. Khanna, R. Abolhassani, Y.K. Mishra, and M. Madsen, "Progress of hybrid nanocomposite materials for thermoelectric applications", *Mater Adv,* vol. 2, pp. 1927-1956, 2021.
[http://dx.doi.org/10.1039/D0MA01030H]

[120] J-F. Li, W-S. Liu, L-D. Zhao, and M. Zhou, "High-performance nanostructured thermoelectric materials", *NPG Asia Mater.*, vol. 2, pp. 152-158, 2010.
[http://dx.doi.org/10.1038/asiamat.2010.138]

[121] Y. Zhang, Y-J. Heo, M. Park, and S-J. Park, "Recent advances in organic thermoelectric materials: principle mechanisms and emerging carbon-based green energy materials", *Polymers (Basel)*, vol. 11, p. 167, 2019.
[http://dx.doi.org/10.3390/polym11010167]

[122] S. Rajput, M. Averbukh, and N. Rodriguez, "Energy harvesting and energy storage systems", *Electronics (Basel),* vol. 11, p. 984, 2022.

[http://dx.doi.org/10.3390/electronics11070984]

[123] G. Zhu, J. Chen, T. Zhang, Q. Jing, and Z.L. Wang, "Radial-arrayed rotary electrification for high performance triboelectric generator", *Nat. Commun.*, vol. 5, p. 3426, 2014.
[http://dx.doi.org/10.1038/ncomms4426]

[124] Y. Zi, J. Wang, S. Wang, S. Li, Z. Wen, H. Guo, and Z.L. Wang, "Effective energy storage from a triboelectric nanogenerator", *Nat. Commun.*, vol. 7, p. 10987, 2016.
[http://dx.doi.org/10.1038/ncomms10987]

[125] X. Pu, C. Zhang, and Z.L. Wang, "Triboelectric nanogenerators as wearable power sources and self-powered sensors", *Natl. Sci. Rev.*, vol. 10, 2023.
[http://dx.doi.org/10.1093/nsr/nwac170]

[126] M. Xie, C. Tian, and P.C. Ooi, "Carbon-enhanced piezoelectric materials and applications", In: *Enhanced Carbon-Based Materials and Their Applications.* Wiley, 2023, pp. 155-183.

[127] X. Pu, L. Li, M. Liu, C. Jiang, C. Du, Z. Zhao, W. Hu, and Z.L. Wang, "Wearable self-charging power textile based on flexible yarn supercapacitors and fabric nanogenerators", *Adv. Mater.*, vol. 28, pp. 98-105, 2016.
[http://dx.doi.org/10.1002/adma.201504403]

[128] K. Cheng, S. Wallaert, H. Ardebili, and A. Karim, "Advanced triboelectric nanogenerators based on low-dimension carbon materials: A review", *Carbon N Y,* vol. 194, pp. 81-103, 2022.
[http://dx.doi.org/10.1016/j.carbon.2022.03.037]

[129] X. Hu, X. Bao, J. Wang, X. Zhou, H. Hu, L. Wang, S. Rajput, Z. Zhang, N. Yuan, G. Cheng, and J. Ding, "Enhanced energy harvester performance by a tension annealed carbon nanotube yarn at extreme temperatures", *Nanoscale,* vol. 14, pp. 16185-16192, 2022.
[http://dx.doi.org/10.1039/D2NR05303A]

[130] S. Rajput, S. Parida, and A. Sharma, *Sonika, Dielectric Materials for Energy Storage and Energy Harvesting Devices.* River Publishers: New York, 2023.

CHAPTER 16

Recent Advancements in Metal Oxide Nanocomposites for Energy Harvesting Applications

Akash Ranjan[1,*], Shaistah Tabassum[2] and Sonika[3]

[1] *Faculty of Education, Banaras Hindu University, Kamachha, Varanasi, Uttar Pradesh, India*
[2] *Department of Education, Tezpur University, Tezpur 784028, Assam, India*
[3] *Department of Physics, Rajiv Gandhi University, Rono Hills, Doimukh, Papumpare 791112, Arunachal Pradesh, India*

Abstract: This chapter examines the latest developments in metal oxide nanocomposites and their uses in energy harvesting devices. The chapter begins with a brief overview of metal oxide nanocomposites, emphasizing their makeup and importance in improving energy conversion processes, including thermoelectric, solar, and piezoelectric ones. A thorough review of current developments exposes cutting-edge synthesis methods and the advantages of nanostructuring, which enhance the performance and characteristics of materials. Several forms of metal oxide nanocomposites are classified in this chapter, emphasizing the widely used materials such as ZnO, TiO_2, and SnO_2, along with novel hybrid systems that blend metal oxides with polymers or other materials for improved performance. A detailed discussion is also included regarding the fabrication processes and characterization methodologies that evaluate the performance of these nanocomposites, highlighting their potential applications in solar energy, waste heat recovery systems, and wearable technology. Additionally, the drawbacks and restrictions of metal oxide nanocomposites, including scalability and material stability, are also addressed in this chapter. Emerging materials and their potential integration into smart energy systems are highlighted below, along with future trends and research directions. A case study shows its practical applications and is included in the chapter's conclusion, along with consideration of these materials' sustainability and effects on the environment in the current situation. By comparing metal oxide nanocomposites with other nanomaterials, we highlight the advantages of these materials as well as potential market trends. This comprehensive study thus aims to shed light on the ground-breaking significance that metal oxide nanocomposites play in the search for durable and efficient energy harvesting technologies.

---

[*] **Corresponding author Akash Ranjan:** Faculty of Education, Banaras Hindu University, Kamachha, Varanasi, Uttar Pradesh, India; E-mail: 1aranjanmedmphil@gmail.com

Keywords: Characterization methodologies, Fabrication processes, Hybrid systems, Metal oxide nanocomposites, Nano structuring, Piezoelectric, Thermoelectric, Wearable technology, Waste heat recovery.

INTRODUCTION

Energy plays a crucial role in the creation and upkeep of almost everything around us. However, the production and consumption of energy account for two-thirds of global greenhouse gas emissions. In order to meet the constantly rising global energy demand, it is vital that we explore clean, affordable, and alternative energy sources, especially due to the impending depletion of fossil fuels like coal, oil, and gas. Additionally, the release of greenhouse gases such as methane and carbon dioxide by the industries, and the re-release of toxic organic pollutants like textile dyes, pharmaceutical contaminants, and antibiotics into the water system, pose serious environmental issues, which further cause water pollution, irreversible global warming, and climate change [1]. At the same time, ensuring access to energy is vital for enhancing economic development and quality of life, particularly in underdeveloped parts of the world. Currently, there is a global shift towards renewable energy resources, such as solar and wind energy, to achieve these particular goals. However, the difficulty in shifting away from fossil fuels lies in the efficient and economical harvesting and storage of energy generated from renewable resources. Unfortunately, renewable energy sources cannot be fully exploited at the time of their production and are intermittent in nature. Therefore, there arises an urgent need for effective energy storage systems in order to make green energy a viable option for the future [2]. Therefore, research is being done on nanocomposites, which are materials that incorporate nanoparticles (usually smaller than 100 nm) into a bulk matrix to improve the composite material's characteristics. These nanomaterials have special properties such as greater surface area, improved mechanical strength, enhanced electrical and thermal conductivity, and superior chemical stability compared to their bulk counterparts. Nanocomposites are very appealing for a variety of applications, especially in energy harvesting technologies, because the addition of nanoparticles to the matrix can result in synergies that improve the material's overall performance. The objective of energy harvesting is to transform ambient energy such as mechanical, thermal, or light energy into useful electrical power. In this context, nanocomposites are extremely useful as they may modify their properties to fit various energy conversion methods, thus improving efficiency and performance [3].

The development of energy harvesting systems, which rely on the efficient conversion of various energy sources, including vibrations, temperature gradients, and sunlight, into electrical power, is greatly aided by nanocomposites. These composites can harvest energy in a variety of ways, especially those having thermoelectric, photocatalytic, or piezoelectric characteristics. For instance, thermoelectric nanocomposites use temperature variations to produce power, whereas piezoelectric nanocomposites can transform mechanical vibrations or pressure into electrical energy. In contrast, photocatalytic nanocomposites are employed for the conversion and storage of solar energy. Considering the size, shape, and composition of the nanoparticles can be precisely adjusted to maximize the energy conversion process, nanocomposites are extremely adaptable for a range of energy harvesting applications [4, 5]. Among the most studied materials for energy harvesting are metal oxide-based nanocomposites because of their superior stability, electrical conductivity, and catalytic qualities. It has been demonstrated that adding metal oxide nanoparticles, such as ZnO, TiO_2, and CuO, to polymer matrices greatly increases the effectiveness of energy conversion devices. Energy harvesting systems perform better in terms of energy efficiency, mechanical durability, and environmental stability due to the cooperation of metal oxides and nanostructured materials [6]. Additionally, improved charge generation, transport, and storage are made possible by the nanoscale contact between the matrix and the nanoparticles, which improves device performance overall.

INTRODUCTION TO NANOCOMPOSITES IN ENERGY HARVESTING

Energy harvesting stands as a crucial innovation in sustainable energy generation, converting ambient environmental energy into usable electrical power. As global energy demands surge and renewable sources gain prominence, these systems are attracting widespread attention. They harness energy from diverse sources motion, vibration, heat, and light to fuel wearable technology, sensors, and compact electronic devices, making them indispensable in the push toward a more efficient and eco-friendly future [7].

Nanocomposites advanced materials that blend nanoparticles with other materials to increase their properties are among the most promising materials for energy harvesting. Nanocomposites are ideal for energy harvesting applications because of their special qualities, which include their high surface area, customized electrical conductivity, and flexibility. Because of their exceptional stability, broad availability, and tunable electronic and optical properties, metal oxides in particular have attracted a lot of interest in the field of nanocomposites. These qualities make them perfect candidates for a variety of energy conversion processes, such as photovoltaic, thermoelectric, piezoelectric, and triboelectric applications [7].

In order to enhance the performance of the base material in particular applications, nanoparticles are incorporated into a matrix material to create nanocomposites. A polymer or metal oxide substance is frequently used as the matrix in energy harvesting, whereas metals, semiconductors, or carbon-based nanomaterials can be used to create the nanoparticles. When these materials are combined, a hybrid material is created that has better mechanical, electrical, and thermal qualities than the constituent parts. Higher energy conversion efficiencies are made possible by nanocomposites' capacity to optimize material properties at the nanoscale, which is essential to their efficacy in energy harvesting. For instance, adding metal oxide nanoparticles to polymer matrices can greatly increase the material's surface area, electrical conductivity, and mechanical strength all of which enhance energy harvesting device performance [7]. The capacity of nanocomposites to efficiently capture various types of environmental energy is what motivates their usage in energy harvesting. Metal oxide nanocomposites, for example, can be employed as light-absorbing materials in photovoltaic applications to transform solar energy into electrical power. By taking advantage of the Seebeck effect, these materials can effectively transform heat into electrical energy for thermoelectric applications. Metal oxide-containing nanocomposites can gather mechanical energy from motion or vibrations and transform it into electrical power in piezoelectric and triboelectric energy harvesting. One of the main benefits of employing nanocomposites in energy harvesting systems is their adaptability in energy conversion methods. Furthermore, the development of extremely effective and adaptable energy harvesting materials is made possible by the capacity to fine-tune the characteristics of metal oxide nanocomposites, such as modifying the particle size, distribution, and composition. Additionally, adding metal oxides to nanocomposites has special benefits in terms of stability and economy. Long-term outdoor energy harvesting applications benefit greatly from the abundance, low cost, and exceptional environmental stability of metal oxides such as Iron Oxide (FeO_3), Zinc Oxide (ZnO), and Titanium Dioxide (TiO_2). Furthermore, when paired with nanoparticles of other materials, metal oxides' desired qualities such as their huge band gaps, strong surface reactivity, and good charge transport characteristics can be improved. For instance, enhanced conductivity and mechanical characteristics can come from the mixing of metal oxides with carbon-based nanoparticles, such as graphene or carbon nanotubes, which can improve energy conversion and storage efficiency [7]. Moreover, advancements in nanocomposite design and fabrication have led to innovative materials that seamlessly integrate into flexible substrates. These breakthroughs have opened doors to cutting-edge energy harvesting technologies that boast both efficiency and adaptability, catering to diverse form factors [5].

NANOCOMPOSITES: STRUCTURE, PROPERTIES, AND SYNTHESIS

To enhance the base material's properties, tiny nanoscale structures nanoparticles, nanofibers, or nanotubes are embedded within a bulk matrix, giving rise to advanced nanocomposites. Unlike their larger-scale counterparts, these minuscule particles, typically ranging from 1 to 100 nanometers, exhibit extraordinary mechanical, chemical, and physical attributes. When the distinctive characteristics of both the bulk matrix and the nanoscale materials merge, the resulting composite boasts superior strength, heightened electrical conductivity, enhanced thermal stability, and refined optical features. These exceptional qualities make nanocomposites highly desirable for a wide array of applications, including energy harvesting. The choice of matrix material and nanoparticle depends on the intended qualities and application [5].

Nanocomposites can be created from a range of matrices, including metals, polymers, ceramics, and even metal oxides. Particularly in the area of energy harvesting, metal oxides can offer stability, resistance to corrosion, and enhanced energy conversion efficiency when utilized as matrices [5]. Nanocomposites' characteristics and effectiveness in energy harvesting applications are greatly influenced by their structure, as depicted in Fig. (1). Due to the nanoparticles' larger surface area and heightened reactivity, interactions between them and the matrix are considerably stronger at the nanoscale. The properties of the composite, including mechanical strength, optical behavior, thermal conductivity, and electrical conductivity, are greatly influenced by these interactions. The synthesis process and conditions determine whether the nanoparticles combine into clusters or are uniformly distributed throughout the matrix to form a homogenous structure. Controlling the synthesis process is crucial to achieving the required material performance since the behavior of the composite can be greatly influenced by the size and distribution of the nanoparticles in the matrix [5]. Furthermore, the overall mechanical and electrical characteristics of the nanocomposite can be affected by the interface between the nanoparticles and the matrix material, sometimes referred to as the matrix-filler interface. This is particularly true in energy harvesting applications where effective charge transfer is crucial. The improved qualities of nanocomposites over their constituent parts are one of their distinguishing characteristics. Numerous enhancements can result from the addition of nanoparticles to a matrix material. For instance, when compared to pure metal oxides, metal oxide-based nanocomposites may show greater stability, increased surface area, and electrical conductivity. These characteristics make them ideal for energy harvesting systems, which often require materials with high surface reactivity, stability in a variety of environmental conditions, and exceptional charge transport capabilities. Improved light absorption in photovoltaic systems is one example of the special optical qualities that metal oxide nanocomposites can display that are not found in the individual metal oxides. The choice of material, the size, shape, and concentration of the nanoparticles, as well as the synthesis technique, all affect the particular characteristics of the nanocomposite [7, 5]. A variety of techniques are used in the synthesis of nanocomposites, each designed to accomplish specific material qualities or capabilities. Top-down and bottom-up methods are the two main ways to synthesize nanocomposites. The top-down method uses mechanical, chemical, or thermal techniques to break down bigger bulk material parts into nanoscale particles. When making nanocomposites out of pre-existing materials, this method is frequently employed [7].

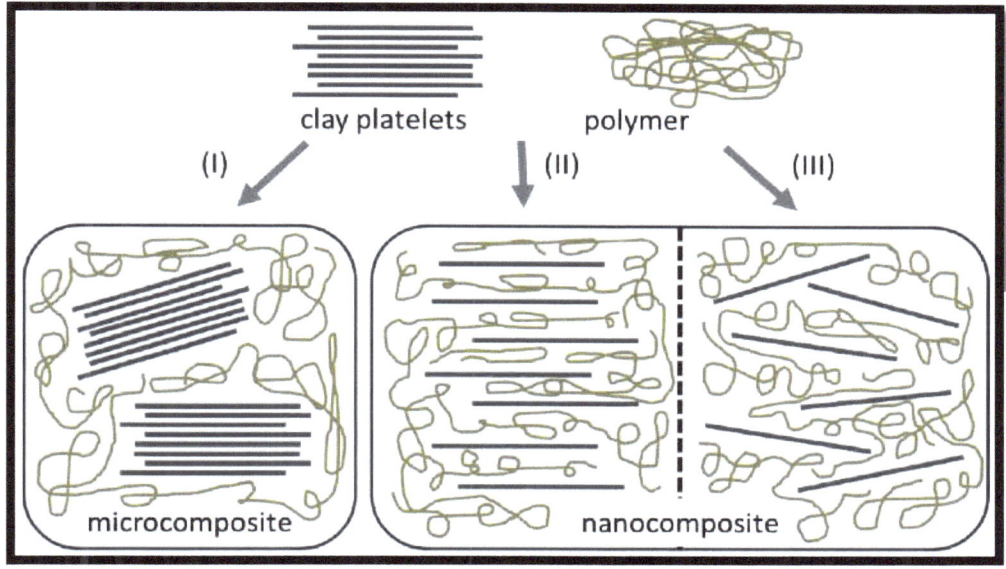

Fig. (1). Different structures of nanocomposites arising from the interaction of layered silicates and polymer. (I) Phase separated micro composite, (II) intercalated nanocomposites, and (III) exfoliated nanocomposites [5].

Examples of top-down techniques for creating nanoparticles or nanostructured components that can be integrated into the matrix include ball milling, laser ablation, and high-energy sputtering. Though it may take more time and energy than bottom-up techniques, the top-down method can be quite successful in regulating the size and shape of the nanoparticles. In contrast, the bottom-up technique uses chemical reactions or self-assembly processes to create nanoparticles from molecular precursors. As it gives exact control over the size, shape, and content of the nanoparticles, this method is frequently utilized for the manufacture of metal oxide nanocomposites. Sol-gel procedures, Chemical Vapor

Deposition (CVD), hydrothermal synthesis, and electrochemical deposition are examples of common bottom-up techniques. The sol-gel procedure, for instance, involves dissolving metal oxide precursors in a solvent to create nanoparticles through chemical reactions. These nanoparticles can subsequently be integrated into a polymer or other matrix material. In order to create well-defined metal oxide nanoparticles that can be combined with polymers or other materials to create nanocomposites, hydrothermal synthesis uses high-pressure water or solvent solutions at high temperatures. Furthermore, hybrid approaches that incorporate both top-down and bottom-up procedures are frequently used in the synthesis of nanocomposite materials. By combining the best features of both techniques, these hybrid techniques can create nanocomposites with incredibly precise characteristics. To further optimize the size or dispersion of the particles in the matrix, a hybrid process might, for instance, use a top-down technique like ball milling after producing metal oxide nanoparticles using a bottom-up technique like sol-gel synthesis. When creating nanocomposites with improved characteristics and a high degree of homogeneity both necessary for effective energy harvesting such hybrid approaches can be especially helpful. The dispersion of nanoparticles inside the matrix is another important consideration in the synthesis of nanocomposite materials. To guarantee that the improved qualities of the constituent components are completely realized in the finished composite, a uniform distribution of nanoparticles must be achieved. The effectiveness of the composite material in energy harvesting applications may be hampered by aggregates or clusters that form as a result of poor dispersion. High-shear mixing, ultrasonic treatment, and the use of stabilizers or surfactants to avoid agglomeration are some of the techniques used to enhance nanoparticle dispersion. To enhance the compatibility between nanoparticles and the matrix, surface functionalization can be employed, ensuring better dispersion and interaction. In energy harvesting applications, nanocomposites particularly those incorporating metal oxides stand out as versatile and compelling materials. By merging bulk matrices with nanoscale particles, they exhibit superior properties compared to their individual components, making them ideal for efficient energy conversion. Precise control over nanoparticle size, shape, and distribution, alongside careful matrix selection, is crucial in the synthesis of these composites. Researchers leverage diverse fabrication techniques to tailor nanocomposites for specific energy harvesting technologies, including photovoltaics, thermoelectrics, and piezoelectric devices. As advancements in synthesis and processing continue, the field is poised to unveil even more cost-effective and high-performance materials for next-generation energy harvesting [7].

ROLE OF METAL OXIDES IN ENERGY HARVESTING

In recognition of their exceptional electrical, optical, mechanical, and chemical characteristics, metal oxides have emerged as a key component of contemporary energy harvesting devices. Given their adaptability, durability, and affordability, these materials which include Iron Oxide (FeO_3), Copper Oxide (CuO), Zinc Oxide (ZnO), and Titanium Dioxide (TiO_2) are frequently used as essential parts in a variety of energy conversion systems. In applications like photovoltaic devices, thermoelectric systems, piezoelectric devices, and triboelectric generators, where they aid in converting ambient energy into useful electrical power, their function in energy harvesting is very important. Metal oxides are perfect for improving the effectiveness and performance of energy harvesting systems because of their inherent qualities, which include large band gaps, high surface area, and exceptional environmental resilience. Furthermore, the capacity to manipulate and alter these characteristics *via* doping and nano-structuring has created new opportunities for the development of high-performance metal oxide-based materials for a range of energy applications [6].

Metal oxides' function as semiconductors in photovoltaic systems is one of their main contributions to energy harvesting. In light of their superior light absorption qualities and capacity to produce charge carriers when exposed to sunlight, metal oxide semiconductors, such as TiO_2 and ZnO, are frequently utilized in solar cells. Specifically, TiO_2 has emerged as one of the most widely utilized metal oxides for the photoanode of Dye-Sensitized Solar Cells (DSSCs). TiO_2's wide band gap guarantees stability in the presence of sunlight, and its vast surface area offers more active sites for photon absorption, which improves charge production. Furthermore, metal oxides like ZnO and CuO have been included in organic-inorganic hybrid solar cells, improving the devices' capacity to absorb light and transfer charges. In addition to increasing solar energy conversion efficiency, these metal oxide-based semiconductors are more affordable and durable than conventional silicon-based solar cells [8].

Metal oxides are essential for thermoelectric energy harvesting, which uses them to turn heat into electricity, in addition to photovoltaic applications. Metal oxide materials can be used to take advantage of the thermoelectric effect, which is the production of electrical voltage in response to a temperature gradient. Due to their high electrical conductivity and low thermal conductivity two essential properties for effective thermoelectric energy conversion certain metal oxides, like Tin Oxide (SnO_2) and Bismuth Telluride (Bi_2Te_3), have demonstrated potential in thermoelectric applications. The main difficulty in thermoelectric energy harvesting is to minimize thermal conductivity while increasing power factor, which is the product of electrical conductivity and the square of the Seebeck coefficient. The efficiency of thermoelectric devices can be increased by optimizing metal oxides to attain these characteristics, especially when they are created at the nanoscale. It has been demonstrated that by lowering thermal conductivity while preserving high electrical conductivity, nano-structuring metal oxide materials for example, by forming thin films or integrating them into nanocomposites significantly improves thermoelectric performance [8].

Another important function of metal oxides in energy harvesting is their participation in piezoelectric and triboelectric energy harvesting. These substances are employed to transform mechanical energy into electrical energy by motion, pressure, or vibration. Since its nanostructures, particularly in the form of nanowires, can effectively transform mechanical strain into electrical charge, Zinc Oxide (ZnO) is one of the most researched piezoelectric materials. Because ZnO nanowires respond so well to mechanical deformation, they are perfect for energy harvesters that use vibrations, human movement, and environmental stressors to gather energy. Wearable electronics, sensors, and self-powered systems are now using metal oxide-based piezoelectric devices, which produce tiny amounts of electricity in response to vibrations in the surroundings or body motions. The performance of Triboelectric Nanogenerators (TENGs), which capture energy from the triboelectric effect created when two materials come into contact and generate charge through friction can also be improved by combining metal oxides with other materials. For TENGs utilized in applications like small-scale energy harvesting from ambient mechanical sources, metal oxide materials are advantageous due to their high surface charge density and durability [6].

Metal oxides offer stability and structural integrity in addition to their electrical characteristics, which greatly improve energy harvesting systems' overall performance. Metal oxides increase the system's mechanical strength and guarantee long-term operational longevity under challenging environmental circumstances, especially when added to nanocomposites. For example, metal oxides are frequently employed to create lightweight and flexible energy harvesting systems where mechanical robustness is crucial, like flexible solar cells or wearable sensors. Metal oxides are ideal for outdoor and long-term applications due to their great resilience in the face of harsh temperatures, humidity, and UV light. Titanium dioxide, for instance, is well-known for its remarkable chemical stability, resistance to deterioration, and capacity to function consistently over long stretches of time, which makes it perfect for use in outdoor energy harvesting systems like solar panels or environmental monitoring equipment. By enhancing charge transfer, mechanical flexibility, and overall device efficiency, metal oxides' involvement in energy harvesting is further enhanced when combined with other nanomaterials, such as carbon-based materials (graphene, carbon nanotubes), or polymers. For instance, it has been demonstrated that adding graphene or carbon nanotubes to metal oxide matrices improves the composite's electrical conductivity, enabling quicker charge transport and more effective energy conversion. The potential of metal oxide nanocomposites, like TiO_2/graphene or ZnO/graphene composites, to improve the efficiency of solar cells, thermoelectric devices, and piezoelectric systems has been investigated. These composites' synergy between metal oxides and carbon-based materials improves the energy harvesting devices' overall performance and increases their capacity to capture and convert energy from a variety of sources [9].

RECENT DEVELOPMENTS IN METAL OXIDE NANOCOMPOSITES FOR ENERGY HARVESTING

Recent advancements in metal oxide nanocomposites for energy harvesting have generated a lot of interest because of their encouraging potential to improve the scalability and efficiency of renewable energy solutions. In a range of energy harvesting applications, metal oxides like Zinc Oxide (ZnO), Titanium Dioxide (TiO_2), Cobalt Oxide (CoO_4), and Nickel Oxide (NiO) have shown impressive improvements when paired with other cutting-edge nanomaterials like graphene, Carbon Nanotubes (CNTs), and conductive polymers. Through improved light absorption, charge transport, and overall device performance, these nanocomposites are being investigated for their potential to improve solar energy harvesting [5, 9]. Metal oxide nanocomposites have demonstrated potential in thermoelectric energy harvesting by enhancing thermoelectric characteristics and striking a balance between thermal control and electrical conductivity. These materials have also shown notable progress in triboelectric and piezoelectric harvesting, which effectively transforms mechanical energy from vibration and motion into useful power. Additionally, metal oxide-based nanocomposites are being used in the field of electrochemical energy storage to enhance the performance of batteries and supercapacitors by providing higher charge/discharge rates, cycle life, and energy density. These metal oxide nanocomposites are anticipated to play a crucial role in the creation of more adaptable, affordable, and sustainable energy harvesting systems as research advances, greatly advancing self-powered gadgets and renewable energy technologies (Table **1**)[9].

Table 1. Advantages of metal oxide nanocomposites against other nanomaterials in key applications [9-11].

Property	Metal Oxide Nanocomposites	Carbon-Based Materials (Graphene, CNTs)	Organic Semiconductors
Solar Cells	High stability, tunable bandgap, effective charge transport	High conductivity, lightweight, flexible	Solution-processable, low-cost, tunable optical properties

(Table 1) cont.....

Property	Metal Oxide Nanocomposites	Carbon-Based Materials (Graphene, CNTs)	Organic Semiconductors
Thermoelectric Devices	Good thermal and electrical properties, stability at high temperatures	High electrical conductivity, low thermal conductivity	Flexible, lightweight, moderate thermoelectric efficiency
Supercapacitors	High redox activity, stability, good capacitance	High surface area, excellent electrical conductivity	Biocompatibility, low toxicity, moderate energy storage efficiency
Cost	Moderate, scalable production	High for pure graphene, CNT synthesis	Low, easy solution-based processing
Flexibility	Limited, but can be integrated into flexible substrates	Excellent flexibility, bendable structures	High flexibility, suitable for wearable applications
Environmental Impact	Stable, recyclable, moderate impact	Concerns over fabrication waste, toxicity of CNTs	Low environmental impact, biodegradable options available

NANOCOMPOSITE METAL OXIDES FOR SOLAR ENERGY HARVESTING

Significant research has been conducted on improved materials for solar energy harvesting due to the increased demand for renewable energy solutions; nanocomposite metal oxides have emerged as viable options. Given their advantageous qualities such as high stability, environmental friendliness, affordability, and tunable electronic and optical characteristics metal oxides, including Copper Oxide (CuO), Zinc Oxide (ZnO), Tin Oxide (SnO_2), and Titanium Dioxide (TiO_2), have long been researched for their potential in solar energy conversion. When compared to their bulk counterparts, these materials' nanoscale structures show noticeably improved characteristics, such as larger surface area, better charge transport, and optimal light absorption all of which are essential for effective energy conversion. Metal oxide nanocomposites are being developed as part of the latest developments in solar energy harvesting technologies to optimize the performance and efficiency of solar cells, especially in devices such as organic-inorganic hybrid solar cells, Dye-Sensitized Solar Cells (DSSCs), and perovskite solar cells. Metal oxide nanoparticles or nanostructures are an essential aspect of the upcoming generation of solar energy harvesting systems since their insertion into composite materials has created new opportunities to improve the stability, scalability, and efficiency of solar cells [10].

The incorporation of nanostructured metal oxides into the photoanode layer of Dye-Sensitized Solar Cells (DSSCs) has been one of the most important advancements in metal oxide-based solar cells. In light of its superior electrical characteristics, large surface area, and capacity to effectively promote charge transport, TiO_2 in particular has been the subject of much research and use in DSSCs. Since TiO_2 can now be fabricated in nanostructured forms such as nanoparticles, nanowires, and nanotubes to enhance surface area for light absorption and boost charge separation efficiency, its application in DSSCs has undergone substantial evolution. Due to the increased dye loading made possible by these nanostructures, more sunlight can be absorbed and transformed into electrical energy. By adding other materials like graphene, carbon nanotubes, or other conductive nanomaterials that enhance the electron transport channels within the photoanode, TiO_2-based nanocomposites have significantly improved the performance of DSSCs. For instance, compared to pure TiO_2, TiO_2-graphene composites have demonstrated enhanced stability and charge transfer capabilities, leading to increased power conversion efficiency. The development of next-generation DSSCs is aided by these nanocomposites' exceptional energy conversion efficiency, which is a result of the special fusion of graphene's conductivity and stability with the high surface area and low recombination rates of TiO_2. Another well-known metal oxide that has drawn interest for its use in solar energy harvesting is Zinc Oxide (ZnO). ZnO is a great option for use in solar cells because, like TiO_2. It has a broad band gap and high electron mobility [10].

In order to produce more effective devices, ZnO nanocomposites are frequently combined with organic molecules or polymers as a photoanode material in DSSCs or organic-inorganic hybrid solar cells. When ZnO is added to these hybrid structures, the total electron transport within the cell is improved, and the light-harvesting efficiency is increased. ZnO-based nanocomposites have demonstrated improved performance in solar energy harvesting, particularly when combined with materials like graphene oxide or carbon nanotubes. ZnO/graphene oxide composites, for example, can greatly enhance charge transfer efficiency, boost photoanode conductivity, and decrease charge recombination, all of which raise power conversion efficiency. In order to optimize the light-harvesting capabilities, ZnO nanostructures, including ZnO nanorods or nanowires, have been integrated into the active layers of solar cells. This is because they provide a greater surface area for photon absorption. In view of these developments, ZnO-based nanocomposites are appealing for use in DSSCs and organic solar cells, where they contribute to improved efficiency and reduced production costs. Another potential field of research is the application of Tin Oxide (SnO_2) in solar energy harvesting. Another wide band gap metal oxide with strong conductivity and stability is SnO_2, which can be used as an Electron Transport Layer (ETL) in organic-inorganic hybrid solar cells and perovskite solar cells. In order to improve its qualities for solar energy conversion, recent research has concentrated on integrating SnO_2 into nanocomposites with other materials, such as metal oxides or carbon-based nanomaterials. To enhance the electron transport characteristics and raise the overall efficiency of perovskite solar cells, for example, SnO_2/graphene composites have been investigated. Better charge collection, decreased energy loss from recombination, and increased electron mobility are the outcomes of combining SnO_2 with conductive carbon-based materials. Due to these advancements, SnO_2-based nanocomposites are appealing for the development of stable, affordable, and high-performing perovskite solar cells, which are quickly becoming one of the most promising next-generation solar technologies [11].

Copper Oxide (CuO), which has drawn interest as a possible material for use in Photoelectrochemical (PEC) cells and hybrid solar cells, is incorporated into metal oxide nanocomposites for solar energy harvesting in another novel way. Compared to other metal oxides, CuO has a smaller band gap, which can be useful for visible spectrum light absorption and make it appropriate for solar energy conversion applications. To increase the effectiveness of composites with other metal oxides and conducting materials, CuO nanostructures such as nanorods, nanowires, and nanoparticles have been included. CuO-based nanocomposites, which blend CuO with TiO_2, ZnO, or carbon-based materials like graphene to improve their electrical conductivity and durability under solar irradiation, have been developed recently. In PEC cells, these composites can act as photoanodes, effectively splitting water to produce hydrogen, a clean energy source. In order to improve charge separation, lower recombination losses, and raise the device's overall power conversion efficiency, CuO nanocomposites can also be utilized in solar cells. Combining the adaptability of nanocomposite formation with the tunability of CuO's electrical properties creates new opportunities for solar energy applications, particularly in the creation of solar fuel and hybrid devices. Integrating metal oxide nanostructures with conductive polymers is a key trend in the creation of metal oxide nanocomposites for solar energy harvesting. These composites are perfect for creating flexible and lightweight solar cells because they provide a special fusion of the flexibility and processability of polymers with the high conductivity and stability of metal oxides. It has been demonstrated that metal oxide nanocomposites, such as ZnO/polymer or TiO_2/polymer, enhance the mechanical characteristics, charge transport effectiveness, and general performance of solar cells. The ability of polymer matrices to carry charge is improved by the addition of metal oxide nanostructures, which is a frequent drawback of purely organic solar cells. These composites present an intriguing path toward the creation of flexible and wearable solar energy harvesting devices by generating flexible solar cells that may be incorporated into a variety of surfaces, including textiles or portable electronics. Additionally, the potential of metal oxide/polymer composites to increase the efficiency of perovskite solar cells is being investigated. In these composites, metal oxides such as SnO_2 or TiO_2 are employed as Electron Transport Layers (ETLs) in conjunction with perovskite materials to improve device stability and charge extraction [10].

THERMOELECTRIC APPLICATIONS OF METAL OXIDE NANOCOMPOSITES

In the fields of waste heat recovery, renewable energy, and portable power generation, thermoelectric materials which use the Seebeck effect to directly convert heat into electricity have attracted a lot of interest. The efficiency of thermoelectric materials is quantified by the dimensionless figure of merit, $ZT = \{S^2 \sigma T\}/ \kappa$, where S is the Seebeck coefficient (thermoelectric voltage), σ\sigma is the electrical conductivity, T is the temperature, and κ\kappa is the thermal conductivity. Low heat conductivity, a high Seebeck coefficient, and high electrical conductivity are necessary for an effective thermoelectric material. It has been difficult to accomplish these qualities in a single material at the same time, nevertheless. A promising way to address these issues is through metal oxide nanocomposites, which present an intriguing way to boost thermoelectric performance by utilizing the special qualities of nano-structuring and combining several materials that can improve the material's overall performance. When combined with other nanomaterials, such as carbon-based nanostructures, polymers, or other metal oxides, metal oxides like Manganese Oxide (MnO_2), Zinc Oxide (ZnO), Tin Oxide (SnO_2), and Titanium Oxide (TiO_2) have demonstrated promise in thermoelectric applications by producing composites with improved thermoelectric qualities [12].

One of the main areas of research has been the creation of nanostructured metal oxides for thermoelectric applications. Due to their large band gaps and the flexibility to modify their electrical and thermal characteristics by nanostructuring, metal oxides particularly TiO_2 and ZnO are naturally excellent choices for thermoelectric materials. Due to phonon scattering at the nanostructures' interfaces, thermal conductivity can be greatly decreased, but electrical conductivity is maintained or even increased by forming nanostructured forms like nanowires, nanorods, or nanoporous materials. For example, the thermoelectric characteristics of ZnO-based nanocomposites have been extensively researched. It has been demonstrated that zinc oxide, when converted into nanorods or nanowires, has a lower heat conductivity than its bulk equivalent, which is an essential property for enhancing thermoelectric efficiency. In order to increase ZnO's electrical conductivity and decrease its lattice thermal conductivity, recent research has concentrated on doping it with different elements, such as gallium or aluminum. By fusing the poor heat conductivity and durability of ZnO nanostructures with the high electrical conductivity and wide surface area of Carbon Nanotubes (CNTs), ZnO/CNT composites in particular have shown notable advances in thermoelectric characteristics. These materials exhibit promise for effective thermoelectric conversion, especially for applications involving waste heat recovery in industrial operations or vehicle exhaust systems. Despite its historically poor thermoelectric performance, Titanium Dioxide (TiO_2) is another metal oxide that has garnered a lot of interest for thermoelectric applications. The creation of nanostructured forms, such as TiO_2 nanotubes, nanowires, and nanoporous structures, has improved the thermoelectric performance of this wide-bandgap semiconductor. By scattering phonons, these changes not only increase the surface area available for heat absorption but also lower thermal conductivity, creating a more conducive environment for thermal management. In order to create nanocomposites that improve its thermoelectric qualities, TiO_2 is frequently combined with other materials. For example, it has been demonstrated that adding graphene or Carbon Nanotubes (CNTs) to TiO_2 increases its electrical conductivity while decreasing its thermal conductivity. Due to its huge surface area and superior electrical conductivity, graphene helps to raise the power factor, which is crucial for enhancing thermoelectric efficiency. The resulting TiO_2/graphene nanocomposites are appropriate for Thermoelectric Generators (TEGs), which transform waste heat into useful electrical power, since they exhibit improved thermoelectric characteristics with a balance of high electrical conductivity and low thermal conductivity [12].

Researchers studying thermoelectrics have also been interested in Tin Oxide (SnO_2) because of its doping capabilities and ease of nanostructure formation. Since SnO_2 is transparent, electrically conductive, and has a high chemical stability, it is perfect for thermoelectric applications, particularly in flexible thermoelectric devices. SnO_2-based nanocomposites have demonstrated encouraging outcomes for thermoelectric applications, especially when paired with conductive polymers or carbon-based materials like graphene. The addition of SnO_2/graphene composites, for example, has resulted in a notable improvement in the thermoelectric properties since graphene enhances charge transport and electrical conductivity, while SnO_2's nanoscale characteristics aid in lowering the total thermal conductivity. SnO_2 has been doped with elements such as lanthanum or antimony in certain experiments to improve its Seebeck coefficient and carrier concentration, thus further optimizing its thermoelectric performance. For thermoelectric power generation applications, these SnO_2-based nanocomposites are being investigated, especially in flexible and portable devices where high-performance and lightweight materials are essential. Another fascinating field of study is the creation of Manganese Oxide (MnO_2) nanocomposites for thermoelectric applications. MnO_2 possesses a special set of characteristics, including comparatively low heat conductivity and the capacity to behave semiconductingly under specific circumstances, that make it appropriate for thermoelectric applications [13].

In order to improve its electronic conductivity and thermoelectric performance, recent research has concentrated on developing MnO_2-based nanocomposites, in which MnO_2 is mixed with carbon nanomaterials like graphene oxide or carbon nanotubes. In addition to increasing charge carrier mobility, the use of carbon-based elements helps regulate the composite's thermal conductivity, increasing thermoelectric efficiency overall. Composites based on MnO_2 have demonstrated potential for energy harvesting from waste heat, and because of their abundance and low cost, they are especially well-suited for usage in large-scale industrial applications. In order to create multifunctional composites that balance charge transport, thermal properties, and stability and result in effective thermoelectric performance for energy conversion systems, researchers have also been looking into MnO_2/metal oxide hybrids, which combine MnO_2 with oxides like TiO_2 or ZnO [13].

To further improve thermoelectric performance, hybrid nanocomposites containing metal oxides and other materials, like organic polymers or other inorganic semiconductors, have also been produced. Polymers like polypyrrole or polyaniline can be added to metal oxide nanocomposites to increase their overall conductivity and provide them with flexibility. These metal oxide-polymer composites are becoming more and more popular in flexible and wearable thermoelectric devices because of their durable, lightweight, and elastic properties, which enable more useful applications in body heat or ambient heat harvesting. To enhance the overall thermoelectric qualities, metal oxides have also been investigated in combination with other inorganic semiconductors like silicon or germanium. By improving the band structure and carrier concentration, these hybrid composites can increase their power factor and thermoelectric efficiency. Enhancing the scalability and sustainability of thermoelectric materials is another significant advancement in thermoelectric metal oxide nanocomposites. Metal oxides are a great option for large-scale, environmentally friendly thermoelectric power generation since they are cheap, plentiful, and non-toxic. The synthesis of metal oxide nanocomposites is being optimized by researchers using cost-effective and scalable methods such as hydrothermal synthesis, sol-gel techniques, and solution-based processing. The thermoelectric metal oxide nanocomposites have the potential to be widely used in waste heat recovery systems in industrial processes, automobiles, and even wearable electronics, where effective energy conversion can greatly lessen reliance on traditional power sources. This is due to their emphasis on environmentally friendly processing techniques and the use of plentiful raw materials [13].

PIEZOELECTRIC AND TRIBOELECTRIC ENERGY HARVESTING WITH METAL OXIDE NANOCOMPOSITES

The development of piezoelectric and triboelectric energy harvesting devices, which can transform mechanical energy such as vibrations, motion, or friction into electrical energy, has attracted a lot of attention as a result of the search for effective, sustainable energy harvesting methods. These techniques hold great promise for use in sensor networks, wearable electronics, and self-sufficient systems that are able to capture mechanical energy from their environment. Owing to their superior mechanical, electrical, and structural qualities which can be created and adjusted at the nanoscale to increase energy conversion efficiency metal oxide nanocomposites are becoming a crucial component for both piezoelectric and triboelectric energy harvesting. When incorporated into nanocomposite structures, these materials such as Barium Titanate ($BaTiO_3$), Zinc Oxide (ZnO), Lead Oxide (PbO), and Titanium Oxide (TiO_2) display advantageous piezoelectric and triboelectric properties that facilitate the effective transformation of mechanical energy into useful electrical power [14].

Materials that produce an electric charge when mechanical stress is applied are the foundation of piezoelectric energy harvesting, the working of which is shown in Fig. (**2**). Some materials, like Titanium Dioxide (TiO_2) and Zinc Oxide (ZnO), are prime candidates for piezoelectric energy conversion due to their intrinsic piezoelectric qualities. Given the increased surface area and the ability to design their crystal structure, these metal oxides can show noticeably improved piezoelectric capabilities at the nanoscale. ZnO nanowires' excellent piezoelectric responsiveness and capacity to provide a significant electrical output under mechanical stress have made them one of the most researched materials for piezoelectric energy harvesting. When exposed to mechanical stress, ZnO nanowires' huge surface-to-volume ratio and one-dimensional structure facilitate effective strain transfer, allowing the material to produce a greater voltage. By increasing mechanical strength, conductivity, and overall energy conversion efficiency, ZnO nanowire-based nanocomposites where ZnO is mixed with graphene, Carbon Nanotubes (CNTs), or polymers have demonstrated significant promise for improving piezoelectric performance. By increasing electron mobility and charge transport inside the nanocomposite, ZnO's integration with carbon-based materials, including graphene oxide, offers further advantages and raises the total power production efficiency in piezoelectric energy harvesting devices. These nanocomposites are especially well-suited for low-power electronics applications, where sensors or tiny electronic devices can be powered by tiny mechanical deformations, like those caused by human motion. Likewise, the well-known and thoroughly researched metal oxide Titanium Dioxide (TiO_2) has demonstrated great potential in piezoelectric energy harvesting. TiO_2 is a good option for piezoelectric devices because of its strong crystalline structure, particularly when it is manufactured into nanostructured forms like nanorods or nanotubes. By mixing TiO_2 nanostructures with other materials to create nanocomposites, their distinctive piezoelectric properties can be improved. For instance, it has been demonstrated that TiO_2/graphene nanocomposites improve the material's electrical and mechanical characteristics, increasing its capacity to transform mechanical energy into electrical power. In addition to enhancing charge transfer and piezoelectric output, graphene's incorporation into TiO_2 increases the composite's flexibility and endurance, which makes it perfect for flexible piezoelectric devices. Wearable technology and energy harvesters for tiny electronics that need low-frequency vibrations or movement to produce energy are just two uses for these TiO_2-based nanocomposites. Additionally, by changing the material's stoichiometry or doping it with other elements, the piezoelectric response of TiO_2 may be further adjusted, providing precise control over its energy conversion efficiency. TiO_2 nanocomposites are very adaptable and scalable for energy harvesting applications because of their material engineering flexibility [14].

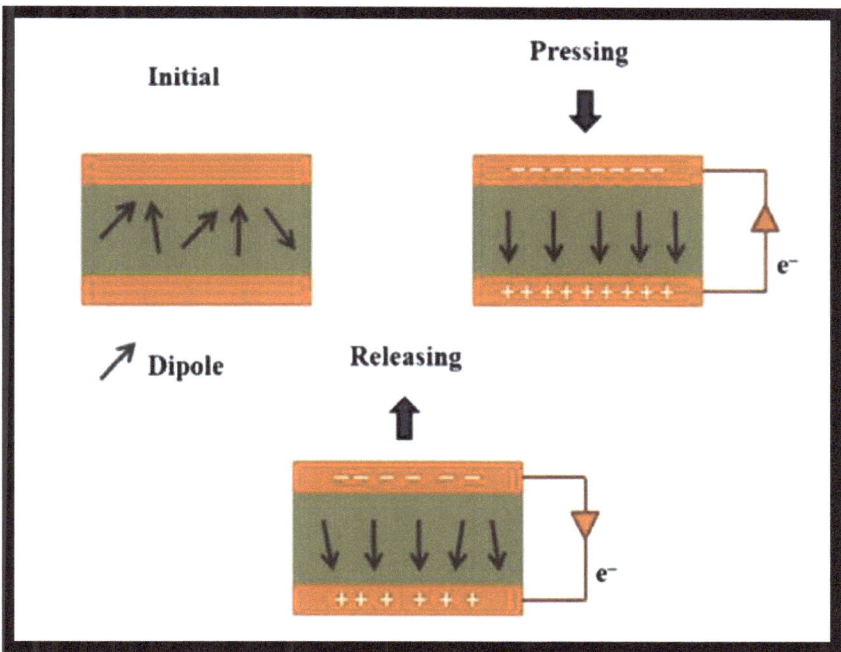

Fig. (2). Schematic working mechanism of PENG [14].

In contrast, triboelectric energy harvesting takes advantage of the contact electrification phenomenon, which occurs when two materials rub against one another and acquire opposing electric charges. There are, however, different modes of operation of the triboelectric energy harvesting, which are depicted in Fig. (3). Using the Triboelectric Effect, the Triboelectric Nanogenerator (TENG) has become a very effective device for transforming mechanical energy into electrical energy. Due to their superior surface qualities, high stability, and capacity to generate a high charge density under friction, metal oxide nanocomposites have been extensively investigated for TENGs. As was previously indicated for piezoelectric applications, Zinc Oxide (ZnO) is also important for triboelectric energy harvesting. Enhancing the triboelectric charge density requires a larger surface area, which is produced by ZnO's capacity to form nanostructures like nanorods, nanosheets, or nanowires. By adding conductive materials like carbon nanotubes or conductive polymers, which can increase triboelectric charge generation and decrease energy loss from charge recombination, ZnO-based nanocomposites can further increase the efficiency of TENGs. These ZnO-based composites are frequently utilized in wearable and flexible triboelectric devices, which are able to capture energy from human motions like bending, walking, and rubbing [15].

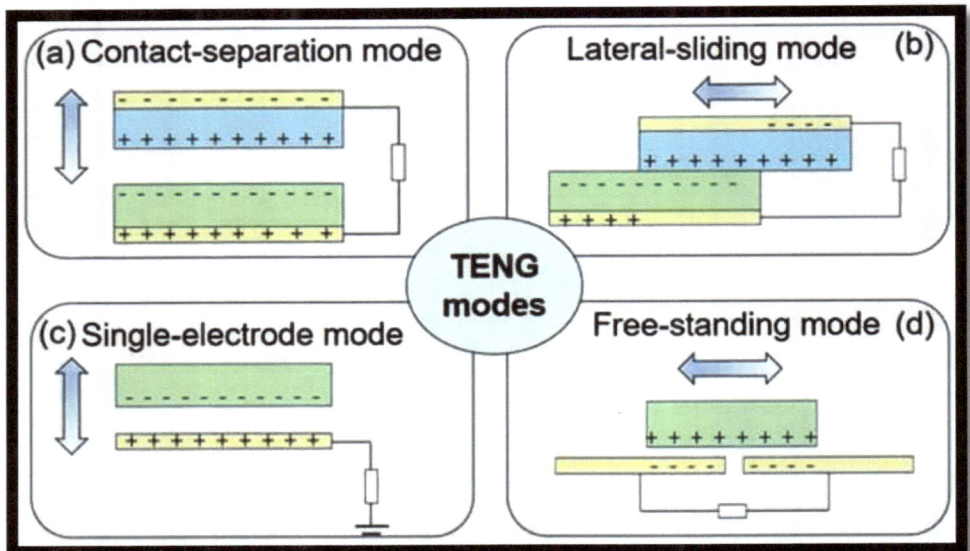

Fig. (3). Schematic illustration of modes of operation of TENG (a) contact and separation mode, (b) lateral sliding mode, (c) single electrode mode, and (d) free standing mode [34].

Triboelectric energy harvesting also relies heavily on Titanium Dioxide (TiO_2) in addition to ZnO. Due to its exceptional dielectric qualities, TiO_2 is a popular choice for producing triboelectric charges. To improve the total energy conversion efficiency, TiO_2 nanostructures, such as nanofibers or nanoparticles, have been included in triboelectric nanogenerators. The triboelectric performance is much improved when TiO_2 is mixed with conductive polymers or carbon-based compounds. For instance, TiO_2's surface charge density and charge retention capacities are increased when graphene oxide or polypyrrole is added, increasing the TENG's total power output. For effective triboelectric charge generation, surface area and dielectric characteristics must be optimized, which is made possible by the ability to design TiO_2 into a variety of nanostructures. Nanocomposites based on TiO_2 have demonstrated great promise in capturing low-frequency mechanical energy from environmental energy sources, human motion, and vibrations [15]. Another intriguing metal oxide that has drawn interest in piezoelectric and triboelectric energy harvesting is Barium Titanate ($BaTiO_3$). Although $BaTiO_3$ is a ferroelectric material with good piezoelectric qualities, its high polarization and dielectric constant make it useful for triboelectric applications as well. For effective energy conversion, $BaTiO_3$ must maintain a high triboelectric charge density, which is made possible by its ferroelectric nature. Nanocomposites based on $BaTiO_3$, especially when paired with conductive nanomaterials like graphene or carbon nanotubes, can greatly enhance triboelectric generator performance. It has been demonstrated that combining $BaTiO_3$ with carbon nanotubes improves the composite's mechanical qualities and raises the charge density, which in turn increases the triboelectric output. Applications requiring high output power and flexibility, such as wearable electronics, portable gadgets, and energy harvesting from body motion, benefit greatly from these nanocomposites. In order to create versatile energy harvesters that can function in a variety of settings and support a sustainable energy generation solution, $BaTiO_3$ is also advantageous. Another area of increasing interest is the synergy between triboelectric and piezoelectric processes in a single device, especially for improving the overall efficiency of energy harvesting. Researchers have created hybrid energy harvesters that can use both mechanical deformation and frictional contact to produce electricity by fusing triboelectric and piezoelectric nanocomposites. For instance, hybrid energy harvesting devices that combine the triboelectric and piezoelectric effects to produce larger power outputs have made use of TiO_2-ZnO nanocomposites. As the triboelectric component can harvest energy from additional mechanical interactions like rubbing or sliding, and the piezoelectric component can provide a continuous energy supply from mechanical deformation, the combination of these two mechanisms may lead to more efficient energy conversion. In applications where both frictional and vibrational pressures are present, such as wearable technology or self-powered sensors, these hybrid devices hold promise for more adaptable and effective energy harvesting systems [14, 15].

ELECTROCHEMICAL ENERGY HARVESTING: METAL OXIDE NANOCOMPOSITES IN SUPERCAPACITORS AND BATTERIES

The efficient storage and release of energy has made electrochemical energy harvesting especially through devices like supercapacitors and batteries a major focus in the field of renewable energy and energy storage technology. Modern energy systems, where there is a growing demand for long-term, high-efficiency, and sustainable storage options, depend heavily on these devices. The enhanced electrochemical characteristics of metal oxide nanocomposites, such as high capacitance, charge-discharge efficiency, cycling stability, and energy density, have made them essential components for improving the performance of supercapacitors and batteries. Due to their varied electronic structures, high specific surface area, and capacity for reversible redox reactions, metal oxides like Manganese Oxide (MnO_2), Cobalt Oxide (Co_3O_4), Iron Oxide (Fe_2O_3), Nickel Oxide (NiO), and Titanium Oxide (TiO_2) are particularly appealing as potential electrochemical energy storage devices. Supercapacitors, sometimes referred to as ultracapacitors, are energy storage devices that use an electrolyte and electrodes with a large surface area to store energy electrostatically. They are perfect for applications needing rapid energy bursts, such as backup energy, pulse power, and regenerative braking systems, because of their extended cycle life, high power density, and quick charge-discharge cycles. The electrode materials used in supercapacitors are crucial to their performance; they must have high specific capacitance and conductivity. As metal oxide nanocomposites can increase charge storage capacity and enable quicker charge/discharge rates, they are very successful at enhancing supercapacitor performance. An appealing material for supercapacitor electrodes is Manganese Oxide (MnO_2), which has long been known for its exceptional electrochemical performance, including high pseudocapacitance and superior stability. However, its total performance is limited due to its comparatively low conductivity. In order to improve mechanical stability and electrical conductivity, scientists have resorted to MnO_2-based nanocomposites, which combine MnO_2 with conductive substances like graphene, polypyrrole, or Carbon Nanotubes (CNTs). By boosting the effective surface area and promoting improved charge transfer, these composites greatly enhance the electrode's overall electrochemical performance. With regard to its high conductivity and mechanical strength, Graphene Oxide (GO) has also been successfully incorporated into MnO_2 to improve cycle life by preventing structural degradation of MnO_2 during charge/discharge cycles [2].

Similarly, because of its favorable redox behavior and high theoretical capacitance, Cobalt Oxide (CoO_4) has garnered a lot of interest for usage in supercapacitors. However, CoO_4, like MnO_2, has drawbacks, including low stability and poor conductivity over extended cycling. Co_3O_4-based nanocomposites, which blend Co_3O_4 with other metal oxides or carbon-based nanomaterials to enhance charge-discharge stability and efficiency, as shown in Fig. (**4**), have been developed to solve these issues. Co_3O_4/graphene composites, for example, display higher electrochemical properties by boosting the overall conductivity and mechanical integrity of the electrode material. By acting as a conductive matrix, graphene supports the structure and enhances charge flow between CoO_4 particles, preventing capacity fading. High-performance supercapacitors with enhanced specific capacitance, enhanced cycle stability, and quicker charge/discharge rates all critical for high-efficiency energy storage systems look promising when using these nanocomposite electrodes. With respect to its abundance, affordability, and environmental friendliness, iron oxide (FeO_3), another potential material, has been extensively studied for supercapacitor applications. However, FeO_3's performance in supercapacitors is limited by its poor cycle stability and relatively low electrical conductivity. FeO_3-based nanocomposites have been created by mixing FeO_3 with high-conductivity substances like graphene, carbon nanotubes, or conductive polymers like Polyaniline (PANI) in order to enhance these properties. These composites improve charge storage and enable quicker charge/discharge cycles by utilizing the large surface area of the carbon-based components. Furthermore, the addition of conductive polymers such as PANI to FeO_3 enhances the composite's mechanical characteristics and provides greater cycle stability. When compared to pure FeO_3, the resulting FeO_3/graphene or FeO_3/PANI nanocomposites show noticeably better electrochemical performance, with increased capacitance, conductivity, and cycling stability. Supercapacitors that need high energy and power densities, such as those found in electric cars, portable gadgets, and renewable energy storage devices, are a good fit for these materials [2]. Another example is Nickel Oxide (NiO), a metal oxide that has shown great promise for use in supercapacitors because of its capacity for high capacitance and reversible redox processes. However, NiO has weak cycle stability and low electrical conductivity, just like other metal oxides. In order to get over these restrictions, NiO-based nanocomposites have been created by mixing NiO with carbon-based substances such as activated carbon, graphene, or carbon nanotubes. As graphene's high conductivity and NiO's high specific capacitance work in collaboration, NiO/graphene composites in particular have demonstrated substantial gains in electrochemical performance. High-performance supercapacitors with exceptional power density and long-term cycling stability can benefit from the addition of graphene, which improves the composite's overall charge transfer and stability throughout repeated charge-discharge cycles. These nanocomposites are also lightweight, flexible, and simple to produce, which makes them perfect for next-generation energy storage applications in flexible electronics, wearable technology, and portable electronics. Metal oxide nanocomposites, in addition to supercapacitors, are essential to the development of batteries because they increase the devices' stability, energy density, and cycle life. The selection of electrode materials has a significant impact on the performance of batteries, including Lithium-Ion Batteries (LIBs) and sodium-ion batteries (SIBs). Conventional battery electrodes, including cobalt oxide for cathodes and graphite for anodes, frequently have problems like sluggish charge/discharge rates, low cycle stability, and restricted capacity. These issues can be resolved by metal oxide nanocomposites, which have a high energy density, quick charge and discharge times, and long-term stability [3].

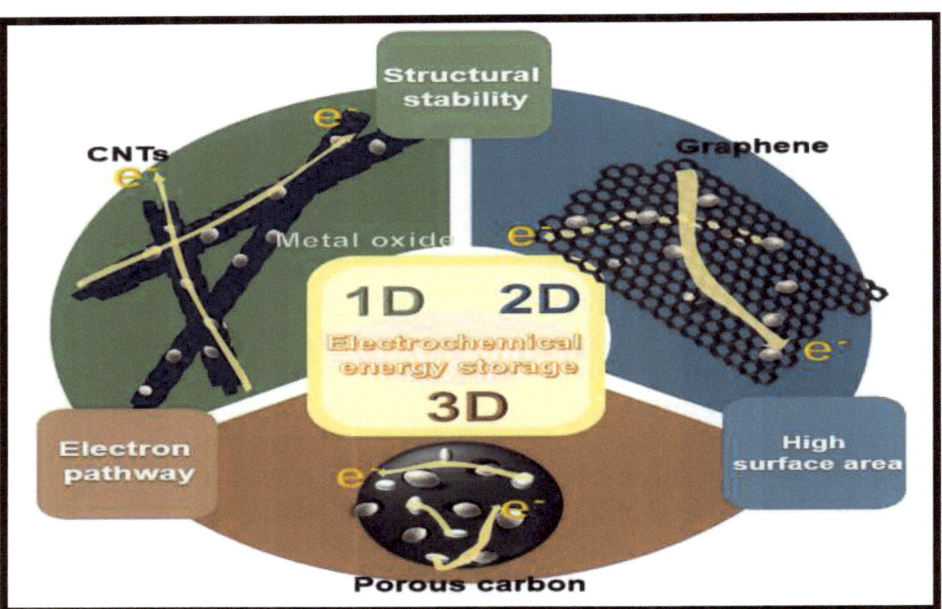

Fig. (4). Carbon/metal-oxide composites of various dimensions for electrochemical energy storage devices [19].

Metal oxide nanocomposites for anode and cathode materials have been developed to help Lithium-Ion Batteries (LIBs), which are widely utilized in electric vehicles, portable gadgets, and grid storage. For example, because of their high theoretical capacity for lithium storage, Iron Oxide (FeO_3) and Tin Oxide (SnO_2) have demonstrated potential as anode materials. However, when cycling, these materials undergo a large volume expansion, which reduces their capacity and shortens their cycle life. In order to solve this problem, conductive carbon elements such as graphene or Carbon Nanotubes (CNTs) have been included in FeO_3-based and SnO_2-based nanocomposites. This helps to reduce volume expansion and enhance the composite's overall conductivity. These nanocomposites are perfect for use in high-performance LIBs because of their noticeably improved cycling stability and capacity retention [3].

Cobalt Oxide (CoO_4) and Manganese Oxide (MnO_2) are frequently utilized on the cathode side due to their high theoretical capacity and good cycling stability. Nevertheless, the limited conductivity of these metal oxides can restrict their rate performance and overall battery efficiency. To increase conductivity, charge/discharge rates, and overall efficiency, CoO_4/graphene or MnO_2/graphene composites have been investigated as cathode materials. Higher power densities and better cycling stability result from the use of graphene, which creates an efficient conductive network that speeds up charge/discharge kinetics and improves charge transfer between metal oxide particles. Since sodium is abundant and inexpensive, these nanocomposites show great promise for application in Sodium-Ion Batteries (SIBs), which are viewed as an alluring substitute for lithium-ion batteries [3]. Apart from their application in LIBs and SIBs, metal oxide nanocomposites are also being investigated in other cutting-edge battery systems, like lithium-sulfur and zinc-ion batteries, which have the potential to provide more sustainable energy storage options and higher energy densities. Metal oxide-based nanocomposites are promising candidates for the next generation of high-performance, environmentally friendly energy storage devices because they help address important issues like the low energy density, poor conductivity, and short cycling life of conventional materials [3].

CASE STUDIES AND REAL-WORLD APPLICATIONS

Recent years have seen a notable increase in interest in the use of metal oxide nanocomposites in energy harvesting devices because of their exceptional capacity to improve power generation, sustainability, and energy conversion efficiency. These cutting-edge materials have the potential to completely transform a number of energy harvesting technologies, as seen by the numerous successful implementations and ongoing development of real-world applications. With an emphasis on solar energy, thermoelectric, piezoelectric, triboelectric, and electrochemical energy harvesting, this section offers a thorough examination of case studies utilizing metal oxide nanocomposites in various energy harvesting systems.

Solar Energy Harvesting with Metal Oxide Nanocomposites

Solar energy harvesting is one of the most promising uses for metal oxide nanocomposites. With respect to their high stability, ease of production, and semiconducting qualities, metal oxides, including Iron Oxide (FeO_3), Zinc Oxide (ZnO), and Titanium Dioxide (TiO_2), have been employed extensively in the construction of solar cells. By adding these elements to nanocomposites containing conductive polymers, carbon nanomaterials, and other metal oxides, researchers have greatly increased their efficiency. For example, the application of a TiO_2/graphene oxide nanocomposite in Dye-Sensitized Solar Cells (DSSCs) was studied. A TiO_2/graphene oxide composite was formed in a study in order to improve electron transport and raise the overall efficiency of DSSCs. The power conversion efficiency of the TiO_2/graphene oxide combination was 8.3%, while that of pure TiO_2-based DSSCs was just 5.7%. The higher performance was attributed to graphene oxide's better charge transport, which assisted in lowering photogenerated electron recombination. The potential of graphene oxide as an electron-transporting material in conjunction with TiO_2 is demonstrated in this case study, allowing for notable increases in solar cell efficiency [16].

Another significant application is the usage of ZnO nanowires in Perovskite Solar Cells (PSCs). ZnO nanowire arrays were integrated into the electron transport layer of perovskite solar cells in a ground-breaking study. This resulted in a power conversion efficiency of 22.1%, which is on par with the most advanced PSCs made of conventional materials. ZnO-based nanocomposites are a great option for effective perovskite solar cells since the ZnO nanowires improved the device's ability to extract charges and transport electrons [17].

Thermoelectric Applications of Metal Oxide Nanocomposites

Utilizing materials that display the Seebeck phenomenon, thermoelectric energy harvesting transforms heat into electrical energy. In light of their strong thermoelectric performance and capacity to retain high stability over a broad temperature range, metal oxide nanocomposites have become attractive options for Thermoelectric Generators (TEGs). The application of Bi_2Te_3/TiO_2 nanocomposites in thermoelectric devices is a noteworthy example. Although Bi_3Te_3 has long been known for having superior thermoelectric qualities, its performance may suffer at high temperatures because of its unstable nature. Researchers greatly increased the thermoelectric material's power factor and thermal stability by adding TiO_2 nanoparticles to the Bi_2Te_3 matrix. With a ZT value (a figure of merit for thermoelectric materials) of 1.2 at 500 K, the Bi_2Te_3/TiO_2 nanocomposite demonstrated a 30% increase in thermoelectric power factor when compared to pure Bi_2Te_3 in a study [18]. The performance of thermoelectric materials for waste heat recovery applications can be significantly improved by metal oxide nanocomposites, as this development showed [19]. Furthermore, thermoelectric generators have made use of Cobalt Oxide (CoO_4) and Nickel Oxide (NiO). The thermoelectric characteristics of NiO/CoO_4 nanocomposites were examined by Li *et al.* (2021), who obtained a ZT value of 0.8 at 300 K. The synergistic effects of the NiO and CoO phases were credited with improving performance by facilitating improved charge transport and decreased thermal conductivity in the nanocomposite material. This case study highlights how metal oxide nanocomposites can be designed to have improved thermoelectric qualities, which makes them appropriate for wearable thermoelectric devices and low-temperature waste heat recovery applications in industrial settings [20].

Piezoelectric Energy Harvesting with Metal Oxide Nanocomposites

Metal oxide nanocomposites have demonstrated significant potential for enhancing the efficiency of piezoelectric energy harvesting systems, which use piezoelectric materials to transform mechanical strain into electrical energy. Zinc Oxide (ZnO) is one of the most researched materials for Piezoelectric Nanogenerators (PENGs) because of its high piezoelectric coefficient and simplicity of nanoscale production [21].

Graphene Oxide (GO) was combined with ZnO nanowire-based Piezoelectric Nanogenerators (PENGs) to improve the energy harvesting capabilities. To enhance charge transmission, GO was applied as a conductive coating over ZnO nanowires that were vertically aligned on a flexible substrate. With a power output of up to 3.5 µW under mechanical deformation, the ZnO/GO nanocomposite-based PENG outperformed pure ZnO nanowire-based PENGs, which only produced 1.1 µW. This study demonstrates how graphene oxide may be used as an efficient conductive matrix to improve metal oxide nanocomposites' piezoelectric performance, which makes them perfect for flexible and wearable electronics that are driven by mechanical vibrations in the surrounding environment [21]. The application of TiO_2 nanorods in piezoelectric energy harvesting is another example. Although TiO_2 has demonstrated promise as a piezoelectric material, it frequently performs less well than other materials, such as ZnO. A TiO_2/graphene oxide nanocomposite was created by someresearchers [22], in which the graphene oxide improved mechanical stability and charge transfer, hence increasing the piezoelectric performance. With an output power density of 1.2 mW/cm^2, this TiO_2/GO nanocomposite demonstrated a notable improvement over pure TiO_2. The potential of TiO_2-based composites for low-frequency vibration energy harvesting applications, such as smart devices and sensor networks, is demonstrated in this case study [23].

Triboelectric Energy Harvesting with Metal Oxide Nanocomposites

The creation of metal oxide nanocomposites has transformed triboelectric energy harvesting, which takes advantage of the triboelectric effect. These materials are perfect for wearable electronics, self-powered sensors, and environmental energy harvesters because they can effectively capture energy from frictional contact between materials. The incorporation of ZnO into Triboelectric Nanogenerators (TENGs) is one of the most effective uses of metal oxide-based triboelectric energy harvesting devices. The usage of ZnO nanowires in the creation of flexible TENGs was investigated in a study [7, 15], which showed a power density of 3.2 W/m^2 and a voltage output of 300 V. The ZnO nanowires' structure was optimized, and they were combined with a Polydimethylsiloxane (PDMS) triboelectric layer to provide this power output. In addition to improving the device's mechanical durability and charge generation, the addition of ZnO nanowires qualified it for use in wearable electronics and energy harvesting from human motion [14].

Additionally, a TiO_2/CNT nanocomposite was designed for triboelectric energy harvesting in a recent study. Under human footstep motion, the TiO_2/CNT nanocomposite demonstrated a power density of 4.8 W/m2, greatly surpassing that of traditional materials. The synergistic effect of CNT's electrical conductivity and TiO_2's high triboelectric charge density was credited with improving performance by facilitating effective charge transfer and energy conversion. The potential of TiO_2-based nanocomposites in self-powered wearables and sensors is highlighted in this case study [23].

Electrochemical Energy Harvesting with Metal Oxide Nanocomposites

Metal oxide nanocomposites have also been used to store and harvest energy in supercapacitors and batteries. The usage of MnO_2/graphene nanocomposites for supercapacitor applications was examined in a recent work, which achieved a specific capacitance of 310 F/g at a high scan rate of 5 mV/s. Compared to pure MnO_2, which had a specific capacitance of 225 F/g, this was a notable improvement. Higher capacitance, improved rate capability, and superior cycling stability were the outcomes of the use of graphene, which created a conductive matrix that improved electron transport within the nanocomposite [24]. Furthermore, the application of ZnO/graphene nanocomposites in zinc-ion batteries has been investigated. A study by Liu *et al.* (2021) demonstrated that ZnO/graphene composites employed as anode materials displayed a high discharge capacity of 1060 mAh/g after 1000 cycles, exceeding typical zinc-based anodes. The ZnO nanoparticles and graphene worked in unison to prevent capacity loss during long-term cycling, which was responsible for the outstanding performance [25].

CHALLENGES AND LIMITATIONS IN METAL OXIDE NANOCOMPOSITE DEVELOPMENT

Although very promising, the development of metal oxide nanocomposites for energy harvesting applications is hampered by a number of important obstacles and constraints. These challenges stem from problems with the materials themselves as well as the difficulties in synthesizing, scaling, and integrating these materials into practical devices. Even though metal oxide nanocomposites have many advantages, such as improved conductivity, mechanical stability, and adjustable electronic characteristics, their development is frequently hampered by a number of important factors, such as difficulties with synthesis, material stability, scalability, cost considerations, and compatibility with current technologies [26].

Synthesis and Fabrication Challenges

The synthesis of metal oxide nanocomposites with exact control over their structural, chemical, and physical properties is one of the main challenges in their development. It can be challenging to disperse nanoparticles or nanostructures uniformly in a composite material, which can result in problems with reproducibility and inconsistent performance. For instance, exact control over the interaction between the metal oxide and the conductive phase is necessary for the creation of metal oxide nanocomposites that incorporate graphene, Carbon Nanotubes (CNTs), or other conductive elements. In applications like energy storage or harvesting, inadequate component integration can result in subpar performance, ineffective charge transfer, and poor mechanical stability. Furthermore, a number of synthesis methods, including Chemical Vapor Deposition (CVD), sol-gel procedures, and hydrothermal synthesis, are intricate, necessitate particular circumstances (such as high temperatures or hazardous solvents), and may result in the production of undesirable byproducts. These difficulties impede the materials' large-scale manufacture for commercial application and make them more difficult to reproduce. Another significant barrier is the scaling up of synthesis techniques for mass production. Although high-quality nanocomposites with superior performance metrics can be produced using lab-scale techniques, bringing these procedures to an industrial scale presents challenges such as variable material quality, trouble sustaining high yields, and higher manufacturing costs. For instance, when attempting to maintain homogeneity over larger amounts of material, the hydrothermal method which is frequently employed to synthesize nanostructured metal oxides may encounter scalability challenges. Finding economical, scalable, and ecologically friendly production techniques is essential for the broad use of metal oxide nanocomposites as demand for them rises [26].

Material Stability and Durability

The stability and longevity of metal oxide nanocomposites in practical settings present a major obstacle to their development for energy harvesting. Despite having great energy conversion qualities, metal oxides frequently experience chemical instability, especially in harsh environments like high temperatures, high humidity, or UV light exposure. For example, Manganese Oxide (MnO_2), which is frequently utilized in solar cells and supercapacitors, can deteriorate over time as a result of electrochemical corrosion or surface oxidation, which can drastically shorten its lifespan and efficiency. In a similar vein, Iron Oxide (FeO_3), despite being widely available and reasonably priced, is susceptible to structural deterioration during charge-discharge cycles in electrochemical devices, which lowers their lifespan and efficiency. Another issue with metal oxide nanocomposites is their mechanical stability, especially in devices like piezoelectric and triboelectric energy harvesters that are subjected to mechanical stress or deformation. The performance of the device may be impacted by problems like fracture or delamination of the composite layers, even though metal oxide-based nanocomposites that incorporate graphene or carbon nanotubes can aid in increasing mechanical strength. Repetitive volume expansion and contraction of the metal oxide particles during charge-discharge cycles can cause structural failure and capacity fading in energy storage devices such as batteries or supercapacitors. Researchers have resorted to coatings or hybrid materials that improve mechanical stability and shield metal oxide from deterioration in an effort to address these problems, but these approaches can complicate and increase the expense of the material development process [27].

Limited Conductivity and Performance

Low electrical conductivity is still a major drawback for many metal oxide nanocomposites, despite their promising qualities. This is especially true for applications involving energy harvesting and storage. Pure metal oxides, like TiO_2, ZnO, and FeO_3, for instance, typically have low electrical conductivity, which restricts their use in solar cells, batteries, and supercapacitors devices that need effective charge transport. The conductivity of metal oxide composites can be greatly increased by adding conductive elements like graphene, Carbon Nanotubes (CNTs), or polypyrrole; however, striking the correct balance between conductivity, mechanical strength, and other characteristics is a challenge. While low conductivity might impede charge collection and lower the device's effectiveness, excessive conductivity can make the material more vulnerable to electronic leakage or excessive self-heating. For example, whereas MnO_2-based nanocomposites can achieve high pseudocapacitance in supercapacitor applications, their poor intrinsic conductivity may limit the pace at which charge is stored and released, hence affecting the device's overall energy efficiency. Similarly, in solar cells, the power conversion efficiency of metal oxide-based materials can be impeded by low charge carrier mobility, leading to significant losses in the photoelectric conversion process. The development of metal oxide nanocomposites is made more complex by the need to overcome these obstacles, which calls for the careful selection of appropriate conductive phases, composite structure optimization, and precise material interface engineering [28].

Compatibility with Existing Technologies and Integration

Another significant challenge is integrating metal oxide nanocomposites into current electronic systems and energy harvesting devices. Numerous metal oxide nanocomposites are engineered to have particular properties (such as improved energy conversion or charge storage), but integrating them into commercial products frequently necessitates further adjustments to guarantee compatibility with current materials and technologies. For instance, in order to guarantee optimal device performance, solar cell topologies frequently combine conductive polymers, metal contacts, and electrolytes that must work with the metal oxide nanocomposite. The total efficiency of the device, however, may be lowered by problems like interface resistance between the metal oxide and other materials. Similarly, the electrolyte-metal oxide interface can slow down the charge-discharge kinetics or create considerable resistance in energy storage devices like batteries and supercapacitors. The electrode-metal oxide interaction is a crucial factor in determining the device's efficiency in the case of piezoelectric and triboelectric energy harvesters. The triboelectric output can be greatly decreased if the electrodes or conductive layers are not properly matched with the metal oxide nanocomposite. Furthermore, metal oxide nanocomposites' mechanical qualities and flexibility must be carefully designed for incorporation into flexible or wearable electronics, which frequently call for materials that can withstand considerable deformation without losing their performance [29].

High Costs and Environmental Impact

Another major obstacle is the cost and environmental impact of creating metal oxide nanocomposites for energy-collecting applications. The production of high-quality metal oxide nanocomposites at a cost-effective scale is challenging since it frequently necessitates the employment of costly chemicals, high-energy procedures, and specialized equipment. Some metal oxide compounds, like iron oxide or zinc oxide, are cheap and widely available, but others, like Manganese Oxide (MnO_2) or Cobalt Oxide (CoO_4), are more expensive, which prevents their widespread use in energy systems. Concerns regarding the environmental effects of nanocomposites' creation and disposal can also arise from the use of hazardous solvents or high-temperature procedures. The total cost of producing metal oxide nanocomposites is still considerable when compared to traditional materials, even with continuous research into green chemistry and other sustainable synthesis techniques. For large-scale applications where cost-effectiveness is essential, such as solar power generation or energy storage in electric cars and grid systems, this is especially challenging. Furthermore, because the buildup of nanoparticles in the environment may have unforeseen ecological consequences, the recyclability and biodegradability of metal oxide-based nanocomposites are issues of concern [30]. Therefore, even though metal oxide nanocomposites have a lot of potential to improve energy harvesting technologies, several obstacles still stand in the way of their advancement. To fully realize the potential of these materials, problems including expensive and time-consuming synthesis procedures, low conductivity, poor material stability, trouble integrating with current technology, and high production costs must be resolved. More investigation into innovative synthesis techniques, the creation of more stable and effective nanocomposites, and the implementation of scalable and ecologically friendly production procedures will all be necessary to overcome these constraints. The use of metal oxide nanocomposites in energy harvesting and storage technologies is anticipated to grow in popularity as these obstacles are overcome, providing effective and sustainable solutions for energy systems of the future [30].

FUTURE PROSPECTS OF METAL OXIDE NANOCOMPOSITES IN ENERGY HARVESTING

Applications including solar energy conversion, thermoelectric power generation, piezoelectric and triboelectric harvesting, and electrochemical storage are just a few of the renewable energy domains that metal oxide nanocomposites are poised to revolutionize in the future. Metal oxide nanocomposites have enormous potential for creating effective and sustainable energy harvesting devices that can function in a range of environmental circumstances because of their distinctive blend of high stability, tunable electrical characteristics, and material design flexibility. In order to develop practical uses that can help lessen dependency on non-renewable energy sources and build self-sustaining systems, these materials must overcome present obstacles in their synthesis, scalability, stability, and performance. Improving solar energy harvesting devices is the most immediate and prospective future use of metal oxide nanocomposites. In solar cells, metal oxides such as Iron Oxide (FeO_3), Zinc Oxide (ZnO), and Titanium Dioxide (TiO_2) have demonstrated significant promise, especially in Dye-Sensitized Solar Cells (DSSCs) and Perovskite Solar Cells (PSCs). In order to increase the effectiveness of charge transport and electron-hole separation in these devices, future research is probably going to concentrate on mixing metal oxides with cutting-edge nanomaterials like graphene, Carbon Nanotubes (CNTs), and quantum dots. Higher energy conversion efficiencies can be achieved by altering the bandgap and surface area of metal oxide nanocomposites, which will enable solar cells to function more effectively even in low light. For instance, it has been demonstrated that Graphene Oxide (GO)-metal oxide composites increase charge carrier mobility and decrease recombination in solar cells, resulting in notable efficiency gains. Additionally, one potential area of research is the incorporation of metal oxide nanocomposites into Perovskite Solar Cells (PSCs). In these cells, the composites can serve as electron transport layers, promoting effective charge transfer and stabilizing the perovskite materials. To increase the scalability, stability, and efficiency of solar energy harvesting systems, research into novel metal oxide combinations that can be used with other next-generation materials, including organometallic perovskites, will be essential. Furthermore, the use of transparent and flexible metal oxide nanocomposites may open the door for flexible solar windows and cells, allowing energy collection in a wider variety of applications, including wearable electronics and building-integrated photovoltaics [27]. Metal oxide nanocomposites are at the forefront of a technology called thermoelectric energy harvesting, which has the ability to transform waste heat into useful electrical power. The intrinsic trade-off between heat and electrical conductivity makes the creation of high-performance thermoelectric materials a major issue. Metal oxide nanocomposites, on the other hand, offer a potential remedy for this issue. Cobalt Oxide (Co_3O_4) and Bismuth Telluride (Bi_2Te_3)-based nanocomposites have demonstrated the capacity to strike a balance between low thermal conductivity and high electrical conductivity, leading to higher figure of merit (ZT) values, which are essential for thermoelectric devices. Future research will probably concentrate on nano-structuring metal oxides to increase phonon scattering and electron mobility, which will lower heat loss and boost energy conversion efficiency. In order to develop compact, high-efficiency Thermoelectric Generators (TEGs) for use in applications such as waste heat recovery from industrial processes, automotive exhaust systems, and even wearable thermoelectric devices that generate power from body heat, Bi_2Te_3/TiO_2 and NiO/ZnO nanocomposites, for instance, are anticipated to be further investigated due to their superior thermoelectric properties. The development of three-dimensional nanostructured materials may result in self-cooling thermoelectric materials, which would transform thermoelectric power generation and increase its viability and efficiency for large-scale uses [26 - 28].

Metal oxide nanocomposites have a very bright future in piezoelectric and triboelectric energy harvesting, especially in wearable electronics, sensor networks, and smart fabrics. Since it has already been shown that metal oxide nanocomposites like ZnO nanowires, TiO_2 nanorods, and CuO may produce power from mechanical vibrations or human motion, they are excellent options for self-powered gadgets. Future developments will probably concentrate on using innovative nano-structuring methods, such as 3D nanostructures and heterostructure composites, to increase the output power and efficiency of these materials [14]. For instance, by increasing the surface charge density and charge transfer efficiency, TiO_2 in conjunction with Graphene Oxide (GO) or Carbon Nanotubes (CNTs) can greatly improve the triboelectric performance. Future research will examine hybrid nanocomposites, which mix triboelectric and piezoelectric materials. This will enable dual-mode energy harvesting, which may be able to absorb energy from frictional contact as well as mechanical vibrations. The development of wearable technology, biological sensors, and environmental monitoring systems that run on mechanical energy or ambient motion will be made possible by these technologies, which will lessen the need for external power sources like batteries. Furthermore, the incorporation of these lightweight, flexible nanocomposites into smart textiles may create new opportunities for body-worn electronics that can capture energy from environmental cues and movement. Triboelectric Nanogenerator (TENG)-powered smart clothes and energy-harvesting shoes may become widely available, boosting wearable technologies and customized healthcare systems. The advancement of energy storage technologies, including supercapacitors, batteries, and hybrid devices, is key to the future of electrochemical energy harvesting employing metal oxide nanocomposites. The capacitance, charge-discharge cycles, and energy density of electrochemical energy storage devices have already been demonstrated to be greatly improved by metal oxide nanocomposites, particularly those based on materials such as Manganese Oxide (MnO_2), Nickel Oxide (NiO), and cobalt oxide (Co_3O_4). For large-scale applications, more reliable, effective, and sustainable supercapacitors and batteries still need to be developed [2].

To further improve the conductivity and structural integrity of electrochemical devices, future research will concentrate on creating nanostructured composites that combine metal oxides with conductive carbon-based materials like graphene and carbon nanotubes. For instance, it has been demonstrated that MnO_2/graphene and TiO_2/CNT composites enhance both electrical conductivity and pseudocapacitance, resulting in superior electrochemical performance in supercapacitors. Furthermore, the development of metal oxide nanocomposite-based anode materials for lithium-ion batteries and zinc-ion batteries has the potential to considerably improve cycle life and charge/discharge efficiency. Another area of intense investigation is hybrid devices, which combine the high energy density of batteries with the high-power density of supercapacitors. Metal oxide nanocomposites can be crucial in developing these hybrid storage systems for a variety of applications, including electric vehicles and grid energy storage. When lightweight and high-performance energy storage systems are needed for wearable electronics, health monitoring devices, and Internet of Things applications, solid-state and flexible energy storage devices based on metal oxide nanocomposites may find greater use in the future. The development of self-sustaining, portable power systems will be made possible by the capacity to customize the electrical characteristics and surface features of metal oxide nanocomposites. This will enable the design of multipurpose devices that can store and harvest energy. Future developments in metal oxide nanocomposites will place a greater emphasis on sustainability and environmental friendliness as the need for energy harvesting technology increases. In order to lessen the environmental impact of these materials, the synthesis of nanocomposites employing green chemistry techniques such as the use of precursors produced from biomass, water-based solvents, and low-energy processes will become more common. Furthermore, metal oxide nanocomposites' capacity to be recycled and biodegraded will be essential to guarantee that the extensive use of these materials does not present serious environmental hazards [26 - 30].

A key component of the future course of energy harvesting technologies will be the creation of environmentally benign metal oxide nanocomposites using non-toxic components and biocompatible materials. For instance, as the demand for environmentally acceptable and sustainable materials in energy harvesting applications increases, iron oxide and zinc oxide both of which are plentiful and non-toxic will probably become even more crucial. To determine the most sustainable production practices and assess the materials' environmental impact, future life cycle assessments will be essential.

Wearable electronics, sensor networks, and smart textiles all depend on metal oxide nanocomposites because of their special mechanical, electrical, and thermal characteristics. Here are some particular instances along with an explanation of how they work: • **Wearable electronics**: To improve conductivity and sensing capabilities, metal oxide nanocomposites, such as those based on ZnO and TiO_2, are incorporated into flexible substrates. By identifying physiological indicators like heart rate, perspiration composition, and temperature, these materials allow for real-time health monitoring.• **Sensor Networks:** Metal oxide nanocomposites are extremely sensitive gas sensors used in biomedical and environmental applications. For example, because of their superior chemoresistive qualities, SnO_2-based nanocomposites are frequently utilized for the detection of harmful gases like CO and NO_2.

Smart Fabrics: Textiles that are responsive to external forces are made by embedding metal oxide nanocomposites into them. CuO and NiO nanocomposites, for instance, are perfect for self-powered wearable technology since they are utilized in thermoelectric fabrics, which use body heat to create power. Exact control over the size, shape, and dispersion of nanoparticles inside the matrix is essential to the efficacy of these nanocomposites. To enhance their performance in energy harvesting, sensing, and wearable applications, researchers are constantly improving synthesis techniques.

CONCLUSION

To sum up, metal oxide nanocomposites have shown themselves to be an extremely diverse and promising class of materials for energy harvesting applications. Their special qualities, which are amplified by nano-structuring and sophisticated synthesis methods, have greatly increased the efficiency of solar energy harvesting, thermoelectric power generation, and piezoelectric energy conversion devices. New hybrid systems and important materials like ZnO, TiO_2, and SnO_2 show improved efficiency and suitability in a variety of energy harvesting devices. However, issues like cost, scalability, and stability continue to stand in the way of their widespread use. Notwithstanding these drawbacks, current studies are looking into novel production techniques and cutting-edge materials that may be able to overcome them. The real-world uses of these nanocomposites, as demonstrated in case studies, demonstrate their potential to be extremely important in the creation of effective and sustainable energy systems. Metal oxide nanocomposites are anticipated to be essential to the upcoming generation of intelligent, renewable energy solutions as the industry develops.

REFERENCES

[1] M.Z. Hussain, Z. Yang, Z. Huang, Q. Jia, Y. Zhu, and Y. Xia,

[2] A. Kumar, H.K. Rathore, D. Sarkar, and A. Shukla, "Nanoarchitectured transition metal oxides and their composites for supercapacitors", *Electrochem. Sci. Adv.,* 2021.

[3] P. H. Patil, V. V. Kulkarni, and S. K. A. Jadhav, "An overview of recent advancements in conducting polymer–metal oxide nanocomposites for supercapacitor application," J. Compos. Sci., 2022.

[4] X. Zhu, "Recent advances in metal oxide-based nanocomposites for energy harvesting applications", *Mater. Sci. Eng. Rep.,* vol. 140, pp. 1-22, 2020.

[5] Y. Li, "Nanocomposites for efficient energy harvesting: Mechanisms, materials, and applications", *Nano Energy,* vol. 80, p. 105514, 2021.

[6] Z.L. Wang, "Triboelectric nanogenerators for self-powered systems", *Nano Energy,* vol. 54, pp. 88-101, 2019.

[7] H. Zhang, "Nanocomposite materials for energy harvesting applications", *Nanomaterials (Basel),* vol. 8, no. 5, p. 276, 2018.

[8] Y. Bai, Z.L. Wang, and X. Zhang, "Metal oxide nanocomposites for energy harvesting and storage", *Adv. Mater.,* vol. 32, no. 10, p. 1905463, 2020.

[9] J. Wang, and S. Li, "Metal oxide nanocomposites for electrochemical energy storage", *Adv. Mater.,* vol. 30, no. 22, p. 1800501, 2018.

[10] S.H. Lee, Y. Kim, and J.H. Lee, "Recent progress in metal oxide-based nanocomposites for high-efficiency dye-sensitized solar cells", *J. Mater. Chem. A Mater. Energy Sustain.,* vol. 7, no. 8, pp. 4112-4133, 2019.

[11] L. Jin, and J. Li, "Recent advances in metal oxide nanocomposites for energy conversion and storage", *Mater. Today Energy,* vol. 10, p. 100323, 2018.

[12]   L. Zhao, and X. Zhang, "Recent advances in thermoelectric properties of metal oxide-based nanocomposites", *J. Mater. Sci. Technol.*, vol. 35, no. 6, pp. 1101-1110, 2019.

[13]   H. Wu, Y. Sun, and X. Wang, "Recent advances in thermoelectric nanocomposites based on metal oxides", *Energy Rep.*, vol. 7, pp. 2229-2241, 2021.

[14]   H. Zhou, T. Liu, and Z. Zhang, "Piezoelectric and triboelectric energy harvesting with metal oxide nanocomposites", *Nano Energy*, vol. 75, p. 104898, 2020.

[15]   W. Yang, H. Guo, and Z.L. Wang, "Triboelectric nanogenerators based on metal oxide nanocomposites for energy harvesting", *Adv. Funct. Mater.*, vol. 30, no. 27, p. 2001611, 2020.

[16]   H. Zhou, J. Zhang, and J. Wang, "Application of TiO2/graphene oxide nanocomposite in dye-sensitized solar cells for improved electron transport and efficiency", *J. Mater. Sci. Technol.*, vol. 34, no. 7, pp. 1123-1130, 2018.

[17]   H. Lee, S. Kim, and J. Lee, "ZnO nanowires in perovskite solar cells for enhanced charge transport and power conversion efficiency", *Adv. Energy Mater.*, vol. 9, no. 4, p. 1800984, 2019.

[18]   F. Hao, T. Xing, P. Qiu, P. Hu, T-R. Wei, D. Ren, X. Shi, and L. Chen, "Enhanced thermoelectric performance in n-type Bi2Te3-based alloys via suppressing intrinsic excitation", *ACS Appl. Mater. Interfaces*, vol. 10, 2018.
[http://dx.doi.org/10.1021/acsami.8b06533]

[19]   Y. Chen, X. Zhang, H. Wang, and J. Li, "Enhanced thermoelectric performance of Bi2Te3 by incorporating TiO2 nanoparticles", *J. Alloys Compd.*, vol. 822, p. 153546, 2020.

[20]   Y. Li, X. Zhang, H. Wang, and J. Li, "Enhanced thermoelectric performance of NiO/CoO4 nanocomposites", *J. Mater. Sci. Mater. Electron.*, vol. 32, no. 18, pp. 14283-14290, 2021.

[21]   X. Wang, Y. Zhou, Y. Zhang, and N. Hu, "Enhanced piezoelectric performance of ZnO nanowire based nanogenerator by graphene oxide coating", *Nano Energy*, vol. 34, pp. 197-204, 2017.

[22]   Y. Xu, Y. Li, Y. Zhang, and J. Liu, "Enhanced piezoelectric performance of TiO2 nanorods by graphene oxide for energy harvesting applications", *Ceram. Int.*, vol. 46, no. 15, pp. 25170-25177, 2020.

[23]   Y. Zhou, Y. Zhang, Z. Wang, and J. Liu, "High-performance triboelectric nanogenerator based on TiO2/CNT nanocomposite for energy harvesting", *Nano Energy*, vol. 107, p. 107758, 2023.

[24]   Y. Yang, X. Zhang, H. Wang, and J. Li, "Enhanced electrochemical performance of MnO2/graphene nanocomposites for supercapacitor applications", *J. Mater. Sci. Mater. Electron.*, vol. 33, no. 20, pp. 17165-17174, 2022.

[25]   Y. Liu, X. Zhang, H. Wang, and J. Li, "ZnO/graphene nanocomposites as anodes for high-performance zinc-ion batteries", *ACS Appl. Mater. Interfaces*, vol. 13, no. 36, pp. 43421-43430, 2021.

[26]   M. Aslam, M.A. Gondal, and A.M. Al-Mayouf, "Challenges and opportunities in the synthesis of metal oxide nanocomposites", *Arab. J. Chem.*, vol. 10, no. 1, pp. 1-10, 2017.

[27]   J. Wang, X. Li, and Y. Yan, "Challenges and opportunities for metal oxide-based nanomaterials in energy harvesting applications", *Adv. Mater.*, vol. 32, no. 20, p. 1906959, 2020.

[28]   S. Bai, Y. Lei, and L. Li, "Improving the electrical conductivity of metal oxide nanocomposites for energy applications", *J. Mater. Chem. A Mater. Energy Sustain.*, vol. 6, no. 28, pp. 13212-13230, 2018. [J.].

[29]   S.K. Lee, and D. Park, "Integration challenges of metal oxide nanocomposites in energy harvesting and storage devices", *ACS Appl. Mater. Interfaces*, vol. 13, no. 12, pp. 14321-14330, 2021.

[30]   M. Kumar, A. Singh, and R.K. Sharma, "Cost-effectiveness and environmental sustainability of metal oxide nanocomposites for energy applications", *Renew. Sustain. Energy Rev.*, vol. 166, p. 111825, 2022.

CHAPTER 17

Current Developments and Future Perspectives of Magnetic Nanocomposites for Advanced Applications

Debasish Banerjee[1], Arindam Banerjee[1], Arbind Prasad[2,*], Bidyanand Mahto[3] and Sudipto Datta[4]

[1] *Mechanical Engineering Department, Omdayal Group of Institutions, Howrah 711316, West Bengal, India*

[2] *Mechanical Engineering Department, Katihar Engineering College (Under Department of Science, Technology and Technical Education, Government of Bihar), Katihar 854109, Bihar, India*

[3] *Government Engineering College, Vaishali (Under Department of Science, Technology and Technical Education, Government of Bihar), Vaishali 844115, Bihar, India*

[4] *Department of Materials Engineering, Indian Institute of Science, Bangalore 560012, Karnataka, India*

> **Abstract:** Magnetic nanocomposites have become increasingly important materials in distinct fields in terms of application and potential, such as biomedical engineering, environmental chemistry, energy storage, and the electronic industry. These materials incorporate characteristics of nanoparticles with the structural benefits of composite matrices, including high thermal stability, magnetic responsivity, and multi-functional properties. Recent advancements have been determined by improvements in synthesis methods, performance, and modified properties for various uses. Advancements in biomedical applications, such as drug targeting, magnetic hyperthermia, and biosensors, point towards a revolutionary future for healthcare. At the same time, they are becoming more involved in clean energy technologies such as supercapacitors and catalysts in the process of energy conversion. Machine learning for property prediction is another area that needs to be focused on, and the development of novel green synthesis methods is the major future research direction. This paper presents an analysis of magnetic nanocomposites and, additionally, future developments that suggest advancements in various fields with various challenges to combating world issues.

---

* **Corresponding author Arbind Prasad:** Mechanical Engineering Department, Omdayal Group of Institutions, Howrah 711316, West Bengal, India; E-mail: arbind.iitg@gmail.com

Keywords: Energy storage, Environmental remediation, Magnetic nanocomposites, Magnetic hyperthermia, Targeted drug delivery.

INTRODUCTION

Magnetic nanocomposites are a group of novel materials that incorporate magnetic nanofillers into different matrix materials, offering a range of functionalities and stimuli-sensitivity for targetted applications. These magnetic nanoparticles mainly include Iron Oxide (Fe_3O_4), cobalt ferrite, or nickel /nickel-based materials, featuring superparamagnetism, high coercivity, or other stronger magnetic responses when the material is in nano-dimensions. These properties become even more desirable when incorporated into polymeric, ceramic, or metallic structures, thereby creating some of the most superior mechanical, thermal, and magnetic composites [1].

The high demand for smart materials with improved characteristics has led to a focus on research on magnetic nanocomposites [2]. These technologies have numerous uses in various industries, such as biomedical engineering for application in drug delivery, magnetic hyperthermia therapy, and biosensors. They act as good adsorbents for water treatment and as promoters of pollutant degradation in environmental science. Furthermore, these materials are increasingly being applied in batteries and supercapacitors, where the electrical conductivity and magnetic nature improve energy density and storage [3].

It is possible for molecules to be adsorbed on the surface or embedded in the coating [4]. Specific developments in synthesis processes, including sol-gel processes, hydrothermal methods, and green synthesis methodologies, enable researchers to control the size, shape, and composition of magnetic nanocomposites [5]. This development and use of artificial intelligence and computational modeling studies on material design help in predicting the best composition of nanoparticles and matrices for changes [6]. Furthermore, the development of new greener synthesis methods that apply bioinspired green chemistry for the formation of materials and devices fits perfectly into the contemporary global shift towards sustainable chemistry in materials science [7, 8].

However, there are some limitations regarding scalability, cost efficiency, and magnetic nanocomposite stability under operating conditions. The future work will be focused on these problems and further development of these prospects in new trends, including flexible electronics, wearable sensors, and advanced energy systems. The following paper presents current advancements in magnetic nanocomposites, focusing on their characteristics, synthesis techniques, and utilization. It also covers the difficulties and trends, emphasizing their indispensable function in sophisticated technologies and environmentally friendly initiatives.

Higher colloidal stability and potential surface modification by functional, reporter, and targeting molecules (such as vitamins, nucleic acids, and antibodies) are typically provided by polymer shells [9].

MAGNETIC NANOCOMPOSITES IN TARGETED DRUG DELIVERY

Target therapy is one of the radical prospects for increasing the efficiency of medications and decreasing unwanted effects, and magnetic nanocomposites are one of the most promising materials within this field [10]. These materials make use of the particle magnetic properties, for instance, iron oxide wrapped within biodegradable gels, to guide the release of the drug under the influence of an external magnetic field [11]. In this context, we are concerned with discussing the potential, mechanisms, and improvements of targeted drug delivery using magnetic nanocomposites, based on several studies.

Nanocomposites are characterized by an ability to target tissues or organs because of their magnetic features. In the presence of an outward magnetic field, the magnetic nanocomposites containing the required drugs can be guided to the desired location and, therefore, concentrate the required doses at the target tissue with minimal exposure to the rest of the organism [12]. It can respond to various stimuli such as pH, temperature, or enzymatic activity once the drug delivery has been localized and controlled [13].

Magnetic nanocomposites are prepared through the incorporation of magnetic nanoparticles, for example, Fe_3O_4, within biocompatible polymers, lipids, or silica matrices. The synthesis methods, such as co-precipitation, solvothermal techniques, and green synthesis, enable this composite to possess high magnetic saturation, stability, and biocompatibility [14]. Moreover, organic tailoring by ligands, antibodies, or peptides can be employed to produce selected interactions with required cells or tissues of the nanocomposites [15].

In oncology, particularly, magnetic nanocomposites have been revealed to hold a lot of potential where there is a need to deliver drugs to tumor regions accurately [16]. For instance, Fe_3O_4-based magnetic nanocomposite systems for the delivery of chemotherapeutic drugs like doxorubicin show enhanced anti-tumor effects with low side effects [17]. Also, magnetic nanocomposites enable magnetic hyperthermia, where a local temperature established by an alternating magnetic field increases the effectiveness of chemotherapy by affecting the cancer cell membrane. The last few years have been marked by the development of the versatile applicability of magnetic nanocomposites [18]. Such carriers of materials can be used as drug delivery vehicles and imaging agents in theranostic systems with treatment and imaging functions [19]. For instance, drug delivery systems using magnetic nanocomposites containing contrast agents for MRI allow immediate tracking of drug release, as well as timely monitoring and control of therapeutic approaches [20].

In addition to medications, MNPs can be used to carry fluorophores, targeting ligands, and responsive components for use in a variety of biomedical applications [19].

MAGNETIC NANOCOMPOSITES IN ENERGY STORAGE AND ENVIRONMENTAL REMEDIATION

Polymeric composites are widely used in biomedical applications such as orthopedics, cardiovascular, skin treatment, wound healing, *etc* [21 - 25]. Ceramic magnetic nanocomposites are widely studied as multifunctional materials for resolving the world's crisis, which centres on energy and environmental issues [26 - 31]. Magnetic nanoparticles promote the performance of functional matrices within new applications for sustainable energy and environmental improvements through their integration [32]. This discussion aims to evaluate these two areas and the key roles and future developments of their contributors.

Magnetic Nanocomposites in Energy Storage

Supercapacitors and batteries are used in energy storage, and these must be materials that effectively possess high conductivity, stability, and energy density. Magnetic nanocomposites meet these criteria because they possess appropriate magnetic and electrochemical characteristics for further energy storage development [33].

Magnetic nanocomposites are used as anode materials in LIBs (as Lithium-Ion Batteries) to improve the performance of the battery. For instance, Fe_3O_4 nanoparticles, when encapsulated within carbonaceous scaffolds, deliver a high value of reversible capacity along with stability, leaving less to worry about problems such as volumetric expansion and the recalcitrant problem of capacity fade during charge/discharge cycles [34]. Moreover, these composites enhance electron transfer and ion transport as well as help increase energy density.

Nanocomposites applied in supercapacitors are known for the integration of both magnetic and conductive characteristics. $CoFe_2O_4$-based nanocomposites with graphene have improved specific capacitance and cyclic stability. They possess high energy density as well as high charge-discharge rates necessary for renewable energy storage [35]. Ceramic and magnetic nanocomposites likewise contribute substantially to hydrogen storage technologies, which are at the vanguard of clean energy systems [36]. Hydrogen storage, both adsorption and desorption, is promoted in nanocomposites with magnetic catalysts, while storage capacities and efficiencies are enhanced [37]. Newer literature highlights that the extra addition of Magnesium Hydride (MgH_2) gives improved kinetics and thermal stability to improve the feasibility of large-scale hydrogen storage [38].

Magnetic Nanocomposites in Environmental Remediation

Nanocomposites qualify as excellent solutions for environmental pollution since they exhibit remarkable magnetic properties that allow them to be used in cleaning up effluent within relatively short times and at low cost [39]. Magnetic nanocomposites act as an adsorbent material for water treatment and other applications, such as the extraction of heavy metals, organic pollutants, and dyes [40]. For instance, iron oxide-based composites with grafted polymer demonstrate very high adsorption for toxic heavy metals like lead and mercury. Furthermore, the separation of magnetic nanocomposites from water is easy by employing a magnetic field; this makes them recyclable [41].

In oil spill cleanup, superhydrophobic and oleophilic magnetic nanocomposites are used to separate oil from water. They are composed of Fe_3O_4 and silica and are light and recyclable, making them suitable for deployment on a large scale for environmental incidents [42]. Magnetic nanocomposites are also used in the elimination of organic contaminants from wastewater [43]. For example, cobalt ferrite-based composites have been found to show appreciable activity in Fenton-like reactions, where even detergents can cause damage to the environment by decomposing hard borate stabilized pollutants into harmless byproducts. Such materials exhibit high efficiency under slightly severe conditions, thus cutting down operating expenses and power consumption [44]. In other air purification methods, magnetic nanocomposites with photocatalysts include titanium dioxide (TiO_2) that breaks down Volatile Organic Compounds (VOCs) and other air contaminants. This makes their recovery and reuse efficient, thus enhancing the development of environmentally friendly air purification systems [45]. The nanocomposite with incorporated magnetism has been among the most notable applications of nanocomposites for environmental cleanup. Magnetic nanocomposites are used in water treatment, oil spill clean-up, and air treatment applications since they allow the easy recovery of pollutants using an external magnetic field, making the process more environmentally friendly [46]. For example, such iron oxide-based magnetic composites, depending on their functional composition, can be particularly effective in capturing heavy metals, including arsenic and lead, in water through the adsorption of these pollutants to the material's surface [50]. The magnetic characteristics enable the enhanced recycling and reuse of the materials, thus making them relatively more environmentally friendly than other conventional filtration mechanisms. In oil spill cleanups, superhydrophobic magnetic nanocomposites can selectively capture oils, which is the focus of oil recovery in an aqueous environment. These materials have high substrate adsorption capacity, which makes it possible to recycle them, hence eliminating the need to discard contaminated materials [47].

MAGNETIC HYPERTHERMIA: A REVOLUTIONARY CANCER TREATMENT APPROACH

Magnetic hyperthermia is one of the most attractive therapeutic approaches based on the application of magnetic nanoparticles that produce heat under the effect of an alternating magnetic field. This heat can be used to destroy cancer cells while leaving adjoining healthy cells relatively unaffected. Self-assembled magnetic nanoparticles with other functional materials, including polymers or biomolecules, are the driving force in this revolutionary therapy.

Mechanism of Magnetic Hyperthermia

Fe_3O_4 and $CoFe_2O_4$ magnetic nanoparticles are commonly applied in magnetic hyperthermia because these nanoparticles can effectively transform the input power of the applied alternating magnetic field into heat. These nanoparticles remain stable when exposed to an external magnetic field; they demonstrate superparamagnetism, the process that enables rapid heating. This localized heat can navigate permanent alteration of the cancer cells, hence killing them [46]. Hyperthermic treatment is commonly sought to improve the efficiency of magnetic nanocomposites [47]. These composites can be developed by the dispersal of magnetic nanoparticles within biocompatible polymeric domains such as polymers, lipids, or hydrogels, which enables improvement in the stability of the composite material and control over drug release and the stimuli-responsive properties of the composites. For example, new magnetic-photosensitive hybrid nanocomposites were designed, demonstrating that light can enhance the hyperthermia effects of these materials [48].

Challenges in Magnetic Hyperthermia

Despite the evident benefits of cancer treatments, magnetic hyperthermia has some challenges. One of the major challenges that remains is the issue of dispersion of magnetic nanocomposites across tumor tissues. The delivered nanoparticles need to specifically and selectively localize and concentrate in tumor tissues, which is an inherently challenging process due to concerns of vascular heterogeneity and altered tumor microcirculation. However, the ability to regulate heating so that the temperature goes high enough to kill tumor cells but low enough to spare the surrounding healthy tissue still poses a great challenge.

Recent Advances in Magnetic Hyperthermia

In the last few years, more attention has been paid to optimizing the design of magnetic nanoparticles for hyperthermia, such as a multifunctional strategy. These particles have been designed not only to produce heat but also to carry chemotherapeutic drugs or improve imaging means, which makes the therapy more complex [49]. Another innovative invention is the application of thermosensitive hydrogels where stimuli-responsive drug delivery can be achieved by heating the magnetic nanoparticles, providing a dual benefit of heat and the heated drug for treating cancer.

MAGNETIC NANOCOMPOSITES AND SUSTAINABILITY

Sustainability is increasingly becoming a critical consideration in the development of new materials, particularly in the context of environmental management and clean energy technologies [50]. Magnetic nanocomposites are emerging as key materials in sustainable practices, as their multifunctionality and magnetic properties contribute to more efficient and eco-friendly solutions in several industries.

Magnetic Nanocomposites in Energy Storage

Magnetic nanocomposites are under investigation for their applications as energy storage material. These materials, especially those that have both magnetic and electronic conductivity characteristics, are used to enhance the performance and reliability of batteries, supercapacitors, and fuel cells. For instance, the combination of carbon-based materials with a limited amount of iron oxide or cobalt ferrite nanoparticles results in improvements in energy storage density and charging-discharging curve, which are crucial in formulating optimal energy storage solutions [44].

Biodegradable and Green Synthesis Approaches

With the increasing interest in eco-friendly products, biodegradable magnetic nanocomposites are encouraged. For instance, the plant extract method or biopolymer synthesis based on plant tissues for magnetic nanoparticles is environmentally friendly compared to the chemical synthesis process. These approaches make the final material less toxic and are cost-efficient to mass-produce the final product [45]. Moreover, the use of biodegradable materials in the formation of magnetic nanocomposites ensures that this material will not pose a threat after its intended usage and can therefore be preferred in both the medical field and the environment. Magnetic nanocomposites are an example of a multi-purpose strategy for today's issues in terms of medicine and ecology. In the case of magnetic hyperthermia, these materials also aid in the local treatment of tumors through heat production by magnetic nanoparticles. However, certain issues, which include targeting, coverage, and temperature, need to be resolved before extensive clinical use. In the field of sustainability, magnetic nanocomposites are extensively useful in environmental cleanup, energy storage, and anti-pollution applications. These properties help pollutants to be attracted or captured quickly, anddue to their flexibility, they can be employed in water treatment, oil spillage, *etc*. Additionally, the continual investigation of biodegradable and green synthesis methods is surely creating a path for more sustainable synthesis and application of magnetic nanocomposites. Owing to these enhancements in magnetic hyperthermia and sustainability, magnetic nanocomposites have a bright future as novel approaches to enhance human health and environmental efficiency, offering innovative, effective, and environmentally friendly.

CHALLENGES AND FUTURE DIRECTIONS

Various issues manifest themselves when magnetic nanocomposites are translated for clinical applications, including toxicity, synthesis procedures, and large-scale production. The environmental decay and longer-term toxicity effects of magnetic nanoparticles, as well as the products of their decay, are still the main areas of concern that have not yet been fully elucidated [31]. Moreover, optimizing the drug loading capacity, drug release profile, and uniform distribution of magnetic nanocomposites in biological systems is another important area of investigation.

New trends have led researchers to develop drug-coated magnetic nanoparticles that respond to changes in pH, temperature, or enzyme milieu. The use of Artificial Intelligence (AI) for real-time control and correction of drug delivery systems is also on the rise. In addition, green synthesis methodologies involving plant extracts or biopolymers aim to improve the biocompatibility and sensory characteristics of magnetic nanocomposites [34].

Further enhancements in nanocomposite structures, like core shells and layered structures of hybrids, are anticipated to improve energy density and stability. Interconnection with RES (Renewable Energy Systems) and smart grids will also improve their scalability factor. Investigations into green synthesis methods and biodegradable magnetic nanocomposite materials are essential to minimize environmental pollution. Furthermore, researchers found that it is beneficial to work with magnetic properties in conjunction with new-age materials, such as MOFs or metal-organic frameworks, for the maximum purification of pollutants.

With the integration of AI and IoT, the functioning of magnetic nanocomposites can be modeled and enhanced for both domains.

CONCLUSION

Magnetic nanocomposites can be classified as a promising class of materials that have revealed great potential for energy storage and environmental applications. In energy storage, they share the role as the enablers of various efficiencies and advancements in batteries, supercapacitors, and hydrogen storage materials critical to renewable energy systems. They also increase energy density, stability, and charge-discharge circulation and promote the construction of new, efficient, and environmentally friendly energy systems. Magnetic nanocomposites have demonstrated great applicability in environmental degradation issues, including water and air pollution, oil spills, and the breakdown of toxic substances globally. Other advantages include a high adsorption capacity, facile magnetically induced separation, and multiple applications, making them feasible for large-scale applications like environmental remediation.

Magnetic nanocomposites hold promising prospects in these areas, given the progress made in designing nanomaterials, surface modification methods, and hybrid structures. Also, AI and IoT technologies are compatible and enhance each other's functionality, offering real-time monitoring and enhanced applications. Some issues like scalability, toxicity, and environmental issues are still unresolved, but people continue to work on them and bring new ideas and innovations. In the future, magnetic nanocomposites can be widely used in renewable energy and environmental care systems. Hence, magnetic nanocomposites carry great potential to address current challenges of the world in both energy storage and environmental management, thus leading to the attainment of sustainable development goals in the future.

REFERENCES

[1] H. Liu, Y. Chen, and Q. Zhang, "Magnetic nanocomposites: Synthesis, properties, and advanced applications", *Adv. Funct. Mater.,* vol. 33, no. 4, p. 2208439, 2023.

[2] Rohith, S., Radhakrishnan, K., Dinesh, A., Sakthivel, S., Patil, R.P., Gnanasekaran, L., Mohanavel, V., Ayyar, M., Iqbal, M., Santhamoorthy, M. and Jaganathan, S.K., 2025. Review on the recent developments in magnetic nanocomposites for energy storage applications. Semiconductors, 59(1), pp.91-114.

[3] Thangaleela, S., Suganya, T., Ali, A., Wang, C.K., Sivamaruthi, B.S. and Chaiyasut, C., 2025. Role of Nanoparticles in Neurological Regeneration and Repair. In Nanoparticles in Modern Neurological Treatment (pp. 131-181). Cham: Springer Nature Switzerland.

[4] Göktürk, I., Güler, K.Ç., Yılmaz, F., Oktar, C., Yılmaz, G.E. and Denizli, A., 2025. Molecularly imprinted polymeric biomaterials in diagnosis and medical practice. Biomedical Materials & Devices, 3(1), pp.299-316.

[5] ROY, A., 2025. Nature-based smart materials. Smart Materials and Applications, p.203.

[6] Kong, T., Zheng, Q., Sun, J., Wang, C., Liu, H., Gao, Z., Qiao, Z. and Yang, W., 2025. Advances in magnetically controlled medical robotics: a review of actuation systems, continuum designs, and clinical prospects for minimally invasive therapies. Micromachines, 16(5), p.561.

[7] Ishaq, M., Ashraf, M.B., Tashkandi, M.A., Ghachem, K. and Kolsi, L., 2025. Artificial neural network analysis of irreversibility in electroosmotic flow of Prandtl–Eyring fluid through an inclined peristaltic channel with ciliated walls. Physics of Fluids, 37(2).

[8] X. Zhao, J. Li, and T. Wang, "Green synthesis of magnetic nanocomposites for sustainable applications", *J. Mater. Chem. A Mater. Energy Sustain.*, vol. 10, no. 12, pp. 5678-5689, 2022.

[9] Emma, P., 2025. Nanoparticle-based drug delivery systems for targeted cancer therapy. Холодная наука, (20), pp.84

[10] Adebayo, D.H., Ajiboye, J.A., Okwor, U.D., Muhammad, A.L., Ugwuijem, C.D., Agbo, E.K. and Stephen, V.I., 2025. Optimizing energy storage for electric grids: Advances in hybrid technologies. management, 10, p.11.

[11] L. Zhang, Y. Xu, and Z. Huang, "Recent developments in magnetic nanocomposites for biomedical applications", *J. Biomed. Mater. Res. A*, vol. 110, no. 3, pp. 897-909, 2022.

[12] Khan, K.A., Ullah, B., Kamal, Z., Esa, M., Riaz, M., Shafique, M., Amanat, M.A., Rahman, A.U., Haq, A.U., Sarfraz, M. and Khan, K.U., 2025. Applications of Au Nanoparticles as drug delivery and theranostic nanoplatforms in cancer therapy: A comprehensive review. Polymer Bulletin, pp.1-68.

[13] Gangadhar, L., Sana, S.S., Mishra, V., Venkatesan, R., Kim, S.C. and Al-Tabakha, M.M., 2025. Recent Trends in Biomedical Applications of Cu2MX4-Based Nanocomposites: An Updated Review. International Journal of Nanomedicine, pp.11895-11939.

[14] S. Chowdhury, Biological smart biomaterials: Materials for biomedical applications.*Applications of Biotribology in Biomedical Systems.* Springer Nature Switzerland, 2024, pp. 313-325.

[15] Singh, S., Das, N., Rathore, M., Nishad, A., Kumari, N. and Guha, R., 2025. Organ-On-A-Chip (OOAC) Technology: Impact on Drug Screening and Personalized Medicine. Regenerative Engineering and Translational Medicine, pp.1-15.

[16] Li, P., Du, Y., Wang, X.D., Shi, Y., Ye, C. and Jin, R., 2025. Preoperative evaluation of endoscopic thyroidectomy via the total areola approach (ETA): a fluid-structure interaction model for predicting lymph node clearance and surgical suitability. Frontiers in Bioengineering and Biotechnology, 13, p.1599770.
[http://dx.doi.org/10.1016/j.matpr.2022.02.498]

[17] Cardoso, B.D., Souza, A., Nobrega, G., Afonso, I.S., Neves, L.B., Faria, C., Ribeiro, J. and Lima, R.A., 2025. Progress in nanofluid technology: from conventional to green nanofluids for biomedical, heat transfer, and machining applications. Nanomaterials, 15(16), p.1242.

[18] Li, B., Tang, Y., Huang, Z., Ma, L., Song, J. and Xue, L., 2025. Synergistic innovation in organ-on-a-chip and organoid technologies: Reshaping the future of disease modeling, drug development and precision medicine. Protein & Cell, p.pwaf058.

[19] D.A. Alromi, S.Y. Madani, and A. Seifalian, "Emerging application of magnetic nanoparticles for diagnosis and treatment of cancer", *Polymers (Basel)*, vol. 13, no. 23, p. 4146, 2021.

[20] Anand, S., Miglani, S. and Anand, R., 2025. AI-optimized cloud architectures for healthcare: enhancing medical data processing and patient care. In Establishing AI-Specific Cloud Computing Infrastructure (pp. 359-378). IGI Global Scientific Publishing.

[21] Packiyadhas, P., Sivaperumal, S.K. and Murugesan, S., 2025. A comprehensive review of food waste: composition, current management, thermal treatment, valorization into bioproducts and sustainable development goals linkages. Journal of Material Cycles and Waste Management, 27(2), pp.777-795.

[22] S. K. Verma, A. Prasad, & V. Katiyar, "State of art review on sustainable biodegradable polymers with a market overview for sustainability packaging". Materials Today Sustainability, 100776, 2024.

[23] Chattopadhyay, A., Suresh Kumar, S., Varol, T. and Perumal, A., 2025. Additive Manufacturing of Soft Materials and Soft Gel for Bio-organs. In Challenges and Innovations in 3D Printed Bio-Organs and Their Materials (pp. 443-473). Cham: Springer Nature Switzerland.

[24] Kalambate, P.K. and Kumar, V., 2025. Decentralized electrochemical biosensors for biomedical applications: From lab to home. Next Nanotechnology, 7, p.100128.

[25] Zahid, A., Leclaire, P., Hammadi, L., Costa-Affonso, R. and El Ballouti, A., 2025. Exploring the potential of industry 4.0 in manufacturing and supply chain systems: Insights and emerging trends from bibliometric analysis. Supply Chain Analytics, 10, p.100108.

[26] A. Prasad, S.M. Bhasney, V. Prasannavenkadesan, M.R. Sankar, and V. Katiyar, "Polylactic acid reinforced with nano-hydroxyapatite bioabsorbable cortical screws for bone fracture treatment", *J. Polym. Res.*, vol. 30, no. 5, p. 177, 2023.

[27] Wu, D., Yan, Z., Ge, B., Dong, Y., Zhao, J., Wu, B., Cao, F., Wang, H., Ma, Z., Li, J. and Zhao, D., 2025. Selective laser melted tantalum three-column porous screw: a novel design achieving both mechanical stability and biological fixation. Discover Applied Sciences, 7(8), p.839.

[28] Puttegowda, M., 2025. Eco-friendly composites: exploring the potential of natural fiber reinforcement. Discover Applied Sciences, 7(5), pp.1-24.

[29] Ngasotter, S., Xavier, K.M., Sagarnaik, C., Sasikala, R., Mohan, C.O., Jaganath, B. and Ninan, G., 2025. Evaluating the reinforcing potential of steam-exploded chitin nanocrystals in chitosan-based biodegradable nanocomposite films for food packaging applications. Carbohydrate Polymers, 348, p.122841.

[30] Park, H., Patil, T.V., Mo, C. and Lim, K.T., 2025. Nanodiamond: a multifaceted exploration of electrospun nanofibers for antibacterial and wound healing applications. Journal of Nanobiotechnology, 23(1), p.285.

[31] Reddy, V.S. and Ramakrishna, S., 2025. Repurposing cement debris into piezo-capacitive sensors for enhanced health monitoring and sustainability. Powder Technology, p.121106.

[32] Meenakshi, M., Bhaskar, R., Kumar, S.A. and Kumar, R.S., 2025. A concise review on magnetic nanoparticles: their properties, types, synthetic methods, and current trending applications. Current Nanoscience, 21(1), pp.2-17.

[33] Asgari, A., Ebrahimnezhad, H. and Sedaaghi, M.H., 2025. Kidney stones CT images classification using graph convolutional network. AUT Journal of Electrical Engineering, 57(2 (Special Issue)), pp.355-368.

[34] Y. Chen, F. Luo, and H. Zhao, "Stimuli-responsive magnetic nanocomposites: Future directions in smart drug delivery systems", *J. Control. Release,* vol. 359, pp. 673-689, 2023.

[35] Gawai, U.P., Kamble, S.D., Kamble, C.D., Kumar, D., Lokhande, R.M. and Chavan, P.G., 2025. Low temperature insulator–semiconductor transition in NiCo2S4/Co9S8 nanocomposites. AIP Advances, 15(9).

[36] J. Li, T. Zhao, and Y. Hu, "Functionalized magnetic nanocomposites for heavy metal removal from water: A review", *J. Environ. Manage.,* vol. 325, p. 116543, 2023.

[37] Esporrín-Ubieto, D., Fraire, J.C., Sánchez-deAlcázar, D. and Sánchez, S., 2025. Engineered plasmonic and fluorescent nanomaterials for biosensing, motion, imaging, and therapeutic applications. Advanced Materials, p.2502171.

[38] A. Kumar, P. Mishra, and N. Singh, "Advanced magnetic nanocomposites for air and water pollution control", *Environ. Sci. Nano,* vol. 10, no. 3, pp. 1234-1246, 2023.

[39] R. Barua, Unleashing human potential: Exploring the advantage of biomechanics in sport performance.*Global Innovations in Physical Education and Health.* IGI Global, 2025, pp. 409-436.

[40] M. Rahman, P. Zhang, and X. Wu, "Superhydrophobic magnetic nanocomposites for oil spill cleanup:

A review of recent advances", *J. Mater. Sci.*, vol. 57, no. 2, pp. 1345-1360, 2022.

[41] Kasala, E.E., Wang, J., Majid, A., Nadege, M.N. and Nyakilla, E.E., 2025. Review and perspectives on enhancing the Hydrogen (H2) storage capacity and stability in geological formations *via* nanoparticle-assisted surfactant/polymer formulations. energy & fuels.

[42] T. Wang, Y. Zhang, and S. Li, "Catalytic performance of magnetic nanocomposites in Fenton-like reactions for wastewater treatment", *Appl. Catal. B,* vol. 332, p. 122563, 2023.

[43] Lv, X., Li, Q. and Wang, K., 2025. State monitoring of lithium-ion batteries based on in situ magnetic techniques: a review. Ionics, pp.1-19.

[44] Manoj, L., Subair, S.M., Varghese, S. and Rakhimol, K.R., 2025. Biocompatible Nanocomposites for Diagnostic Imaging. In Biocompatible Nanocomposites: From Synthesis to Applications (pp. 297-329). Singapore: Springer Nature Singapore.

[45] Safavi, F.S., Ebrahimipour, S.Y., Fatemi, S.J., Mohammadi, P. and Shamspur, T., 2025. Green synthesis of silver nanoparticles and their immobilization on magnetic biochar for the removal of tetracycline and enrofloxacin. Biomass Conversion and Biorefinery, pp.1-15.

[46] J. Li, T. Zhao, and Y. Hu, "Functionalized magnetic nanocomposites for heavy metal removal from water: A review", *J. Environ. Manage.,* vol. 325, p. 116543, 2023.

[47] M. Rahman, P. Zhang, and X. Wu, "Superhydrophobic magnetic nanocomposites for oil spill cleanup: A review of recent advances", *J. Mater. Sci.,* vol. 57, no. 2, pp. 1345-1360, 2022.

[48] X. Sun, X. Wang, and L. Zhang, "Multifunctional magnetic nanocomposites for combined photothermal and hyperthermia therapies", *Nanomedicine (Lond.),* vol. 18, no. 3, pp. 1267-1282, 2023.

[49] T. Wang, Y. Zhang, and S. Li, "Catalytic performance of magnetic nanocomposites in Fenton-like reactions for wastewater treatment", *Appl. Catal. B,* vol. 332, p. 122563, 2023.

[50] Q. Zhao, C. Luo, and T. Zhang, "Magnetic nanoparticles for hyperthermia-based cancer therapy", *Nano Res.,* vol. 16, no. 4, pp. 1099-1116, 2023.

SUBJECT INDEX

A

Activated carbon (AC) 164, 165, 169, 276, 277, 293, 337
Active layer 26, 125, 194, 287, 288, 327
Actuators 31, 41, 55, 67, 94, 119, 125, 128, 149, 227
Adaptability 2, 178, 183, 187, 225, 249, 253, 260, 319, 323, 328
Adhesion 6, 26, 30, 41, 138, 144, 276
Adsorption 111, 112, 169, 203, 345
Alignment 2, 34, 43, 94, 119, 126, 130, 131, 282, 303
Alumina 124, 240
Ammonia 112, 113, 227
Anode 144, 146, 147, 148, 149, 150, 152
Aqueous solution 40, 60, 111, 139, 141, 142
Atmospheric pressure 26, 62

B

Band gap 110, 111, 134, 164
Bio-based materials 276, 277
Biocompatibility 276
Biomass 55, 61, 71, 72, 73, 76, 110, 114, 115, 116, 141, 147, 168, 184, 199, 276, 277, 290, 292, 293
Bismaleimide 30
Butanediol 140

C

Capacitance 148, 149, 165, 169, 170, 171
Carbon 2, 4, 7, 16, 23, 24, 26, 31, 34, 42, 56, 63, 64, 110, 114, 119, 120, 122, 124, 125, 130, 131, 135, 145, 146, 147, 148, 149, 150, 152, 164, 165, 166, 167, 168, 169, 170, 171, 172, 173, 174, 175, 178, 203, 207, 208, 209, 210, 211, 213, 214, 215, 216, 225, 227, 233, 234, 235, 276, 277, 282, 287, 293, 297, 298, 300, 303, 319, 323, 332, 337, 346
  black (CB) 130, 131, 165, 166, 168, 170, 172
  fiber reinforced polymer (CFRP) 2, 7, 23, 26, 42, 63, 64, 165, 174, 233, 235
  fibers (CFS) 2, 4, 7, 24, 26, 42, 63, 165, 166, 167, 173, 174, 233, 235
  nanotubes (CNTS) 2, 4, 31, 34, 42, 64, 119, 122, 124, 125, 130, 131, 147, 148, 165, 167, 169, 170, 172, 173, 174, 207, 208, 209, 210, 211, 213, 214, 215, 216, 225, 227, 233, 234, 235, 277, 282, 287, 297, 303
  quantum dots (CQDS) 166
Catalyst 112, 113
Cellulose 61, 62, 114, 115, 134, 135, 136, 137, 139, 140, 141, 142, 168, 170, 184, 199, 203, 276, 277, 282, 292, 293
Chitosan 64, 139, 141, 142, 170
Composite materials 1, 2, 3, 4, 5, 7, 8, 10, 11, 13, 14, 16, 17, 19, 21, 23, 24, 25, 26, 29, 31, 32, 33, 34, 35, 36, 37, 38, 39, 41, 42, 45, 46, 55, 56, 57, 58, 62, 64, 66, 67, 68, 71, 73, 76, 95, 110, 119, 122, 123, 125, 126, 127, 128, 129, 130, 131, 134, 135, 136, 138, 142, 143, 144, 145, 147, 149, 150, 151, 152, 164, 165, 166, 168, 172, 173, 174, 175, 178, 184, 186, 187, 194, 197, 199, 200, 203, 209, 214, 227, 233, 234, 235, 236, 239, 240, 243, 249, 253, 254, 260, 276, 277, 281, 282, 287, 288, 290, 292, 293, 297, 298, 300, 303, 305, 306, 319, 323, 327, 328, 332, 337, 345, 346, 350
Conductive 110, 116, 125, 130, 164, 172, 173, 282

Sushil Kumar Verma, Sonika & Arbind Prasad (Eds.)
All rights reserved-© 2026 Bentham Science Publishers

Subject Index

fibers 164, 172, 173
polymer 110, 116, 125, 130, 172, 282
Conductivity 2, 4, 6, 7, 8, 17, 26, 30, 31, 33, 35, 41, 42, 43, 56, 58, 63, 64, 66, 72, 94, 116, 119, 120, 122, 124, 125, 126, 127, 128, 129, 130, 131, 134, 135, 144, 145, 147, 148, 150, 164, 165, 166, 167, 168, 169, 170, 171, 172, 173, 174, 175, 203, 209, 210, 227, 234, 235, 249, 254, 276, 277, 282, 293, 297, 298, 303, 328, 337
Copolymers 16, 35, 38, 40, 42, 57, 58, 60, 61, 138, 141
Coupling agent 4, 41
Crystallinity 30, 35, 57, 59, 61, 62, 63, 116, 126, 139, 141, 142, 209, 210, 214, 234

D

Delamination 24, 243
Dielectric 41, 59, 122, 124, 126, 127, 128, 129, 130, 131, 164, 166, 167, 169
  constant 59, 169
  loss 59, 169
  materials 122, 164, 166
  properties 41, 122, 124, 126, 127, 128, 129, 130, 131, 166, 167
Dimethyl 30, 40, 115
  acetamide (DMAC) 40
  formamide (DMF) 30, 40
  sulfoxide (DMSO) 40, 115
Dispersibility 4, 34, 43, 122, 147, 173, 282

E

Elasticity 209, 210, 227
Electrical conductivity 6, 7, 8, 17, 26, 30, 31, 33, 41, 42, 43, 56, 58, 63, 64, 66, 72, 94, 116, 119, 120, 122, 124, 125, 126, 127, 128, 129, 130, 131, 134, 135, 144, 145, 147, 148, 150, 164, 165, 166, 167, 168, 169, 170, 171, 172, 173, 174, 175, 203, 209, 210, 227, 234, 235, 249, 254, 276, 277, 282, 293, 297, 298, 303, 328, 337
Electrochemical performance 116, 144, 146, 147, 150, 152
Electrolyte 144, 146, 147, 148, 150, 151, 152, 164, 165, 168, 170, 171
Electron mobility 26, 125, 287
Electrospinning 2, 60, 61, 116, 130, 131, 147, 170, 173, 211, 214, 233
Energy 2, 26, 31, 35, 55, 67, 72, 94, 110, 114, 116, 119, 125, 128, 130, 131, 134, 143, 144, 145, 146, 147, 148, 149, 150, 151, 152, 164, 165, 166, 167, 168, 169, 170, 171, 172, 173, 174, 175, 178, 183, 187, 194, 207, 225, 227, 233, 249, 253, 260, 276, 277, 287, 288, 293, 297, 303, 305, 306, 319, 323, 327, 328, 332, 337, 345, 346, 350
  conversion 26, 31, 110, 114, 116, 134, 164, 166, 172, 227, 287
  density 144, 146, 148, 149, 150, 152, 165, 171, 293
  harvesting 2, 31, 55, 67, 72, 94, 110, 119, 125, 128, 130, 131, 166, 178, 183, 187, 194, 207, 225, 227, 233, 249, 253, 260, 276, 287, 288, 297, 303, 305, 306, 319, 323, 327, 328, 332, 337, 345, 346, 350
  storage 2, 26, 31, 35, 55, 67, 110, 114, 116, 134, 143, 144, 145, 146, 147, 148, 149, 150, 151, 152, 164, 165, 166, 167, 168, 169, 170, 171, 172, 173, 174, 175, 225, 227, 276, 277, 287, 293, 297, 337, 345, 346, 350
Environmental impact 1, 2, 4, 32, 35, 55, 61, 71, 73, 76, 134, 143, 178, 183, 187, 194, 197, 199, 200, 203, 225,

236, 243, 249, 253, 260, 269, 270, 276, 277, 287, 290, 292, 293, 297, 298, 300, 303, 305, 306, 319, 323, 327, 328, 332, 337, 345, 346, 350
Extrusion 24, 60, 61, 214, 235

F

Friction 119, 124, 126, 127, 129, 227
Fuel cell 110, 113, 114, 144
Fungal decomposition 141

G

Glass fiber reinforced polymer (GFRP) 2, 7, 23, 24, 42, 63, 64, 233, 235, 236
Graphene 4, 31, 34, 42, 64, 119, 122, 125, 126, 127, 128, 129, 130, 131, 147, 148, 165, 166, 167, 168, 170, 172, 173, 174, 214, 235, 277, 282, 297, 303
Gravimetric capacitance 148, 149, 170

H

Heterojunction 111, 112
Hydrogel 138, 141, 142, 270
Hydrogen bond 4, 38, 58, 115, 141, 142, 150, 282
Hydrophobic 34, 41, 115, 124, 127, 168

I

Impedance spectroscopy 170
Ionic conductivity 150

L

Lactide 57, 58, 60, 61

Lignocellulosic biomass 115, 141
Lithium-ion battery (LIB) 144, 146, 147, 148, 149, 150, 151, 152
Load transfer 4, 24, 42

M

Macromolecular chains 30, 57, 58, 59, 61
Microporous 168, 169
Microstructure 4, 17, 26, 30, 34, 43, 57, 60, 122, 125, 126, 130, 131, 139, 142, 144, 147, 150, 166, 172, 173, 209, 214, 233, 282, 298, 303, 328
Monofilament 235
Morphology 4, 6, 17, 30, 34, 43, 57, 60, 111, 115, 116, 122, 125, 126, 130, 131, 139, 142, 144, 147, 150, 166, 172, 173, 209, 214, 233, 282, 298, 303, 328

N

Nano- 2, 4, 31, 34, 42, 60, 61, 64, 110, 111, 112, 113, 114, 116, 119, 122, 124, 125, 126, 127, 128, 129, 130, 131, 134, 135, 136, 137, 138, 139, 140, 141, 142, 147, 148, 164, 165, 166, 167, 168, 169, 170, 171, 172, 173, 174, 175, 207, 208, 209, 210, 211, 213, 214, 215, 216, 225, 227, 233, 234, 235, 277, 282, 287, 297, 303, 328
  hybrid 164, 165, 166, 167, 168, 169, 170, 171, 172, 173, 174, 175
  cellulose 134, 135, 136, 137, 139, 140, 141, 142, 168, 170
  crystals 135, 136, 137, 138, 140
  fibers 2, 60, 61, 116, 130, 131, 147, 170, 173, 211, 214, 233
  materials 4, 110, 111, 112, 113, 114, 116, 119, 122, 124, 125, 126, 127, 128, 129, 130, 131, 134, 135, 147,

Subject Index

164, 165, 166, 167, 168, 169, 170, 171, 172, 173, 174, 175, 227, 282, 287, 297, 303, 328
particle 110, 111, 112, 113, 114, 116, 134, 135, 147, 165, 166, 168, 170, 172, 173, 282
sheets 122, 127
tubes 2, 4, 31, 34, 42, 64, 119, 122, 124, 125, 130, 131, 147, 148, 165, 167, 169, 170, 172, 173, 174, 207, 208, 209, 210, 211, 213, 214, 215, 216, 225, 227, 233, 234, 235, 277, 282, 287, 297, 303
Natural fibers (NFS) 2, 4, 7, 8, 10, 23, 24, 41, 55, 60, 61, 62, 63, 64, 66, 72, 73, 76, 114, 115, 134, 135, 141, 142, 178, 184, 186, 187, 194, 197, 199, 200, 203, 225, 233, 235, 236, 239, 240, 243, 249, 253, 254, 260, 276, 277, 281, 282, 287, 290, 292, 293, 297, 298, 300, 303, 305, 306, 319, 323, 327, 328, 332, 337, 345, 346, 350

O

Oligomers 57, 60, 61, 138, 139
Organic light-emitting diode (OLED) 31, 67, 119, 125, 128, 149, 227
Oxidation 26, 31, 41, 111, 116, 126, 147, 173

P

Packing density 43, 131, 214
Perovskite solar cell (PSC) 26, 125, 287, 288, 327
Phase transition 210, 214
Photocatalyst 110, 111, 112, 113, 114
Photocatalytic hydrogen evolution (PHE) 112, 113

Photovoltaic (PV) 1, 2, 26, 31, 67, 110, 119, 125, 128, 134, 149, 178, 183, 187, 194, 203, 225, 227, 249, 253, 260, 287, 288, 297, 303, 319, 323, 327, 328, 332, 337, 345, 346
Piezoelectricity 119, 120, 121, 122, 124, 125, 126, 127, 128, 129, 130, 131, 227
Polyaniline (PANI) 116, 173
Polycaprolactone (PCL) 58, 60, 61
Polyethylene oxide (PEO) 59, 150
Polyhydroxybutyrate (PHB) 58, 61
Polyimide (PI) 124, 125, 126, 127, 129, 210, 211, 214, 215, 216, 227
Polylactic acid (PLA) 35, 57, 58, 59, 60, 61, 62, 63, 64, 66, 138, 139, 270
Polymer composites 1, 2, 4, 5, 7, 8, 10, 11, 13, 14, 16, 17, 19, 21, 23, 24, 25, 26, 29, 31, 32, 33, 34, 35, 36, 37, 38, 39, 41, 42, 45, 46, 55, 56, 57, 58, 62, 64, 66, 67, 68, 71, 73, 76, 95, 110, 119, 122, 123, 125, 126, 127, 128, 129, 130, 131, 134, 135, 136, 138, 142, 143, 144, 145, 147, 149, 150, 151, 152, 164, 165, 166, 168, 172, 173, 174, 175, 178, 184, 186, 187, 194, 197, 199, 200, 203, 209, 214, 227, 233, 234, 235, 236, 239, 240, 243, 249, 253, 254, 260, 276, 277, 281, 282, 287, 288, 290, 292, 293, 297, 298, 300, 303, 305, 306, 319, 323, 327, 328, 332, 337, 345, 346, 350
Porous structure 111, 116, 168, 169

R

Renewable energy 1, 2, 26, 31, 35, 55, 67, 110, 114, 116, 119, 125, 128, 134, 143, 144, 149, 164, 166, 172, 178,

183, 187, 194, 203, 225, 227, 233,
249, 253, 260, 276, 277, 287, 288,
297, 303, 319, 323, 327, 328, 332,
337, 345, 346

S

Scanning electron microscopy (SEM) 6, 17, 30, 43, 60, 116, 130, 142, 147, 166, 173, 209, 214, 234, 282
Single-walled carbon nanotube (SWCNT) 174, 211, 213, 214, 215
Spectroscopy 8, 17, 30, 43, 60, 116, 130, 142, 147, 166, 173, 209, 214, 234, 282
Spin coating 26, 111, 287
Supercapacitors 2, 116, 134, 144, 147, 148, 149, 150, 152, 164, 165, 166, 167, 168, 169, 170, 171, 172, 174, 175, 293

T

Tensile strength 7, 24, 42, 63, 64, 139, 209, 233, 234
Thermal conductivity 4, 7, 30, 42, 58, 63, 66, 124, 127, 130, 131, 165, 203, 235, 240, 249, 253, 260
Thermogravimetric analysis (TGA) 6, 43, 60, 142, 147, 235
Thermoplastic polymers 57, 58, 61
Triboelectric nanogenerator (TENG) 119, 120, 121, 122, 124, 125, 126, 127, 128, 129, 130, 131, 227
Twistron 207, 208, 212, 213, 215

U

Ultraviolet (UV) 30, 42, 111, 112, 134, 138, 141, 142, 164

W

Weaving 214, 234
Wet spinning 209, 210, 214

X

X-ray photoelectron spectroscopy (XPS) 8
X-ray diffraction (XRD) 2, 6, 11, 19, 21

Y

Yarn 209, 213
   bias angle 213
   density 213
   porosity 213
   strength 209
Young's modulus 59, 209, 233

Z

Zeta potential analysis 9
Zinc oxide 110, 111, 148, 149, 152, 227, 287, 297, 332

www.ingramcontent.com/pod-product-compliance
Lightning Source LLC
Chambersburg PA
CBHW041455280526
45792CB00004B/1022